AF813177

Special Issue: Age and Growth of Chondrichthyan Fishes: New Methods, Techniques and Analysis

Developments in environmental biology of fishes 25

Series Editor
DAVID L.G. NOAKES

Special Issue: Age and Growth of Chondrichthyan Fishes: New Methods, Techniques and Analysis

Editor:

Carlson, John K. (Volume editor), National Marine Fisheries Service, Panama City, USA
Goldman, Kenneth J. (Volume editor), Alaska Department of Fish and Game, Homer, USA

Reprinted from *Environmental Biology of Fishes*, Volume 77 (3-4) 2006

Springer

A C.I.P catalogue record for this book is available from the Library of Congress

ISBN-10 1-4020-5569-2
ISBN-13 978-1-4020-5569-0

Published by Springer,
P.O. Box 17, 3300 AA Dordrecht, The Netherlands

Sold and distributed in North, Central and South America
By Springer,
101 Philip Drive, Norwell, MA 02061, USA

In all other countries, sold and distributed
By Springer,
P.O. Box 322, 3300 AH Dordrecht, The Netherlands

http://www.springer.com
Printed on acid-free paper

All Rights Reserved
© 2006 Springer
No part of the material protected by this copyright notice may be reproduced or utilized in any form or by any means, electric or mechanical, including photocopying, recording or by any information storage and retrieval system, without written permission from the copyright owner.

Printed in the Netherlands

CONTENTS

Special Issue: Age and Growth of Chondrichthyan Fishes: New Methods, Techniques and Analysis
Guest Editors: Carlson, John K. & Goldman, Kenneth J.

Environ Biol Fish (2006) 77:209–210
DOI 10.1007/s10641-006-9138-9

Acknowledgment of referees

Published online: 17 October 2006
© Springer Science+Business Media B.V. 2006

The guest editors would like to acknowledge the following referees for taking the time out of their schedules to review manuscripts for this volume.

Referee	Institution
Allen Andrews	Moss Landing Marine Laboratories
Andre Punt	University of Washington
Andrew Piercy	Florida Museum of Natural History
Ashley Williams	James Cook University
Brian Gervelis	National Marine Fisheries Service
Christina Conrath	Florida Museum of Natural History
Colin Simpfendorfer	Mote Marine Laboratory
Craig Kastelle	National Marine Fisheries Service
David Ebert	Moss Landing Marine Laboratories
David Kulka	Department of Fisheries & Oceans-Canada
David Milton	CSIRO Marine Research-Australia
R. Dean Grubbs	Virginia Institute of Marine Science
Enric Cortes	National Marine Fisheries Service
Greg Skomal	Massachusetts Division of Marine Fisheries
Henry Mollet	Monterey Bay Aquarium
Ivy Baremore	University of Florida
Referee	Institution
Jack Musick	Virginia Institute of Marine Science
James Gelsleichter	Mote Marine Laboratory
James Hobbs	University of California, Davis
James Sulikowski	Florida Museum of Natural History
Jason Romine	Virginia Institute of Marine Science
Jody Spence	University of Victoria
John Hoenig	Virginia Institute of Marine Science

Referee	Institution
Joshua Loefer	South Carolina Department of Natural Resources
Julie Neer	National Marine Fisheries Service
Kate Siegfried	University of California, Santa Cruz
Linda Lombardi	National Marine Fisheries Service
Lisa Kerr	University of Maryland
Lisa Natanson	National Marine Fisheries Service
Lori Hale	National Marine Fisheries Service
Malcolm Francis	National Institute of Water and Atmospheric Research-New Zealand
Malcolm Smale	Port Elizabeth Museum-South Africa
Marta Nammack	National Marine Fisheries Service
Michael Gallagher	Irish Sea Fisheries Board
Peter Sheridan	National Marine Fisheries Service
Rick Officer	Fisheries Science Services Marine Institute-Ireland
RICC Francis	National Institute of Water and Atmospheric Research-New Zealand
Rich Beamish	Department of Fisheries & Oceans-Canada
Rory McAuley	Western Australian Fisheries and Marine Research Laboratories
Sarah Gaichas	National Marine Fisheries Service
Sarah Irvine	Deakin University-Australia
Scott Meyer	Alaska Department of Fish and Game

Referee	Institution
Steve Campana	Bedford Institute of Oceanography-Canada
Todd Gedamke	Virginia Institute of Marine Science
Warren Joyce	Bedford Institute of Oceanography-Canada
Wade Smith	Moss Landing Marine Laboratories
Yiota Apostolaki	Center for Environment, Fisheries and Aquaculture Science-Lowestoft, England

Acknowledgment of financial support

This publication and travel to the symposium for some participants was supported in part by the National Sea Grant College, Program of the U.S. Department of Commerce's National Oceanic and Atmospheric Administration under NOAA Grant # NA04OAR4170038, project # W05-23PD, through the California Sea Grant College Program; and in part by the California State Resources Agency. Funding was also provided by the American Elasmobranch Society, the National Marine Fisheries Service-Southeast Fisheries Science Center, and the Florida Aquarium. The views expressed herein do not necessarily reflect the views of any of those organizations.

Special acknowledgements

The guest editors extend their appreciation to George Burgess, Jeff Carrier, John Morrissey, Julie Neer, Ilze Berzins, Russell Moll, Alex Chester, Nancy Thompson, Michelle Heupel, and Colin Simpfendorfer for helping with various aspects of the symposium. We also thank Lori Hale and Dana Bethea for assisting with the edits to all the manuscripts. Finally, our appreciation goes to Suzanne Mekking, Martine van Bezooijen, and David Noakes for working with us on getting this volume published in Environmental Biology of Fishes.

Environ Biol Fish (2006) 77:211–228
DOI 10.1007/s10641-006-9105-5

ELASMOBRANCHS

Age and growth studies of chondrichthyan fishes: the need for consistency in terminology, verification, validation, and growth function fitting

Gregor M. Cailliet · Wade D. Smith · Henry F. Mollet · Kenneth J. Goldman

Received: 2 June 2006 / Accepted: 22 June 2006 / Published online: 29 September 2006
© Springer Science+Business Media B.V. 2006

Abstract Validated age and growth estimates are important for constructing age-structured population dynamic models of chondrichthyan fishes, especially those which are exploited. We review age and growth studies of chondrichthyan fishes, using 28 recent studies to identify areas where improvements can be made in describing the characteristics of ageing structures (both traditional and novel) utilized to estimate ages of sharks, rays, and chimaeras. The topics identified that need consistency include the: (1) terminology used to describe growth features; (2) methods used to both verify and validate age estimates from chondrichthyan calcified structures, especially edge and marginal increment analyses; and (3) the functions used to produce and describe growth parameters, stressing the incorporation of size at birth (L_0) and multiple functions to characterize growth characteristics, age at maturity and longevity.

G. M. Cailliet (✉) · W. D. Smith · H. F. Mollet
Moss Landing Marine Laboratories, 8272 Moss Landing Road, Moss Landing, CA 95039, USA
e-mail: cailliet@mlml.calstate.edu

K. J. Goldman
Alaska Department of Fish and Game, 3298 Douglas Place, Homer, AK 99603, USA

Keywords Age validation · Precision analysis · Chondrichthyes · Growth · Longevity · Vertebrae

Introduction

In recent years, there have been many advances in the quantitative study of age and growth of chondrichthyan fishes (Cailliet and Goldman 2004). Several new hard parts have been shown to provide valid assessments of age in some species, and new techniques for validation (e.g. bomb carbon) are becoming more widely known and applied. Moreover, the importance of assessing the precision and accuracy of counts on ageing structures, and the differences in growth models and their fits to data, are becoming more widely recognized. The book chapter cited above in a book entitled "Biology of Sharks and their Relatives," edited by Carrier et al. (2004), reviewed the field of age and growth in this group of fishes up to 2003. Although it has been less than two years since this publication, numerous papers on these subjects have been published, in addition to several papers not covered in the 2004 review chapter.

Since then, we have found five papers that were missed and 23 new publications that covered age and growth of chondrichthyan fishes. These papers covered the life histories of three species of chimaeras (Francis and Maolagáin 2000, 2001, 2004; Moura et al. 2004), three species of rays (Coelho and Erzini 2002; Neer and Thompson 2005; White and Potter 2005), nine species of skates (Gallagher et al. 2004; Henderson et al. 2004; Francis and Maolagáin 2005; Gedamke et al. 2005; Sulikowski et al. 2005a, b), two species of mackerel shark (Malcolm et al. 2001; Campana et al. 2005), and ten species of ground or requiem sharks (Yamaguchi et al. 1998; Lombardi-Carlson et al. 2003; Oshitani et al. 2003; Cruz-Martinez et al. 2004; Ivory et al. 2004; Joung et al. 2004, 2005; Lessa et al. 2004; Neer and Thompson 2004; Santana and Lessa 2004; Carlson and Baremore 2005; Manning and Francis 2005; Neer et al. 2005).

We reviewed the 28 new or missed papers mentioned above, analyzing the approaches that were taken in them to help identify key problems, if any, still existing in the methods involved in chondrichthyan age and growth studies. We were specifically interested in determining how these authors handled issues like the terminology of growth patterns, verification and validation techniques (focusing on edge characteristics and marginal increment analyses), and growth function fitting (how the von Bertalanffy (1938) growth function was fit and what other functions might have also been useful). At the end of each section, we provide recommendations that will hopefully guide researchers on how to proceed with each type of growth-related analysis.

Cailliet and Goldman (2004) and Goldman (2004) provided many guidelines on how to approach the subjects mentioned above. However, since many of these recent papers used a variety of different approaches, we have chosen to use the variability encountered in them to suggest ways to unify the field so that future papers might use suitable, and hopefully similar, approaches in their assessment of chondrichthyan life history parameters. Considerable variability and inconsistency were found in the: (1) terminology used for growth patterns in these structures; (2) methods used to verify and validate age estimates, including whether or not statistical analyses were applied; and (3) growth function fitting and parameter estimation.

Calcified structures and terminology used for chondrichthyan growth studies

Most age and growth studies of sharks, rays, and chimaeras utilize growth patterns in vertebral centra, dorsal fin spines (especially for those species which do not have suitably calcified centra and/or live in deep-sea habitats), and more recently in skates, caudal thorns (Cailliet and Goldman 2004). These structures tend to accumulate calcified growth material as they age, thus producing concentric areas that often have characteristics reflecting the time of year (season) in which this material is being deposited. Of the papers we recently reviewed, all used calcified structures, including vertebral centra (23 of the studies), dorsal spines (four), and caudal thorns (two).

In the field of fish ageing, there have been several attempts to synthesize the terminology used to describe growth features so that it is consistent among studies, one of the earliest being Wilson et al. (1987). Recently, Panfili et al. (2002) provided a comprehensive review in their "Manual of Fish Sclerochronology." However, this excellent review, with its glossary, is focused more on bony fish ageing, involving the use of otoliths and scales, which are not appropriate for chondrichthyan fishes.

It is important to distinguish between growth patterns that reflect seasonal growth and those, when combined, that may reflect annual (yearly) growth. Therefore, we first need to distinguish the length of time a particular growth pattern reflects. Panfili et al. (2002), for example, discussed daily growth rings, a phenomenon that has not yet been found in chondrichthyans. For these fishes, the first distinction is whether a term reflects a season (i.e., summer or winter patterns; but this may not always be the same in all species) or a year (i.e., an annual pattern, which requires some sort of validation, discussed later). As Panfili et al. (2002) noted, the term *annulus* "has traditionally been used to designate yearly marks even though the term is derived from the Latin "anus," meaning ring, not from "annus," which means year." For the seasonal growth pattern, Panfili et al. (2002) synonymized the words *band*, *ring*, *increment*, and *mark*, something that we feel confuses researchers.

We would prefer to have a standardized terminology that all, or at least the majority of chondrichthyan researchers, should follow.

Cailliet and Goldman (2004) and Goldman (2004) tried to be consistent with their terminology, suggesting that "band" be used for seasonal periods (e.g. opaque bands tending to be deposited in summer and translucent bands tending to be deposited in winter months) and either "ring" or "annulus" be used for those growth patterns demonstrated or assumed to represent a year's period. Cailliet and Goldman (2004) stated that "the most commonly distinguishable banding pattern in sectioned centra when viewed microscopically is one of wide bands separated by distinct narrow bands," and also that the "terms opaque and translucent are commonly used to describe these bands." An additional characterization of chondrichthyan growth bands was applied by Officer et al. (1996, 1997) based upon their relative extent of mineralization; these were identified as "hypermineralized bands."

Although there is often regularity in the width of bands, this can still be a potentially misleading generalization. For broader discussions, we feel that it is important to modify such statements saying "there is often a consistency in the wide/narrow pattern." The width of these opaque and translucent bands can be particularly exaggerated during the early years, and later, as growth slows, widths of these bands become more similar to each other. In fact, opaque and translucent bands may be narrower than translucent bands and/or vice versa in "older" fish. In addition, the relative widths of these bands may not remain consistent throughout the life of the animal. It is the deposition of opaque and translucent bands that is usually more consistent seasonally. Therefore, bands should be described and identified for their optical qualities, rather than dimensions such as band widths, which can be highly variable.

While reviewing the papers published since Cailliet and Goldman (2004), we found that many terms and combinations of terms were used. In the 28 studies reviewed, seasonal patterns were termed "band" (18 studies), "ring" and "zone" (4 each), "increment" (1), and in one publication, not defined at all. The terms used to represent annual patterns in these studies were "band" (7 studies), "band pair" (5), "annulus" (plural annuli), "ring" (in one of these, "growth ring") and "increment" (3 each), while 7 studies did not provide a definition. This supports our assertion that there is a need for consistency in the future use of terminology.

Indeed, years ago Cailliet et al. (1985) suggested counting *band pairs*, defined as one opaque and one translucent *band* combined, in their study of white shark, *Carcharodon carcharias*, growth. Martin and Cailliet (1988) added the term *rings*, which referred to the fine features within and making up either opaque or translucent bands (see Fig. 1 for a recent example). These fine rings have rarely been reported in chondrichthyans. Officer et al. (1996, 1997) identified these features as "minor increments" or "fine check marks" in gummy, *Mustelus antarcticus*, and school shark, *Galeorhinus galeus*, vertebrae, and similar rings were found in vertebral centra of the blue stingray, *Dasyatis chrysonota* (Cowley 1997), smooth hound, *M. mustelus* (Goosen and Smale 1997), and sandtiger sharks, *Carcharias taurus* (Goldman et al. 2006).

Recommendations

For clarity and consistency, we suggest that chondrichthyan fish agers use the following

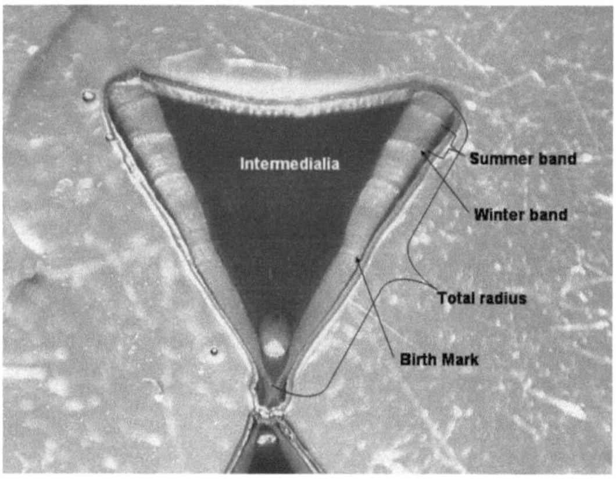

Fig. 1 A thin-sectioned vertebral centrum from an estimated 3.5+ year old spinner shark (*Carcharhinus brevipinna*) is shown (from Carlson and Baremore 2005). Centrum features, including the birthmark, opaque and translucent bands, band pairs, are identified. Also notable is the marginal increment of the ultimate band and the finer rings within the structure

terminology: (1) "Opaque" or "translucent bands" (following Cailliet et al. 1983); (2) "band pairs" (often referred to as annuli and/or rings, sensu Cailliet and Goldman (2004) and Goldman (2004)), comprising one opaque and one translucent band; and (3) "increments" which are measurements of partial to complete growth bands or band pairs (which should be specifically defined by authors). These terms should not be confused with other terminology such as "checks" or "discontinuous bands" (Panfili et al. 2002), although these also appear as translucent and opaque features. We remind investigators that the method of preparation and examination of an ageing structure (e.g. stained vs. unstained, radiographed vs. microphotographed, and viewed using reflected vs. transmitted light) alter the optical properties of calcified structures. Therefore, features characterized as opaque or translucent may vary depending upon methodology. Finally, we propose that future studies ascertain whether bands classified as "opaque" are hyper- or hypomineralized.

Verification and precision analysis

Panfili et al. (2002) defined "verification" as confirming "the consistency of the interpretation of age, i.e., the repeatability and/or precision of a numerical interpretation that may be independent of age." They further define "precision" as "the closeness of repeated measurements of the same quantity." They then pointed out that this can be between or within readers or laboratories. The techniques commonly used to verify age estimates were presented by Campana (2001) for fishes in general and by Cailliet and Goldman (2004) and Goldman (2004) for use on chondrichthyans.

Most of the 28 recently reviewed studies presented evidence that verified or assessed precisions of age estimates. The Index of Average Percent Error (Beamish and Fournier 1981, sometimes also including D and V of Chang 1982) was presented in 13 of these papers, while percentage agreement (Beamish and Fournier 1981; Cailliet et al. 1990; Kimura and Lyons 1991; Campana 2001; Cailliet and Goldman 2004) and age–bias curves (Campana et al. 1995) each were reported in six papers. Combinations of the various verification and precision assessments were common (11 studies). This approach of combining various assessments is a good one because when more than one method produces similar results it gives additional strength to the conclusions.

However, Hoenig et al. (1995) demonstrated that there can be differences in precision that APE indices obscure because they assume that the variability among observations of individual fish can be averaged over all age groups and that this variability can be expressed in relative terms. Additionally, APE indices do not result in values that are independent of the age estimates, do not test for systematic differences, do not distinguish all sources of variability (such as differences in precision with age) and do not take experimental design among studies into account (i.e., number of times each sample was read in each study.

Within a given ageing study, APE indices may serve as good relative indicators of precision within and between readers provided that each reader ages each vertebra the same number of times. However, even this appears only to tell us which reader was less variable, not which one was better or if either were biased. Bias is a more critical issue than precision, particularly in long-lived chondrichthyan fishes. We prefer also using Goldman's (2004) method for assessing precision, in which the percent agreement within and between readers is calculated, with individuals divided into appropriate length or disc width groups (e.g., 5–10 cm increments). This can be done with sexes separate and/or combined. Biases can then be assessed using contingency table methods (Bowker 1948; Hoenig et al. 1995). We feel that there is validity in using percent agreement with individuals grouped by length as a test of precision because it does not rely on ages (which have been estimated or assessed), but rather on empirical length measurements. Of course, age could be used if, and only if, validation of absolute age for all available age classes had been achieved.

Recommendations

Chondrichthyan life history researchers should continue to apply precision analyses in their ageing studies, and should, whenever possible, use multiple methods. In addition, within- or between-reader age–bias curves should be employed, including frequencies and levels of agreements superimposed on these curves (Fig. 2), and contingency tables.

Validation analysis

Panfili et al. (2002) defined "accuracy" as "the closeness of the estimate of a quantity (measured or computed value) to its true value." Thus, to document or test accuracy is to validate that the growth zones being counted represent some temporal unit such as season or year. Again, the techniques commonly used to validate age estimates were presented by Campana (2001) for fishes in general and by Cailliet and Goldman (2004) and Goldman (2004) for use on chondrichthyans.

Campana (2001) included at least eight approaches, all clearly summarized and listed, in order of choice. These were: (1) release of known age and marked fish; (2) bomb radiocarbon; (3) mark-recapture of chemically tagged fish; (4) radiochemical dating; (5) discrete length modes sampled for age structures; (6) natural date-specific markers; (7) marginal increment analysis; and (8) captive rearing (with and without oxytetracycline or OTC). While radiochemical dating has proven to be quite useful for bony fish otoliths (see Andrews et al. 1999, 2005; Stevens et al. 2004 for examples), its assumptions are invalid for cartilaginous chondrichthyan skeletons and it cannot be used on this group of fishes (Welden et al. 1987).

According to Cailliet and Goldman (2004), the techniques most commonly used on chondrichthyan fishes were marginal increment analysis, size frequency modal analysis, release of known-age, marked fish, mark-recapture of chemically tagged fish, and captive rearing. Also, one study

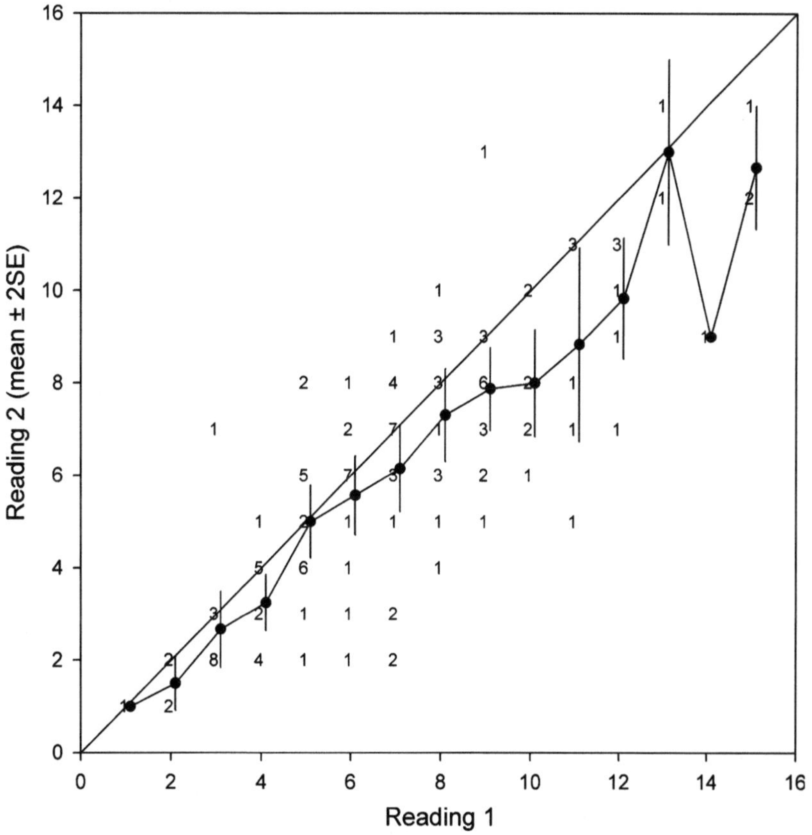

Fig. 2 An intra-reader age–bias plot that also incorporates age-specific agreements from a contingency table of thorn band counts for *Amblyraja georgiana* (from Francis and Maolagáin 2005). Numbers represent number of skates, and dots with error bars are the mean counts of reading 2 (±2 standard errors) relative to reading 1 (offset by +0.1 bands for clarity) for 119 readings. The diagonal line indicates a one-to-one relationship

was published by Campana et al. (2002) utilizing bomb radiocarbon age validation techniques for the porbeagle, *Lamna nasus*, and one vertebral centrum of the shortfin mako, *Isurus oxyrinchus*. While this exciting new validation approach is quite promising, it is also highly technical and expensive, thus no new papers have been published to date. However, three papers were presented at this symposium, which appear in these proceedings (see Ardizzone et al. 2006; Campana et al. 2006; Kerr et al. 2006)

In our review of the recent literature (Cailliet and Goldman 2004), including the more recent 28 studies, validation studies have not been very common for chondrichthyan fishes. This is mainly because of their limited accessibility, large size and mobility, and the difficulty of obtaining monthly, or even seasonal, samples. As a result, almost half of the 28 studies (13) did not report any age validation results, and the rest used a variety of tools.

Edge analysis

As discussed by Cailliet and Goldman (2004) and Goldman (2004), edge analysis characterizes the margin of a structure used for ageing over time in many different individuals to discern seasonal changes in growth. These structures have traditionally been vertebral centra, but this approach could equally apply to spines, thorns and neural arches. Edge analysis involves qualitatively characterizing the margin of the calcified structure as opaque or translucent, light or dark, wide or narrow, or a combination of these features.

In our examination of the recent literature we found that 6 of the 13 studies applied centrum edge analysis as a validation method. One approach was categorizing the edges simply as opaque or translucent (four studies), while the others categorized the edges as one of three grades (two studies).

The use of edge analysis was introduced by Holden and Vince (1973), who determined the timing of band deposition and validated the annual formation of one opaque and translucent band (one band pair) in whole vertebral centra of *Raja clavata* in conjunction with OTC mark recapture. However, they warned that the timing of opaque band formation did not necessarily coincide with the time that they become visible at the edge of the centrum. In their study, recognition of centrum edge types was commonly obscured by remaining vertebral connective tissue. Prompted by this earlier study, Tanaka and Mizue (1979) sectioned vertebral centra to enhance band clarity and determine the periodicity of band formation. Three grades of band development were classified from centrum edges. These grades were based on the optical qualities and width of the ultimate band (I: dark; II: light, narrow; III: light, broad) in relation to the month of capture.

Identifications of edge types may be influenced by many factors. The optical qualities of an ageing structure vary with preparation (e.g. thickness of section), species, its dimensions, and lighting methods. Edge types of stained vertebrae may be more difficult to interpret because of the accumulation of stain at the sample–resin interface. Inter-annual environmental variation may also alter the pattern of band formation and reduce the resolution of the technique. The experience level of those estimating ages and inconsistent criteria for assigning edge grades may further introduce variability and subjectivity into the analysis. It is therefore critical to include only samples of good condition and clarity and to carefully and consistently examine the edges of an ageing structure.

Despite the subjectivity associated with this approach, edge analysis has frequently been used in chondrichthyan ageing studies. The percent frequency of opaque and translucent bands has been compared with month or season of specimen capture (e.g. Roussouw 1984; Kusher et al. 1992; Wintner et al. 2002), and Tanaka and Mizue's (1979) approach has been adopted in numerous studies (e.g. Yudin and Cailliet 1990; Carlson et al. 1999). Following modified edge analysis methods introduced in teleost ageing studies (Anderson et al. 1992; Vilizzi and Walker 1999), Smith (2005) classified four distinct edge categories: narrow translucent, broad translucent, narrow opaque and broad opaque. The width of the forming band (broad/narrow) was determined based on proportional development in relation to the previous like band. Although this approach may not be well suited for small-bodied species or may become more complicated among the largest/oldest specimens, the consideration of four general edge types can provide enhanced details pertaining to seasonal patterns of band formation. When combined with additional techniques, such as Marginal Increment Analysis (MIA), edge analysis can provide valuable corroborative evidence to validate the periodicity of band formation.

Marginal increment analysis

MIA provides a useful, semi-direct (Panfili et al. 2002) method of validating the periodicity of band formation. It is the most commonly employed validation technique among chondrichthyan age and growth studies (Cailliet and Goldman 2004; Goldman 2004). Like edge analysis, MIA requires the recognition and identification of the band type forming on the outer edge of an ageing structure. Typically, the width of the ultimate, developing band (or band pair) is compared to the width of the last fully formed band pair and mean values of these ratios are related to the month of capture. Trends in the periodicity of band formation can be compared by size class, pooled age classes, select age classes (e.g. White et al. 2001; Sulikowski et al. 2005a), or season (Neer and Thompson 2005), but should ideally be restricted to individual age classes (Campana 2001). Specimens estimated to be age 0 cannot be included in MIA because they lack fully formed band pairs.

Campana (2001) identified MIA as one of the most difficult and likely to be abused methods of validation. However, Parsons (1993) successfully established the applicability and resolution of MIA in chondrichthyan growth studies by validating the annual deposition of a single band pair within the vertebral centra of *Sphyrna tiburo* using MIA in conjunction with captive, known-age and OTC-injected recaptured specimens. Although the incorporation of MIA into elasmobranch ageing studies has increased markedly since Parsons' (1993) study, the technique had previously been applied for many years. In his pioneering work, Ishiyama (1951) was the first to present a formula for MIA and examine ratios of ultimate and penultimate marginal widths between months of capture to determine the season of band formation. This attempt, however, was largely overlooked.

More recent authors (e.g. Killam and Parsons 1989; Simpfendorfer 1993; Natanson et al. 1995; Loefer and Sedberry 2003; Santana and Lessa 2004; Goldman and Musick 2006) have applied MIA as a validation technique, but few have provided examples of the formulae used to calculate these values or explicit details of this technique. Consequently, ambiguous and inconsistent terminology associated with MIA may have restricted the effective use, interpretation, and comparative value of these analyses among many elasmobranch ageing studies.

Four publications are commonly cited in association with chondrichthyan MIA and each offer seemingly different approaches and terminology:

(1) Natanson et al. (1995): MIR = $(VR–R_n)/(R_n–R_{n-1})$, in which MIR is the Marginal Increment Ratio, *VR* is the vertebral radius, R_n is the radius of the ultimate band or band pair, and R_{n-1} is the radius of the next to last complete band pair;
(2) Conrath et al. (2002): MIR = *MW*/*PBW*, in which MIR remains as previously defined, *MW* is the margin width, and *PBW* is the previous band pair width;
(3) Lessa et al. (2004): MI = $VR–R_n$, in which MI is termed the marginal increment, *VR* is the vertebral radius, and R_n is the radius of the last complete band or band pair.
(4) Branstetter and Musick (1994) apply the term "relative marginal increment analysis," but did not provide a formula or figure to describe the calculation. In their description of MIA, the authors' definitions of the terms "band" and "ring" were unclear and they proceeded to use them interchangeably making it somewhat difficult to interpret the features to which they were referring. Ambiguity associated with only the presentation of text and terminology may result in differing interpretations as to what features (e.g. bands or band pairs, opaque bands or translucent bands, broad or narrow bands, etc.) should be measured and compared.

Of these four methods, the one detailed by Natanson et al. (1995) has been the most widely cited and originated from Hayashi's (1976) study of marginal increment formation in the otoliths of the red tilefish, *Branchiostegus japonicus*. Each of the techniques described by Branstetter and Musick (1994), Natanson et al. (1995) and Conrath et al. (2002) calculate relative MIRs because the width of the outermost band pair (or band) is divided by the last fully formed band pair, making the marginal increment proportional to the previous growth band, but not necessarily to other fish of different ages. Alternatively, Santana and Lessa (2004) presented a variation that reports the mean relative MIRs by expressing absolute marginal increments as a percentage following Crabtree and Bullock (1998). In contrast, Lessa et al.'s (2004) approach does not provide values of the marginal width. Instead, their formula provides a secondary estimate of vertebral radius minus the ultimate band pair. This approach should not be used as a semi-direct validation method.

If the MIA methods of Branstetter and Musick (1994), Natanson et al. (1995) and Conrath et al. (2002) are interpreted and calculated correctly, they will provide the same result. These methods are not distinct and reflect the most commonly applied form of MIA in teleost ageing studies. The modification presented by Conrath et al. (2002) provides a simplification that directly compares the widths of the ultimate and penultimate band pairs. Secondarily determining width of the penultimate band pair (or band) by subtracting measurements from the vertebral radius introduces additional measurement error into calculations. The percent marginal increment applied by Santana and Lessa (2004) generates an interesting assessment of increment patterns but may inhibit the ability to assess the significance of these trends using most common statistical methods because values are expressed as a percentage. Therefore, we feel using the simplification of MIA as described by Conrath et al. (2002) is the most appropriate technique for validating the temporal periodicity of band deposition among chondrichthyans.

When considering preparation techniques for structures used in age determination and validation, especially MIA, we caution against the use of whole vertebrae. This is mainly because there can be error measuring straight lines on a concave surface. Many authors (e.g. Kusher et al. 1992) have discussed the potential drawbacks of ageing whole vertebrae and stressed the advantages of sectioning these structures so as to more easily discern the growth zones, especially from older fishes.

MIA typically tests the null hypothesis that a single band pair is deposited annually within the ageing structure of a study species. Given that initial assumption, it is imperative that measurements incorporated into these analyses consist of the last fully-formed band pair (one translucent and one opaque band) and the ultimate forming band or band pair. Measurement of opaque or translucent bands alone would not adequately test this null hypothesis. If more than one band pair is formed each year, or no pattern is evident whatsoever, it will be revealed in such analyses.

Following MIA, statistical analyses should be applied to the resulting data set to determine if significant differences exist among months. Too frequently, authors have relied on visual assessments of potential trends in marginal increment formation based solely on graphical representation of the data. Adequate statistical analyses include parametric single factor ANOVA (e.g. Carlson et al. 2003) and non-parametric Kruskal–Wallis one-way ANOVA (e.g. Simpfendorfer et al. 2000). The required assumptions for parametric analyses (i.e., equality of variances and normality) should be tested to determine if ANOVA is appropriate. As a result, transformation of mean marginal increment data may be necessary to perform ANOVA (e.g. Neer et al. 2005). Likewise, power analysis should be applied to assess the adequacy of sample size and potential for statistical error. Few authors have tested (or reported testing) their marginal increment data to ensure that parametric analyses were appropriate. Because non-parametric approaches may be more robust in cases of unequal sample sizes, inequality of variances, or departures from normality (e.g. Zar 1996), Kruskal–Wallis tests on ranks may be particularly well suited for marginal increment data.

Although rarely included, post-hoc tests should be applied to determine the source and extent of variation if significant differences are detected in mean marginal increment ratios among months. Tukey and Newman–Keuls are the most commonly used parametric tests for this purpose and modifications of each are available in the common event of unequal sample sizes (Zar 1996; Santana and Lessa 2004). Equivalent procedures are available for non-parametric evaluations, including the Nemenyi and Dunn tests (Zar 1996; Smith 2005). The approach of Dunn (1964) may be more applicable as it does not require equal sample sizes. The ability to identify which monthly or seasonal mean marginal increment ratios are significant from one another may enhance conclusions on the timing of band deposition and the environmental or biological factors that are associated with these events.

In our review of the recent 28 studies, authors used a version of marginal increment analysis in 12 studies. However, of those studies using MIA, only four attempted any statistical analyses to determine if observed variation in the mean marginal increment ratios differed significantly among months or seasons. These statistics included one factor ANOVA (3 studies), the Tukey test (1 study), and non-parametric ANOVA or the Kruskal–Wallace test (1 study).

Recommendations

For edge analysis, researchers should consider using several grades based upon the optical qualities and width of the ultimate band. It is also essential that only samples of good condition and clarity are used. Structures should be sectioned for use with both edge and marginal increment analyses because characterization and measurement of the critical areas of the margin will be more precise. We also feel that researchers need to develop and apply statistical analyses to categorical edge data, perhaps including log-likelihood ratios, Kolmogorov–Smirnov goodness of fit tests, and frequency distribution analysis, among other possibilities (e.g. Zar 1996; Cappo et al. 2000).

For both edge and marginal increment analysis, we support recent recommendations that trends in the periodicity of band formation be analyzed separately by size class, pooled age classes, selected age classes, and seasons. We also support combining edge analysis with other techniques (e.g. MIA) to strengthen the interpretation of band formation periodicity (Fig. 3).

The simplification of MIA as described by Conrath et al. (2002) is the most appropriate technique for validating the temporal periodicity of band deposition among chondrichthyans. Statistical analyses for MIA are necessary to insure that the edge dimensions really vary significantly with season. These should include tests to determine whether parametric or non-parametric statistics would be most appropriate for a given study. If significant differences among months or seasons are detected, appropriate post-hoc tests should be applied to identify the temporal source of this variation.

L_0 vs. t_0 and other aspects of the von Bertalanffy growth function (VBGF)

The VBGF (von Bertalanffy 1934, 1938, 1960) is the most commonly used growth function in chondrichthyan age and growth studies:

$$L(t) = L_\infty - (L_\infty - L_0)\,\mathrm{e}^{-kt},$$

where $L(t)$ is length as a function of time (t), L_∞ is the theoretical asymptotic length, L_0 is the size at birth, and k is the rate constant. The function that has consistently been presented as von Bertalanffy's (1934) growth function (e.g. Ricker 1979; Gulland 1983; Hilborn and Walters 1992; Haddon 2001) represents a modification of the original formula and is:

$$L(t) = L_\infty\left(1 - \mathrm{e}^{-k(t-t_0)}\right),$$

where t_0 is the theoretical time at zero length and the other parameters are as previously defined. Von Bertalanffy (1934) obtained his growth function by integrating the differential equation:

$$\mathrm{d}w/\mathrm{d}t = \eta w^{2/3} - \kappa w,$$

where η (eta) is the build-up (anabolic), κ (kappa) is the break-down (catabolic) physiological

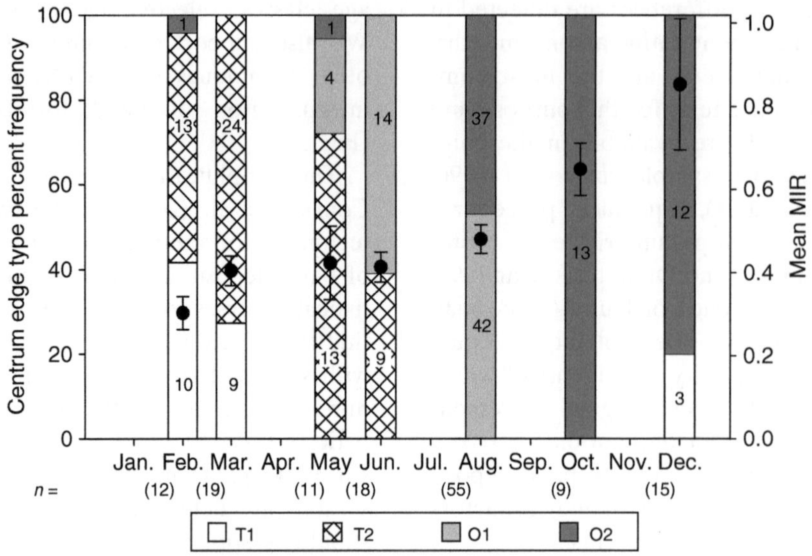

Fig. 3 Monthly variation among four centrum edge types (n = 205) and mean monthly marginal increment ratios (MIR) ±1 standard error (n = 139) determined from pooled sexes and size classes of the diamond stingray, *Dasyatis dipterura* (from Smith 2005). Values within the histogram represent the number of samples included in monthly centrum edge analyses. Sample sizes incorporated into the marginal increment analysis are listed in parentheses below the x-axis

parameter, and w is mass (weight). The constant of integration is determined by the value of $w(t)$ at time zero (y-axis intercept) or $L(t)$, and not some imaginary, negative time when $w(t) = 0$ or $L(t) = 0$ (x-axis intercept). His differential equation applied to mass, but the first step in this integration is the substitution method $y = w^{1/3}$ and this first mathematical step can be interpreted biologically to produce a differential equation for length if we substitute with $L = qw^{1/3}$ where q is a constant. The integration of the differential equation also shows that it is convenient to use the parameter $L_\infty = 3\ q\ (\eta/\mathrm{k})$ which is inversely proportional to the rate constant k (where $k = \kappa/3$). The steady state value $L\ (t = \infty) = L_\infty$ is determined by both η and k, while the time it takes to reach the final length from birth is determined by k alone.

It was Beverton (1954) who first used t_0 instead of L_0 as the third parameter in the VBGF. He mathematically transformed the VBGF with the parameters L_∞, k, and L_0 to an equation with the parameters L_∞, k, and t_0 to simplify yield calculations. He stated in Lecture 9 on p. 43: "It must be remembered that the constant t_0 is largely artificial, insofar as it defines the age at which the organism would be of zero length if it grew throughout life with the same pattern of growth as in the post-larval phase." The VBGF with t_0 as the third parameter was also used in Beverton and Holt (1957), and they also stated on p. 34: "In practice, the constant t_0 must be regarded as quite artificial." Nevertheless, this led to widespread but unfortunate use of the VBGF with t_0 as the third parameter in age and growth studies. Holden (1974) incorrectly assumed that t_0 had biological meaning for elasmobranchs (i.e., gestation period), but it does not (Pratt and Casey 1990; Van Dykhuizen and Mollet 1992).

We also note that the rate constant k has units of reciprocal time and is difficult to interpret. It is easier to interpret k in terms of half-lives ($\ln 2/k$) with units of time. The time it takes to reach the fraction x of L_∞ is given by:

$$t_x = 1/k \ln\left[(L_\infty - L_0)/(L_\infty(1 - x))\right].$$

If, for example, we use $x = 0.95$ (Ricker 1979), we also need to specify L_0 (for example $0.2L_\infty$), then we can interpret $t_{0.95}$ as a longevity estimate as given by:

$t_{0.95} = 2.77/k = 4.0 \ln 2/k (4 \text{ half} - \text{lives}).$

Fabens (1965) defined longevity based on $x = 0.9933$, assumed $L_0 = 0$, and obtained:

$t_{0.99} = 5/k = 7.21 \ln 2/k,$

for 7.2 half-lives. This shows that there is a considerable range for the definition of longevity depending on the value of x postulated.

It does not matter which 3-parameter VBGF is used for fitting length vs. age data as they are mathematically equivalent. However, the equation with L_0 as the third parameter has major advantages. Size at birth of elasmobranchs is often well defined and known. It is therefore easy to judge whether the fitted L_0 is a reasonable value. If the parameters L_∞, k, and t_0 are used as fitting parameters for elasmobranch age and growth studies, then L_0 should at least be calculated ($L_0 = L_\infty(1 - e^{kt_0})$). If the calculation of L_0 is omitted, one cannot evaluate how reasonable the fitted t_0 might be; they often are quite excessive (i.e., a t_0 value that is far too large or far too small) and this is an indication that an unreasonable calculated L_0 will result. Despite the advantages of using L_0 instead of t_0 as the third parameter in the VBGF, few papers dealing with elasmobranch growth have used it (e.g. Aasen 1963; Cailliet et al. 1992; Van Dykhuizen and Mollet 1992; Mollet et al. 2002). We are not sure why t_0 remained the preferred third parameter in most publications over the last 20 or more years, except that it is convenient.

Some authors have dealt with the lack of biological reality involved with estimating t_0 by fixing or anchoring the VBGF with an estimate of L_0 from known or estimated size-at-birth values (e.g. Van Dykhuisen and Mollet 1992; Neer and Thompson 2005). This modification can often significantly alter the other VBGF parameters. For example, it could decrease L_∞ and also increase the mean square error (MSE) and the standard error of the estimate (SEE) of the function. The 2-parameter VBGF with L_0 fixed as only one value can ignore what is often highly variable and sometime rapid early juvenile growth rates. Thus, all known values of L_0 should be used.

In our review of the 28 most recent chondrichthyan growth studies, almost all estimated growth parameters using the VBGF, and most (25 studies) also used the 3-parameter solution solving for k, L_∞, and t_0 but not L_0). However, three studies (Carlson and Baremore 2005; Neer et al. 2005; Santana and Lessa 2004) also used a two-parameter solution, using a fixed (or average) L_0 to anchor the model, solving only for k and L_∞.

Recommendations

Our recommendation is to use L_0 instead of t_0, whenever possible, because it can be biologically meaningful (Cailliet and Goldman 2004). We suggest that t_0 should never be used to estimate meaningful life history parameters of chondrichthyans (e.g. gestation period). If a three-parameter fit for the VBGF is used that incorporates t_0, researchers should check to see whether the resulting, calculated L_0 value crosses the y-axis within the range of observed length at birth.

Multiple growth functions: biological relevance, quality of fit, and convenience

It is often important and even necessary, to use more than one growth function to adequately characterize the growth of a given species. Yet, as previously stated, a single form of the VBGF (after Beverton 1954) has primarily been applied in chondrichthyan ageing studies. However, serious limitations and reservations have been identified with the growth function (e.g. Knight 1968; Roff 1980; Moreau 1987), including a limited ability to reflect early growth (Gamito 1998). Some of the criticisms applied to the VBGF are also relevant to many growth functions in general (e.g. assumption of asymptotic growth). Appropriate models should be selected on the indication of biological reality, statistical basis of their fit, convenience (Moreau 1987), and, as models of increased complexity are applied, parsimony (e.g. Burnham and Anderson 2002; Spiegelhalter et al. 2002; Guthery et al. 2005). If an investigator's objective is to express the growth characteristics of a species in quantitative terms, it is imprudent and may be counter-productive to base this description on a single, exclusive model.

Numerous models have been developed to describe growth characteristics based on size-at-age estimates or mark-recapture data (e.g. Ricker 1979; Baker et al. 1991; Haddon 2001). The VBGF itself has been modified, including two-parameter fits based on known size-at-birth (e.g. Van Dykhuizen and Mollet 1992), weight-at-age estimates (Fabens 1965), a generalized four-parameter form (Pauly 1979), or "near-linear" reparamaterizations developed to improve statistical properties of the model (e.g. Ratkowsky 1986; Hernandez-Llamas and Ratkowsky 2004). Polynomial functions have been suggested and applied as alternatives to the VBGF, but the resulting parameters provide no correlate for biological evaluation (Knight 1968; Chen et al. 1992). Flexible models, such as Richards (1959) and Schnute (1981), provide formulations that are capable of expressing more than one model form. Growth models have also been developed that incorporate the influences of ontogenetic or strong seasonal changes in growth trajectories (Soriano et al. 1992; Porch et al. 2002). However, it is outside the scope of this review to summarize each of the many available growth models and detail their characteristics. Instead, we emphasize that a single universal model is unlikely to adequately describe the growth of all chondrichthyans and encourage the fitting of multiple functions to enhance descriptions of growth.

Moreau (1987, p. 81) stated that "the main criteria for choosing a growth curve are quality of fit and convenience." Goodness of fit is best evaluated using several criteria. Coefficients of determination (r^2) have been the primary and often sole measure of model fit among chondrichthyan ageing studies. However, this approach may not be well suited for non-linear models (e.g. Kvålseth 1985). Recommended methods of evaluating model performance include the lowest residual MSE (also referred to as residual variance) or SEE, examination or comparison of residuals, and level of significance (e.g. $P < 0.05$) (Ratkowsky 1983; Neter et al. 1996). These measures used separately or in combination, are valuable whether considering single or multiple models. Although the potential for misinterpreting the quality of fit based on analysis of residuals alone increases when sample sizes are relatively small, plots of standardized residuals vs. predicted age allow a rapid means of identifying outliers within a dataset. Such outliers could, in turn, disproportionately influence estimates of MSE or SEE (Ratkowsky 1983). Because standardized residuals are normalized by their standard deviation, these plots and related analyses provide useful means of comparing fit between models generated from differing size-at-age (e.g. total length, disc width, weight) variables. Convenience is also an important factor in model selection. Specific models may be preferred for comparison with other studies, application to additional fishery models, or for indirect estimation of mortality and other life history correlates (e.g. Jensen 1996).

Regardless of the quality of fit and need for convenient models, the extent to which a given growth function produces reasonable biological estimates must remain a primary factor in model selection. Goodness of fit, used alone, could lead to choosing an inappropriate growth function. Using a combination of fit and a biological interpretation of one or more of the parameters, such as L_0 (from t_0 if necessary), longevity (from k), and L_∞ (directly), may ensure that the most biologically meaningful growth function is chosen.

In our review of the 28 most recent chondrichthyan growth studies, most applied only the VBGF (25 studies). However, four studies (Carlson and Baremore 2005; Neer et al. 2005; Neer and Thompson 2005; Santana and Lessa 2004) also fit their data to alternate growth functions (Ricker 1979), including the Gompertz (3 studies), logistic (3 studies), modified VBGF using a fixed L_0 based on a known size-at-birth (2 studies), Richards (2 studies), and Schnute (1 study) models (e.g. Winsor 1932; Ricker 1979; Schnute 1981). In each of these four studies, goodness of fit was evaluated by one or more measures other than the coefficients of determination (r^2).

Alternative models to the VBGF have been demonstrated to provide improved fits or generate more biologically reasonable representations of chondrichthyan growth in some studies. Gompertz and logistic models have been reported to produce significantly better fits to weight-at-age estimates than other model forms for *Rhinoptera bonasus* (Neer and Thompson 2005) and *Carcharhinus limbatus* (Killam and Parsons 1989), respectively. A logistic model fit to total length-at-age was presented as the most appropriate descriptor of growth for *Raja binoculata* (Zeiner and Wolf 1993). In some instances (Neer and Cailliet 2001), alternative growth functions provided the best fit to observed size-at-age data but were not reported because of the convenience and recognition of using the VBGF. Although the traditional VBGF may be an unsuitable descriptor of growth for species which do not attenuate toward an asymptote with increasing age, data quality, sample size, and dispersion of data across size-classes influence model performance, and subsequent selection. These examples illustrate the value of evaluating alternative models.

Recommendations

Although the VBGF may often provide a suitable description of growth, we encourage the use of multiple growth models to evaluate the growth characteristics of a given species. We also recommend that one should certainly consider convenience (i.e., the ability to compare parameters between sexes and among studies, locations, or species) and fit (i.e., using numerous growth functions, statistically examining them, and choosing those that best fit the actual size-at-age data) when characterizing growth for a given species. This approach is needed, considering that not all species follow the same growth function and different stages of their lives may undergo different characteristic growth patterns (Moreau 1987; Prince et al. 1991; Soriano et al. 1992; Hernandez-Llamas and Ratkowsky 2004). Finally, we encourage authors to consider and fit alternate metrics of body size for use with various growth models. For example, it may be more relevant to use girth, disc width, and/or weight rather than total length for angel sharks (Natanson and Cailliet 1990) and many species of batoids.

Summary and conclusions

Since validated age and growth estimates are important for constructing age-structured population dynamic models of chondrichthyan fishes, we have reviewed the field of age and growth on these fishes, briefly summarizing 28 recent studies either missed or new since the publication of the summary chapter on chondrichthyan ageing by Cailliet and Goldman (2004). We used these recent studies to identify areas where improvements can be made in describing the characteristics of ageing structures (both traditional and novel) utilized to estimate ages of sharks, rays, and chimaeras. The topics identified that we believe would be improved through greater consistency include the: (1) terminology used to describe growth features, promoting the use of the terms bands and band pairs; (2) methods used to both verify and validate age estimates from chondrichthyan calcified structures, especially edge and marginal increment analyses, and including statistical analyses; (3) the functions used to produce and interpret growth model parameters, stressing the incorporation of size at birth (L_0); and (4) use of multiple functions to characterize chondrichthyan growth, age at maturity, and longevity. Finally, we also strongly urge chondrichthyan agers to consult, review and incorporate established and novel methods used in age and growth studies of other organisms, including bony fishes (e.g. Campana 2001; Panfili et al. 2002).

Acknowledgements We dedicate this paper to all our graduate students and colleagues who have made it possible for us to keep up with the studies of fish age and growth, especially those who helped generate and evaluate ageing techniques. We really appreciate the efforts of John Carlson and Ken Goldman in putting together the slate of symposium speakers (and contributors to this volume) at the Joint Meeting of Ichthyologists and Herpetologists and American Elasmobranch Society Annual Meeting in 2005 entitled "Age and Growth of Chondrichthyan Fishes: New Methods, Techniques, and Analyses" in Tampa, Florida, 6–11 July, 2005. We appreciate the constructive comments on this manuscript by Colin Simpfendorfer, Jack Musick and several other anonymous reviewers. We acknowledge John Carlson, Ivy Baremore, Malcolm Francis, and C.O. Maolagáin for allowing us to use their figures. This study was supported by funds from NOAA/NMFS to the National Shark Research Consortium.

References

Aasen O (1963) Length and growth of the porbeagle (*Lamna nasus*, Bonnaterre) in the North West Atlantic. Rep Norwegian Fish Mar Investig 13(6):20–37

Anderson JR, Morison AK, Ray D (1992) Age and growth of the Murray cod, *Maccullochella peeli* (Perciformes: Percichthyidae), in the Lower Murray-Darling Basin, using thin-sectioned otoliths. Aust J Mar Freshw Res 43:983–1013

Andrews AH, Cailliet GM, Coale KH (1999) Age and growth of the Pacific grenadier (*Coryphaenoides acrolepis*) with age estimate validation using an improved radiometric ageing technique. Can J Fish Aquat Sci 56:1339–1350

Andrews AH, Burton EJ, Kerr LA, Cailliet GM, Coale KH, Lundstrom CC, Brown TA (2005) Bomb radiocarbon and lead-radium disequilibria in otoliths of bocaccio rockfish (*Sebastes paucispinis*): a determination of age and longevity for a difficult-to-age fish. Proceedings of the 3rd international symposium on otolith research and application. Marine Freshw Res 56:517–528

Ardizzone, D, Cailliet GM, Natanson LJ, Andrews AH, Kerr LA, Brown TA (2006) Application of bomb radiocarbon chronologies to shortfin mako (*Isurus oxyrinchus*) age validation. In: Carlson JK, Goldman KJ (eds) Age and growth of chondrichthyan fishes: new methods, techniques and analyses. Special Volume from symposium of the American Elasmobranch Society, July 2005. Environ Biol Fish (in press)

Baker TT, Lafferty R, Quinn II TJ (1991) A general growth model for mark-recapture data. Fish Res 11:257–281

Beamish RJ, Fournier DA (1981) A method for comparing the precision of a set of age determinations. Can J Fish Aquat Sci 38:982–983

von Bertalanffy L (1934) Untersuchungen ueber die Gesetzlichkeit des Wachstums. Wilhelm Roux' Arch Entwick Organ 131:613–652

von Bertalanffy L (1938) A quantitative theory of organic growth (inquiries on growth laws II). Hum Biol 10:181–213

von Bertalanffy L (1960) Principles and theory of growth. In: Wowinski WW (ed) Fundamental aspects of normal and malignant growth. Elsevier, Amsterdam, pp 137–259

Beverton RJH (1954) Notes on the use of theoretical models in the study of the dynamics of exploited fish populations. United States Fishery Laboratory, Beaufort, North Carolina, Miscellaneous Contribution (2), 159 pp

Beverton RJH, Holt SJ (1957) On the dynamics of exploited fish populations. United Kingdom Ministry of Agriculture and Fisheries, Fisheries Investigations (Series 2) 19, 533 pp

Bowker AH (1948) A test for symmetry in contingency tables. J Am Stat Assoc 43:572–574

Branstetter S, Musick JA (1994) Age and growth estimates for the sand tiger in the northwestern Atlantic ocean. Trans Am Fish Soc 123:242–254

Burnham KP, Anderson DR (2002) Model selection and multimodel inference: a practical information-theoretic approach, 2nd edn. Springer-Verlag, New York, 488 pp

Cailliet GM, Goldman KJ (2004) Age determination and validation in chondrichthyan fishes. In: Carrier J, Musick JA, Heithaus MR (eds) Biology of sharks and their relatives. CRC Press LLC, Boca Raton, FL, pp 399–447

Cailliet GM, Martin LK, Kusher D, Wolf P, Welden BA (1983) Techniques for enhancing vertebral bands in age estimation of California elasmobranchs. In: Prince ED, Pulos LM (eds) Proceedings international workshop on age determination of oceanic pelagic fishes: tunas, billfishes, sharks. NOAA Tech. Rep. NMFS 8, pp 157–165

Cailliet GM, Natanson LJ, Welden BA, Ebert DA (1985) Preliminary studies on the age and growth of the white shark, *Carcharodon carcharias*, using vertebral bands. S Calif Acad Sci Mem 9:49–60

Cailliet GM, Yudin KG, Tanaka S, Taniuchi T (1990) Growth characteristics of two populations of *Mustelus manazo* from Japan based upon cross-readings of vertebral bands. In: Pratt HL Jr, Gruber SH, Taniuchi T (eds) Elasmobranchs as living resources: advances in the biology, ecology, systematics, and the status of the fisheries. NOAA Technical Report (90), pp 167–176

Cailliet GM, Mollet HF, Pittenger GG, Bedford D, Natanson LJ (1992) Growth and demography of the pacific angel shark (*Squatina californica*), based upon tag returns off California. Aust J Mar Freshw Res 43:1313–1330

Campana SE (2001) Accuracy, precision, and quality control in age determination, including a review of the use and abuse of age validation methods. J Fish Biol 59:197–242

Campana SE, Annand CM, McMillan JI (1995) Graphical and statistical methods for determining the consistency of age determinations. Trans Am Fish Soc 124:131–138

Campana SE, Natanson LJ, Myklevoll S (2002) Bomb dating and age determination of large pelagic sharks. Can J Fish Aquat Sci 59:450–455

Campana SE, Marks L, Joyce W (2005) The biology and fishery of shortfin mako sharks (*Isurus oxyrinchus*) in Atlantic Canadian waters. Fish Res 73:341–352

Campana SE, Jones C, McFarlane GA, Myklevoll S (2006) Bomb dating and age determination of spiny dogfish. In: Carlson JK, Goldman KJ (eds) Age and growth of chondrichthyan fishes: new methods, techniques and analyses. Special Volume from symposium of the American Elasmobranch Society, July 2005. Environ Biol Fish (in press)

Cappo M, Eden P, Newman SJ, Robertson S (2000) A new approach to validation of periodicity and timing of opaque zone formation in the otoliths of eleven species of *Lutjanus* from the central Great Barrier Reef. Fish Bull 98:474–488

Carlson JK, Baremore IE (2005) Growth dynamics of the spinner shark (*Carcharhinus brevipinna*) off the United States southeast and Gulf of Mexico coasts: a comparison of methods. Fish Bull 103:280–291

Carlson JK, Cortes E, Bethea DM (2003) Life history and population dynamics of the finetooth shark (*Carcharhinus isodon*) in the northeastern Gulf of Mexico. Fish Bull 101:281–292

Carlson JK, Cortes E, Johnson AG (1999) Age and growth of the blacknose shark, *Carcharhinus acronotus*, in the eastern Gulf of Mexico. Copeia 1999:684–691

Carrier J, Musick JA, Heithaus MR (eds) (2004) Biology of sharks and their relatives. CRC Press LLC, Boca Raton, FL, 596 pp

Chang WYB (1982) A statistical method for evaluating the reproducibility of age determination. Can J Fish Aquat Sci 39:1208–1210

Chen Y, Jackson DA, Harvey HH (1992) A comparison of von Bertalanffy and polynomial functions in modelling fish growth data. Can J Aquat Sci 49:1228–1235

Coelho R, Erzini K (2002) Age and growth of the undulate ray, *Raja undulata*, in the Algarve (southern Portugal). J Mar Biol Assoc UK 82:987–990

Conrath CL, Gelsleichter J, Musick JA (2002) Age and growth of the smooth dogfish (*Mustelus canis*) in the northwest Atlantic Ocean. Fish Bull 100:674–682

Cowley PDS (1997) Age and growth of the blue stingray *Dasyatis chrysonota chrysonota* from the Southern Eastern Cape coast of South Africa. S Afr J Mar Sci 18:31–38

Crabtree RE, Bullock LH (1998) Age, growth, and reproduction of black grouper, *Mycteroperca bonaci*, in Florida waters. Fish Bull 96:735–753

Cruz-Martinez A, Chiappa-Carrara S, Arenas-Fuentes V (2004) Age and growth of the bull shark, *Carcharhinus leucas*, from southern Gulf of Mexico. E-J Northw Atlant Fish Sci 35(13):1–10

Dunn OJ (1964) Multiple contrasts using ranks sums. Technometrics 6:241–252

Fabens AJ (1965) Properties and fitting of the von Bertalanffy growth curve. Growth 29:265–289

Francis MP, Maolagáin CÓ (2000) Age and growth of ghost sharks. Final Research Report for Ministry of Fisheries Research Project GSH1999/01 Objective 1, National Institute of Water and Atmospheric Research, pp 1–27

Francis MP, Maolagáin CÓ (2001) Development of ageing techniques for dark ghost shark (*Hydrolagus novaezelandiae*). Final Research Report for Ministry of Fisheries Research Project MOF2000/03C, National Institute of Water and Atmospheric Research, pp 1–18

Francis MP, Maolagáin CÓ (2004) Feasibility of ageing pale ghost sharks (*Hydrolagus bemisi*). Final Research Report for Ministry of Fisheries Research Project GSH2002/01 Objective 2, National Institute of Water and Atmospheric Research, pp 1–11

Francis MP, Maolagáin CÓ (2005) Age and growth of the Antarctic skate, (*Amblyraja georgiana*), in the Ross Sea. Commission for the Conservation of Antarctic Marine Living Resources (CCAMLR) Science 12:183–194

Gallagher MJ, Nolan CP, Jeal F (2004) Age, growth and maturity of the commercial ray species from the Irish Sea. E-J Northw Atlant Fish Sci 35(10):22

Gamito S (1998) Growth models and their use in ecological modeling: an application to a fish population. Ecol Model 113:83–94

Gedamke T, DuPaul WD, Musick JA (2005) Observations on the life history of the barndoor skate, *Dipturus laevis*, on George Bank (western North Atlantic). E-J Northw Atlant Fish Sci 35(10):13

Goldman KJ (2004) Age and growth of elasmobranch fishes. In: Musick JA, Bonfil R (eds) Elasmobranch fisheries management techniques. Asia Pacific Economic Cooperation, Singapore, 370 pp, pp 97–132

Goldman KJ, Branstetter S, Musick JA (2006) A re-examination of the age and growth of sand tiger sharks, *Carcharias taurus*, in the western North Atlantic: the importance of ageing protocols and use of multiple back-calculation techniques. In: Carlson JK, Goldman KJ (eds) Special volume from symposium of the American Elasmobranch Society, July 2005. Environ Biol Fish (in press)

Goldman KJ, Musick JA (2006) Growth and maturity of salmon sharks in the eastern and western North Pacific, with comments on back-calculation methods. Fish Bull 104:278–292

Goosen AJJ, Smale MJ (1997) A preliminary study of age and growth of the smoothhound shark *Mustelus mustelus* (Triakidae). S Afr J Mar Sci 18:85–91

Gulland JA (1983) Fish stock assessment. A manual of basic methods. FAO/Wiley Ser Food Agric 1:86–97

Guthery FS, Brennan LA, Peterson MJ, Lusk JJ (2005) Information theory in wildlife science: critique and viewpoint. J Wildlife Manage 69(2):457–465

Haddon M (2001) Chapter 8: Growth of individuals. In: Modeling and quantitative measures in fisheries. Chapman and Hall/CRC, Boca Raton, FL, pp 187–246

Hayashi Y (1976) Studies on the growth of the red tilefish in the east China Sea – 1. A foundational consideration for age determination from otoliths. Bull Jpn Soc Sci Fish 42(11):1237–1242

Henderson AC, Arkhipkin AI, Chtcherbich JN (2004) Distribution, growth and reproduction of the white-spotted skate *Bathyraja albomaculata* (Norman, 1937) around the Falkland Islands. E-J Northw Atlant Fish Sci 35(1):1–10

Hernandez-Llamas A, Ratkowsky DA (2004) Growth of fishes, crustaceans and molluscs: estimation of the von Bertalanffy, Logistic, Gompertz and Richards curves and a new growth model. Mar Ecol Prog Ser 282:237–244

Hilborn R, Walter CJ (1992) Quantitative fisheries stock assessment: choice, dynamics and uncertainty. Chapman and Hall, New York, 570 pp

Hoenig JM, Morgan MJ, Brown CA (1995) Analyzing differences between two age determination methods by tests of symmetry. Can J Fish Aquat Sci 52:364–368

Holden MJ (1974) Problems in the rational exploitation of elasmobranch populations and some suggested solutions. In: Jones EH (ed) Sea fisheries research. Logos, London, pp 187–215

Holden MJ, Vince MR (1973) Age validation studies on the centra of *Raja clavata* using tetracycline. J Conseil Int Explor Mer 35:13–17

Ishiyama R (1951) Studies on the rays and skates belonging to the family Rajidae, found in Japan and adjacent regions. 2. On the age-determination of Japanese black skate *Raja fusca* Garman (Preliminary Report). Bull Jpn Soc Fish 16(12):112–118

Ivory P, Jeal F, Nolan CP (2004) Age determination, growth and reproduction in the lesser-spotted dogfish, *Scyliorhinus canicula*. E-J Northw Atlant Fish Sci 35(2):20

Jensen AL (1996) Beverton and Holt life history invariants result from optimal trade-off of reproduction and survival. Can J Fish Aquat Sci 54:987–989

Joung SJ, Liao YY, Chen CT (2004) Age and growth of sandbar shark, *Carcharhinus plumbeus*, in northeastern Taiwan waters. Fish Res 70:83–96

Joung SJ, Liao YY, Liu KM, Chen CT, Leu LC (2005) Age, growth and reproduction of the spinner shark, *Carcharhinus brevipinna*, in the northeastern waters of Taiwan. Zool Stud 44(1):102–110

Kerr LA, Andrews AH, Cailliet GM, Brown TA, Coale KH (2006) Investigations of radiocarbon and stable carbon and nitrogen ratios in vertebrae of white shark (*Carcharodon carcharias*) from the eastern North Pacific Ocean. In: Carlson JK, Goldman KJ (eds) Age and growth of chondrichthyan fishes: New methods, techniques and analyses. Special volume from symposium of the American Elasmobranch Society, July 2005. Environ Biol Fish (in press)

Killam KA, Parsons GR (1989) Age and growth of the blacktip shark, *Carcharhinus limbatus*, near Tampa Bay, Florida. Fish Bull 87:845–857

Kimura DK, Lyons JJ (1991) Between-reader bias and variability in the age-determination process. Fish Bull 89:53–60

Knight W (1968) Asymptotic growth: an example of nonsense disguised as mathematics. J Fish Res Board Can 25(6):1303–1307

Kusher DI, Smith SE, Cailliet GM (1992) Validated age and growth of the leopard shark, *Triakis semifasciata*, with comments on reproduction. Environ Biol Fish 35:187–203

Kvålseth TO (1985) Cautionary note about r^2. Am Stat 39(4):279–285

Lessa R, Santana FM, Hazin FH (2004) Age and growth of the blue shark *Prionace glauca* (Linnaeus, 1758) off northeastern Brazil. Fish Res 66:19–30

Loefer JK, Sedberry GR (2003) Life history of the Atlantic sharpnose shark (*Rhizoprionodon terraenovae*) (Richardson, 1836) off the southeastern United States. Fish Bull 101:75–88

Lombardi-Carlson LA, Cortes E, Parsons GR, Manire CA (2003) Latitudinal variation in life-history traits of bonnethead sharks, *Sphyrna tiburo*, (Carcharhiniformes, Sphyrnidae), from the eastern Gulf of Mexico. Mar Freshw Res 54:875–883

Malcolm H, Bruce BD, Stevens JD (2001) A review of the biology and status of white sharks in Australian waters. Report to Environment Australia, Marine Species Protection Program, CSIRO Marine Research, Hobart, 81 pp

Manning MJ, Francis MP (2005) Age and growth of the blue shark (*Prionace glauca*) from the New Zealand exclusive economic zone. New Zealand Fisheries Assessment report 2005/26, 52 pp

Martin LK, Cailliet GM (1988) Age and growth of the bat ray, *Myliobatis californica*, off central California. Copeia 1988(3):762–773

Mollet HF, Ezcurra JM, O'Sullivan JB (2002) Captive biology of the pelagic stingray, *Dasyatis violacea* (Bonaparte, 1832). Mar Freshw Res 53:531–541

Moreau J (1987) Mathematical and biological expression of growth in fishes: recent trends and further developments. In: Summerfelt RC, Hall GE (eds) Age and growth of fish. Iowa State University Press, Ames, pp 81–113

Moura T, Figueiredo I, Bordalo Machado P, Serrano Gordo L (2004) Growth pattern and reproductive strategy of the holocephalan *Chimaera monstrosa* along the Portuguese continental slope. J Mar Biol Assoc UK 84:801–804

Natanson LJ, Cailliet GM (1990) Vertebral growth zone deposition in Pacific angel sharks. Copeia 1990(4):1133–1145

Natanson LJ, Casey JG, Kohler NE (1995) Age and growth estimates for the dusky shark, *Carcharhinus obscurus*, in the western North Atlantic Ocean. Fish Bull 93:116–126

Neer JA, Cailliet GM (2001) Aspects of the life history of the Pacific electric ray, *Torpedo californica* (Ayres). Copeia 2001(3):842–847

Neer JA, Thompson BA (2004) Aspects of the biology of the finetooth shark, *Carcharhinus isodon*, in Louisiana waters. Gulf Mexico Sci 2004(1):108–113

Neer JA, Thompson BA (2005) Life history of the cownose ray, *Rhinoptera bonasus*, in the northern Gulf of Mexico, with comments on geographic variability in life history traits. Environ Biol Fish 73:321–331

Neer JA, Thompson BA, Carlson JK (2005) Age and growth of *Carcharhinus leucas* in the northern Gulf of Mexico: incorporating variability in size at birth. J Fish Biol 66:1–14

Neter J, Kutner MH, Nachtsheim CJ, Wasserman W (1996) Applied linear statistical models: a unified practical approach, 4th edn. Irwin, Chicago, IL, 1408 pp

Officer RA, Gason AS, Walker TI, Clement JG (1996) Sources of variation in counts of growth increments in vertebrae from gummy shark, *Mustelus antarcticus*, and school shark, *Galeorhinus galeus*: implications for age determination. Can J Fish Aquat Sci 53:1765–1777

Officer RA, Day RW, Clement JG, Brown LP (1997) Captive gummy sharks, *Mustelus antarcticus*, form hypermineralised bands in their vertebrae during winter. Can J Fish Aquat Sci 54:2677–2683

Oshitani S, Nakano H, Tanaka S (2003) Age and growth of the silky shark *Carcharhinus falciformis* from the Pacific Ocean. Fish Sci 69:456–464

Panfili J, de Pontual H, Troadec H, Wright PJ (eds) (2002) Manual of fish schlerochronology. Ifremer-IRD coedition, Brest, France, 464 pp

Parsons GR (1993) Age determination and growth of the bonnethead shark *Sphyrna tiburo*: a comparison of two populations. Mar Biol 117:23–31

Pauly D (1979) Gill size and temperature as governing factors in fish growth: a generalization of von Bertalanffy's growth formula. Institut für Meereskunde an der Universität Kiel No. 63, 156 pp

Porch CE, Wilson CA, Nieland DL (2002) A new growth model for red drum (*Sciaenops ocellatus*) that accommodates seasonal and ontogenic changes in growth rates. Fish Bull 100:149–152

Pratt HL Jr, Casey JG (1990) Shark reproductive strategies as a limiting factor in directed fisheries, with a review of Holden's method of estimating growth parameters. In: Pratt HL, Gruber SH, Taniuchi T (eds) Elasmobranchs as living resources: advances in the biology, ecology, systematics, and the status of the fisheries. NOAA Technical Report NMFS 90, pp 97–109

Prince ED, Lee DW, Zweifel JR, Brothers EB (1991) Estimating age and growth of young Atlantic blue marlin *Makaira nigricans* from otolith microstructure. Fish Bull 89:441–459

Ratkowsky DA (1983) Nonlinear regression modeling: a unified practical approach. Statistics: textbooks and monographs, vol 48. Marcel Dekker, Inc., New York, 276 pp

Ratkowsky DA (1986) Statistical properties and alternative parameterizations of the von Bertalanffy growth curve. Can J Fish Aquat Sci 43:742–747

Richard FJ (1959) A flexible growth function for empirical use. J Exp Bot 10(29):290–300

Ricker WE (1979) Growth rates and models. In: Hoar WS, Randall DJ, Brett JR (eds) Fish physiology, vol VIII. Bioenergetics and Growth, pp 677–743

Roff DA (1980) A motion for the retirement of the von Bertalanffy function. Can J Fish Aquat Sci 37:127–129

Roussouw GJ (1984) Age and growth of the sand shark, *Rhinobatos annulatus*, in Algoa Bay, South Africa. J Fish Biol 25:213–222

Santana FM, Lessa R (2004) Age determination and growth of the night shark (*Carcharhinus signatus*) off the northeastern Brazilian coast. Fish Bull 102:156–167

Schnute J (1981) A versatile growth model with statistically stable parameters. Can J Fish Aquat Sci 38:1128–1140

Simpfendorfer CA (1993) Age and growth of the Australian sharpnose shark, *Rhizoprionodon taylori*, from north Queensland, Australia. Environ Biol Fish 36(3):233–241

Simpfendorfer CA, Chidlow J, McAuley R, Unsworth P (2000) Age and growth of the whiskery shark, *Furgaleus macki*, from southwestern Australia. Environ Biol Fish 58:335–343

Smith WD (2005) Life history aspects and population dynamics of a commercially exploited stingray, *Dasyatis* dipterura. MS Thesis, Moss Landing Marine Laboratories and San Francisco State University, 221 pp

Soriano M, Moreau J, Hoenig JM, Pauly D (1992) New functions for the analysis of two-phase growth of juvenile and adult fishes, with application to Nile perch. Trans Am Fish Soc 131:486–493

Spiegelhalter DJ, Best NG, Carlin BP, van der Linde A (2002) Bayesian measures of model complexity and fit. J Roy Stat Soc B 64(4):583–639

Stevens MM, Andrews AH, Cailliet GM, Coale KH, Lundstrom CC (2004) Radiometric validation of age, growth, and longevity for the blackgill rockfish, *Sebastes melanostomus*. Fish Bull 102:711 722

Sulikowski JA, Kneebone J, Eizey S, Jurek J, Danley PD, Howell WH, Tsang PCW (2005a) Age and growth estimates of the thorny skate (*Amblyraja radiata*) in the western Gulf of Maine. Fish Bull 103:161–168

Sulikowski JA, Tsang PCW, Howel WH (2005b) Age and size at sexual maturity for the winter skate, *Leucoraja ocellata*, in the western Gulf of Maine based on morphological, histological, and steroid based analysis. Environ Biol Fish 72:429–441

Tanaka S, Mizue K (1979) Studies on sharks – XV. Age and growth of Japanese dogfish, *Mustelus manazo* Bleeker in the East China Sea. Bull Jpn Soc Sci Fish 45:43–50

Van Dykhuizen G, Mollet HF (1992) Growth, age estimation and feeding of captive sevengill sharks, *Notorynchus cepedianus*, at the Monterey Bay Aquarium. Aust J Mar Freshw Res 43:297–318

Vilizzi L, Walker KF (1999) Age and growth of the common carp, *Cyprinus carpio*, in the River Murray, Australia: validation, consistency of interpretation, and growth models. Environ Biol Fish 54:77–106

Welden BA, Cailliet GM, Flegal AR (1987) Comparison of radiometric with vertebral band age estimates in four California elasmobranchs. In: Summerfelt RC, Hall GE (eds) The age and growth of fish. The Iowa State University Press, Ames, pp 301–315

White WT, Platell ME, Potter IC (2001) Relationship between reproductive biology and age composition and growth in *Urolophus lobatus* (Batoidea: Urolophidae). Mar Biol 138:135–147

White WT, Potter IC (2005) Reproductive biology, size and age compositions and growth of the batoid, *Urolophus* paucimaculatus, including comparisons with other species of the Urolophidae. Mar Freshw Res 56:101–110

Wilson CA, Chairman, Glossary Committee (1987) Glossary. In: Summerfelt RC, Hall GE (eds) Age and growth of fish. Iowa State University Press, Ames, pp 527–530

Winsor CP (1932) The Gompertz curve as a growth curve. Proc Natl Acad Sci 18(1):1–8

Wintner SP, Dudley SFJ, Kistnasamy N, Everett (2002) Age and growth estimates for the Zambezi shark, *Carcharhinus leucas*, from the east coast of South Aftica. Mar Freshw Res 53:557–556

Yamaguchi A, Taniuchi T, Shimizu M (1998) Goegraphic variation in growth of the starspotted dogfish *Mustelus manazo* from five localities in Japan and Taiwan. Fish Sci 65:732–739

Yudin KG, Cailliet GM (1990) Age and growth of the gray smoothhound, *Mustelus californicus*, and the brown smoothhound, *M. henlei*, sharks from central California. Copeia 1990:191–204

Zar JH (1996) Biostatistical analysis, 3rd edn. Prentice Hall, NJ, 662 pp

Zeiner SJ, Wolf P (1993) Growth characteristics and estimates of age at maturity of two species of skates (*Raja binoculata* and *Raja rhina*) from Monterey Bay, California. NOAA Technical Report NMFS 115:87–99

Environ Biol Fish (2006) 77:229–239
DOI 10.1007/s10641-006-9113-5

ORIGINAL PAPER

Age and growth of the sandbar shark, *Carcharhinus plumbeus*, in Hawaiian waters through vertebral analysis

Jason G. Romine · R. Dean Grubbs · John A. Musick

Received: 16 June 2006 / Accepted: 6 July 2006 / Published online: 6 September 2006
© Springer Science+Business Media B.V. 2006

Abstract Age and growth estimates were determined for the sandbar shark, *Carcharhinus plumbeus*, from Oahu, Hawaii in the central Pacific Ocean. Age estimates were obtained through vertebral centra analysis of 187 sharks. We verified our age estimates through marginal increment analysis of centra and oxytetracycline marking methods of at liberty sandbar sharks. Sizes of sampled sharks ranged from 46 to 147 cm pre-caudal length. Four growth models were fitted to length-at-age data; two forms of the von Bertalanffy growth model, the Gompertz growth model, and a logistic growth model. Males and females exhibited statistically significant differences in growth, indicating that females grow slower and attain larger sizes than males. Growth parameter estimates revealed slower growth rates than previously estimated (based on captive specimens) for Hawaiian sandbar sharks. The von Bertalanffy growth model using empirical length-at-birth provided the best biological and statistical fit to the data. This model gave parameter estimates of $L_\infty = 138.5$ cm PCL and $k = 0.12$ year^{-1} for males and $L_\infty = 152.8$ cm PCL, $k = 0.10$ year^{-1} for females. Male and female sandbar sharks mature at approximately 8 and 10 years of age, respectively.

J. G. Romine (✉) · R. D. Grubbs · J. A. Musick
Virginia Institute of Marine Science, 1208 Greate Rd, Gloucester Point, VA 23062, USA
e-mail: jromine@vims.edu

Keywords Age · Growth · Sandbar shark · Hawaii

Introduction

The sandbar shark, *Carcharhinus plumbeus*, is a common large-coastal shark that inhabits temperate and subtropical waters world-wide and attains lengths greater than two meters (Bigelow and Schroeder 1948; Compagno 1984). In Hawaiian waters, sandbar sharks most frequently occur between depths of 10 and 50 m (Wass 1973). The species is not commercially important in the central Pacific, which provides a unique opportunity to examine age and growth of a late maturing carcharhiniform shark that has not been greatly affected by fishing mortality.

Male and female sandbar sharks in Hawaiian waters have historically been shown to reach maximum sizes of 132 and 146 cm pre-caudal length (PCL), respectively (Wass 1973). The sandbar shark is viviparous via yolk-sac placenta, giving birth to well-developed live young following a gestation period of approximately 9–12 months (Springer 1960; Clark and von Schmidt 1965; Wass 1973; Lawler 1976). In Hawaiian waters, pups are approximately 47 cm PCL at birth and litter sizes average 5.5 pups per litter (Wass 1973). Wass (1973) estimated maturity to occur at 110 cm PCL for males and at 115 cm PCL for females.

The age and growth of the sandbar shark off Hawaii has been previously investigated. However, dissimilarities in some parameter estimates such as growth rates and age-at-maturity exist. Using data from captive sharks, Wass (1973) reported very fast growth rates ($k = 0.4015$ year^{-1} for males, $k = 0.3745$ year^{-1} for females) and indicated that sandbar sharks in Hawaii reached maturity at 3 years of age. Conversely, growth rate estimates obtained from tooth replacement calculations suggested maturity occurred at 13 years of age. The discrepancy between the two methods may be due to the use of captive animals, which may not be indicative of growth rates in the Hawaiian wild population. Furthermore, the results based on tooth replacement rates are comparable to other sandbar shark populations around the world. For example, sandbar sharks in the Northwest Atlantic Ocean attain maturity between 12 and 15 years of age (Casey et al. 1985; Sminkey and Musick 1996). Additionally, Joung et al. (2004) estimated age-at-maturity to be between 7.5 and 8.2 years of age for females and 8.2 years of age for males for sandbar sharks in Taiwanese waters. Given the variability in growth estimates calculated by Wass (1973), we used vertebral centra from wild sandbar sharks to re-estimate growth-rates in the Hawaiian population.

Materials and methods

Sample collection and preparation

We collected sandbar sharks using demersal longlines outside of Kaneohe Bay, Hawaii at depths between 70 and 100 m (Fig. 1). Longlines were set perpendicular to the shoreline and baited with sardines, *Sardinops sagax*, chub mackerel, *Scomber japonicus*, yellowfin tuna, *Thunnus albacares*, skipjack tuna, *Katsuwonus pelamis*, barracuda, *Sphyraena barracuda*, and mahi-mahi, *Coryphaena hippurus*. Gangions consisted of a stainless-steel snap-clip attached to 3 m of monofilament followed by a 1-m stainless-steel leader that was attached to a circle hook. We used two sizes of gangions. Smaller gangions included 250 kg monofilament, 1.6 mm stainless-steel leaders, and 14/0 galvanized circle hooks, whereas large gangions included 410 kg monofilament, 2.2 mm stainless-steel leaders, and 18/0 stainless-steel circle hooks. Hooks were allowed to fish for 3 h before being retrieved. Captured sharks were landed, measured, and euthanized if needed for samples. At least five male and female sharks within each 5 cm size class between 45 cm PCL and 150 cm PCL were euthanized and vertebral samples were removed from below the first dorsal fin. Once the required vertebral samples had been collected, subsequently caught sharks were injected with oxytetracycline (OTC, 25 mg kg body weight^{-1}), tagged with Hallprint dart tags, and released for age-validation purposes.

Vertebral samples were frozen after collection. We cleaned the thawed vertebrae of excess tissue and stored five centra from each specimen in 75% ETOH. Using a Beuhler Isomet rotary diamond saw, we sectioned vertebral centra sagitally through the focus of the centrum. Sections were then dried for 24 h. Once dry, samples were mounted on a microscope slide via mounting medium. Samples were polished using a Metaserv 2000 grinder polisher until light was readily transmitted through the samples and rings were distinguishable using a dissection microscope. Vertebrae of sharks that were recaptured and sacrificed were examined under ultraviolet light for OTC marks.

Maturity

We determined maturity of males and females using macroscopic methods. Male sharks were classified as mature if claspers were deemed fully calcified (i.e. hard) and could be rotated forward (Clark and von Schmidt 1965; Driggers et al. 2004). Females were classified as mature if they were pregnant or had enlarged oviducal glands and well developed uteri (Castro 1993).

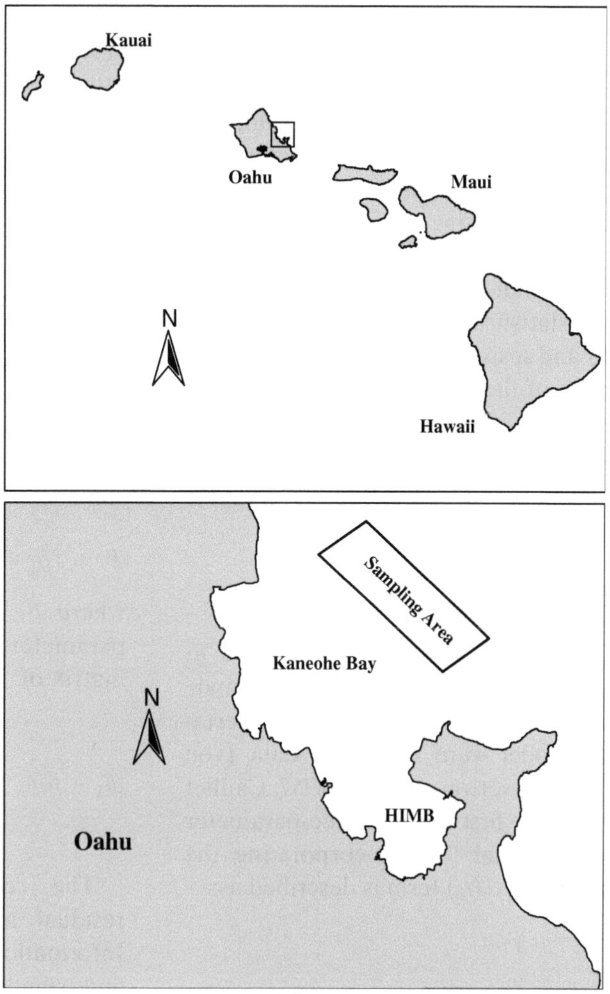

Fig. 1 Sampling area on the windward coast of Oahu. All sharks used for age and growth were captured within the rectangle noted as sampling area

Age assignment and validation

The rings or annuli counted for age estimates were defined as a band pair consisting of an opaque zone combined with a wider translucent zone in the intermedialia, which continued on to the corpus calcareum (Casey et al. 1985; Sminkey and Musick 1995). The birthmark was determined as the first band that intersected the inflection of the corpus calcareum. If annuli were not readily distinguishable, samples were stained with a 0.01% crystal violet solution to enhance readability.

Mounted vertebral sections were examined for age using a dissecting microscope and the Optimas video imaging system. The principal author and another reader conducted multiple blind readings of all vertebrae. Once all vertebrae were read, Hoenig's (1995) and Evans and Hoenig's (1998) tests of symmetry were conducted to test the hypothesis that age estimates between readers did not differ significantly and were due to random error.

Age estimates for vertebrae that were not consistent between readers were reexamined by both readers until a consensus was reached. The consensus estimate was used in the final analysis. If a consensus age estimate could not be reached the sample was removed from the study (Cailliet and Goldman 2004).

A relative marginal increment analysis was conducted to determine periodicity of ring formation (Branstetter and Musick 1994; Natanson et al. 1995; Goldman and Musick 2006). The Marginal Increment Ratio (MIR) is defined as:

$$\mathrm{MIR} = (\mathrm{VR} - R_n)/(R_n - R_{n-1}),$$

where VR = centrum radius, R_n = distance from the focus to the last complete narrow band, and R_{n-1} = the distance to the penultimate complete narrow band. All measurements were made along the corpus calcareum using an Optimas imaging system. We plotted monthly mean MIR values to determine the periodicity of band pair formation and tested for statistically significant differences for all months and seasons via one-way analysis of variance. Young-of-the-year sharks were not used in MIR analyses as they have no fully formed rings. We used centrum radius measurements to estimate the relationship between radius and precaudal length.

Growth models

Following Carlson and Baremore (2005), we fit four growth models to length-at-age data for male and female sharks. Two forms of the von Bertalanffy growth model were fit to the data (von Bertalanffy 1938; Beverton and Holt 1957; Cailliet et al. this issue). The first form, a three-parameter von Bertalanffy model (VB) incorporating the theoretical age-at-zero (t_0) term is described as:

$$L_t = L_\infty(1 - \mathrm{e}^{-k(t-t_0)}),$$

where t_0 = age or time when length theoretically equals zero. The second form of the model (VB2) used the length-at-birth intercept rather than a theoretical age at zero length and is described as:

$$L_t = L_\infty - (L_\infty - L_0)\mathrm{e}^{-kt},$$

where L_t = length at time t, L_∞ = theoretical asymptotic length, k = coefficient of growth, and L_0 = mean length-at-birth (47 cm PCL). Length-at-birth was estimated from observed at-term embryos and free-swimming young-of-the-year during this study as well as previously reported data by Wass (1973). A modified version of the Gompertz growth model (Ricker 1975) was also fitted to the data:

$$L_t = L_0(\mathrm{e}^{G(1-\mathrm{e}(kt))}),$$

where $G = \ln(L_\infty/L_0)$ (vonBertalanffy 1938). These lengths were determined from empirical data from this study and confirmed by Wass (1973). Finally, a logistic model (Ricker 1975) was fitted to the data:

$$L_t = L_\infty/(1 + \mathrm{e}^{-k(t-t_0)}).$$

All model parameters were estimated using the Marquardt least-squares nonlinear (NLIN) procedure in SAS statistical software (SAS V.9, SAS Institute, Inc). Growth parameter estimates for males and females were compared for statistically significant differences following the methods of Bernard (1981), Quinn and Deriso (1999), and Wang and Milton (2000) with a generalized T^2-statistic:

$$T^2 = (\beta_1 - \beta_2)'V^{-1}(\beta_1 - \beta_2),$$

where β_1 and β_2 are vectors of growth-model parameter estimates, V is the variance–covariance matrix of

$$[\beta_1 - \beta_2] : \beta_1 - \beta_2 = \begin{bmatrix} L_{\infty(1)} - L_{\infty(2)} \\ k_{(1)} - k_{(2)} \\ t_{0(1)} - t_{0(2)} \end{bmatrix}.$$

The coefficient of determination (r^2), residual mean square error (MSE), Akaike's Information criteria (AIC) (Akaike 1973), and standard deviation of the residuals were used as measures of goodness-of-fit for all models. A Shapiro–Wilks test and a normal probability plot of the residuals were used to test for normality, excessive skew or excessive kurtosis using the univariate procedure in SAS statistical software (SAS V.9, SAS Institute, Inc).

Results

Sample collection

We captured a total of 320 sandbar sharks as a part of this study. Vertebral samples were obtained from 194 sharks (Fig. 2) while the remainder of sharks were measured, tagged, injected with OTC and released. Size ranges for females and males captured were 46–147 cm PCL

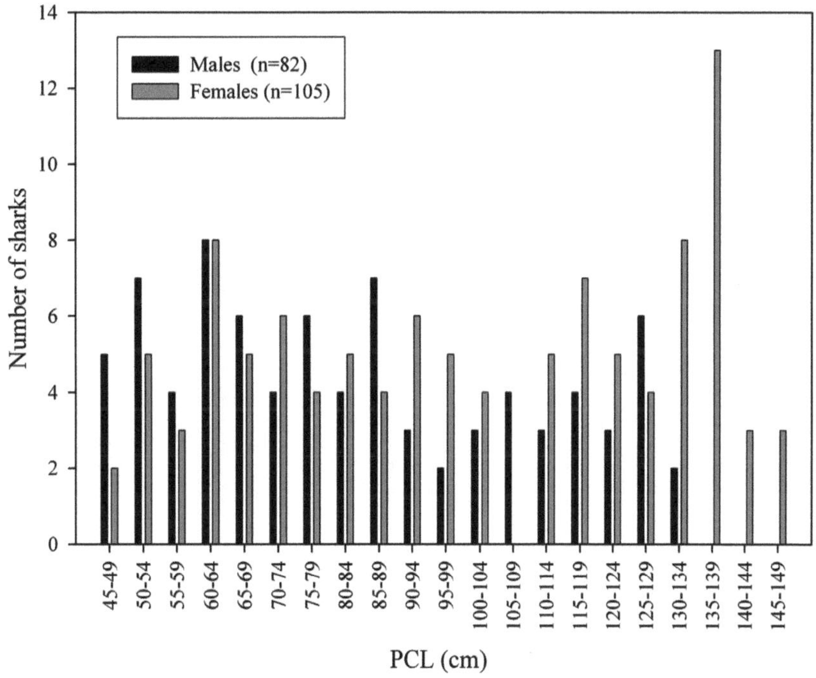

Fig. 2 Size frequency of sandbar sharks used for age and growth in this study

and 46–132 cm PCL, respectively. Sharks were captured in all months except March.

Vertebral radius and length analysis

The relationship between vertebral radius and shark length (PCL = 12.0VR + 8.12) significantly correlated (n = 148, r^2 = 0.97, Fig. 2). No significant difference between males and females was found (Z = 0.109, P = 0.55), thus vertebral radius measurements were combined to estimate the regression.

MIR analysis

For combined sexes, MIR analysis suggests a single growth band pair is formed annually with the narrow opaque band being formed in winter months. Marginal increment ratios increased from spring to winter (Figs. 3, 4). Differences in monthly marginal increment ratios were not significant between all months in which samples were collected, (ANOVA, n = 120, F = 0.64, df = 8, P = 0.74, Fig. 4). The periodicity of band formation was also supported from one recaptured shark. This shark measured 60 cm PCL at its release on 26 June 2004 and 62 cm PCL at its recapture on 20 January 2005. An OTC stained opaque growth band was present at the very margin of the centra, suggesting ring formation had recently begun during the winter months.

Age estimation

After our initial readings, we reached consensus age estimates for 187 (105 females, 82 males) samples. Consensus could not be reached for seven samples which were removed from all analyses. Agreement between blind readings of readers was reached 43.1% of the time. Reader agreement was 71.2% within one band and 84.3% within two bands. We could not reject the hypothesis of symmetry between ages assigned by both readers (X^2 = 38.64, df = 39, P = 0.488, Hoenig 1995) indicating that differences between readers were due to random error.

Growth models

We found significant differences between male and female von Bertalanffy growth-model parameters, when using the form of the model that incorporated the theoretical age at zero length ($T^2 = 8.48 > T_0^2 = 8.11$, $P < 0.05$). Therefore, all models were subsequently fitted to male

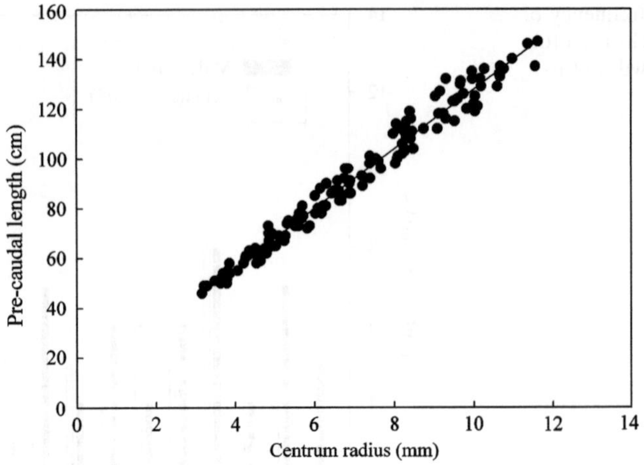

Fig. 3 Regression of pre-caudal length and centrum radius ($r^2 = 0.97$, $n = 148$) for males and females combined

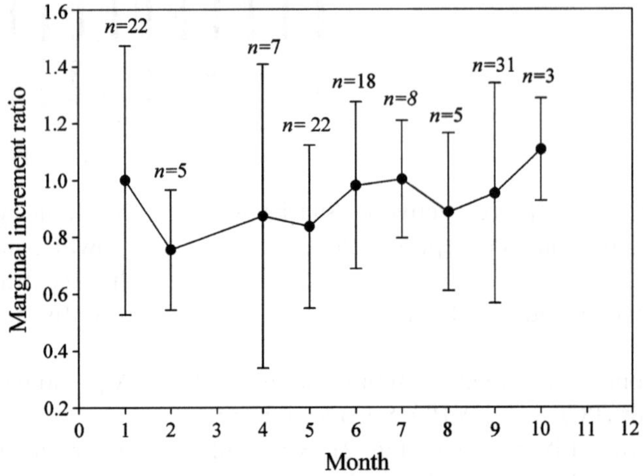

Fig. 4 Mean marginal increment ratio monthly values and standard deviation of the monthly means for both sexes combined

and female length-at-age data for each sex separately.

All growth models fitted to observed length-at-age data were significant ($P < 0.0001$, Table 1, Figs. 5, 6). Coefficients of determination were all greater than 0.94. The residuals of all models were normally distributed and no excessive skew or kurtosis was detected.

Table 1 Estimates of model parameters and goodness of fit statistics for models fitted to length-at-age data for male and female sandbar sharks.

	Model	L_∞ (cm PCL)	k (year^{-1})	t_0	L_0 (cm PCL)	AIC	r^2	MSE	SD of residuals
Males	VB	151.1(± 16.3)	0.09(± 0.03)	−5.01(± 1.02)	na	2639	0.951	33.75	5.74
	VB2	138.5(± 9.7)	0.12(± 0.02)	na	47	3381	0.994	42.74	6.31
	Gompertz	130.4(± 6.6)	0.19(± 0.03)	na	47	3780	0.993	47.8	6.58
	Logistic	134.3(± 7.5)	0.19(± 0.03)	1.98(± 0.66)	na	2740	0.995	35.05	5.85
Females	VB	164.9 (± 14.9)	0.08(± 0.02)	−5.26(± 1.08)	na	5307	0.943	51.97	7.14
	VB2	152.8(± 8.8)	0.10(± 0.02)	na	47	6406	0.994	62.16	7.72
	Gompertz	143.5(± 5.6)	0.17(± 0.02)	na	47	6976	0.995	67.69	7.98
	Logistic	146.4(± 6.4)	0.17(± 0.02)	2.66(± 0.61)	na	5278	0.995	51.69	7.12

All length values are for pre-caudal length (PCL) in cm. Values in parentheses are 95% confidence intervals. (VB = von Bertalanffy three-parameter model with t_0 term, VB2 = von Bertalanffy two-parameter model with empirical length-at-birth (L_0), L_0 = average measured length-at-birth used in VB2 and Gompertz growth models only, AIC = Aikake's Information Criteria, MSE = Mean Square Error, SD = standard deviation, na = not applicable)

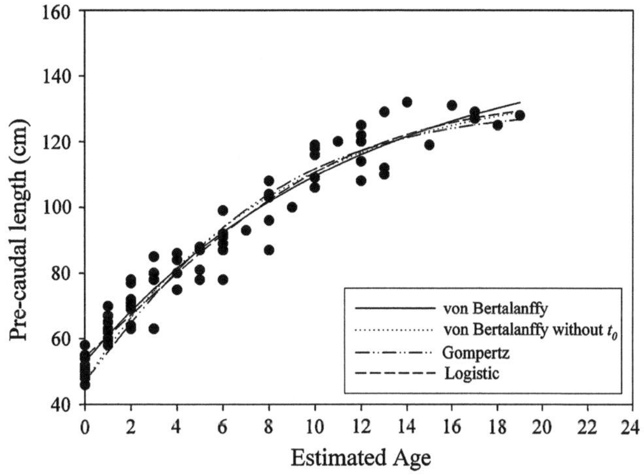

Fig. 5 Length-at-age estimates and growth models fitted to data for male sandbar sharks (n = 81)

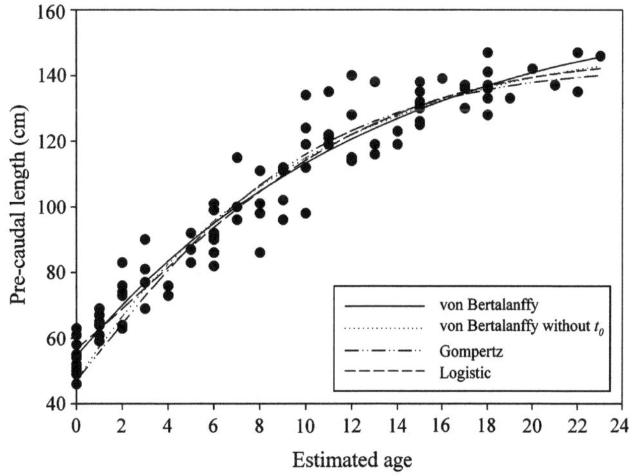

Fig. 6 Length-at-age data and fit of growth models to the data for female sandbar sharks (n = 105)

The von Bertalanffy growth model that included the theoretical t_0 term provided the best statistical fit to the observed size-at-age data for male sandbar sharks. This model had the lowest residual mean square error (MSE), and the lowest AIC values (Table 1). The von Bertalanffy growth model that included the theoretical t_0 term and the logistic model provided the best statistical fits to the observed size-at-age data for female sandbar sharks. These models had the lowest AIC and MSE values, respectively.

The three-parameter von Bertalanffy model produced the highest estimates for asymptotic maximum length for both males (151.1 cm PCL) and females (164.9 cm PCL, Table 1). The Gompertz model produced the lowest estimates for asymptotic length for males (130.4 cm PCL) and females (143.5 cm PCL). Observed maximum lengths for males (132 cm PCL) and females (147 cm PCL) fell within the 95% confidence intervals of all models. The three-parameter von Bertalanffy model produced the lowest estimates of the growth coefficient (k) for males (0.09 year^{-1}) and females (0.08 year^{-1}). The Gompertz and logistic models produced the highest estimates of the growth coefficient for males (0.19 year^{-1}) and females (0.17 year^{-1}).

Six sharks were recaptured over the time period of this study. Time at liberty ranged from 7 to 526 days. Lengths of recaptured sharks ranged from 60 to 77 cm PCL at release. The average growth rate was 6.97 cm year^{-1} for the five sharks that were at liberty for more than 100 days. This value agreed with growth rates for sharks in this size range estimated from vertebral analyses.

Mean length-at-age estimates determined from the three-parameter von Bertalanffy model differed between males and females (Table 2). Females were generally larger at a given age and attained older ages than males. Males attained maturity between 100 and 110 cm PCL and females attained maturity between 110 and 120 cm PCL. These sizes correspond to 8 and 10 years of age for males and females, respectively, as determined by the two-parameter von Bertalanffy model. Maximum observed age was 19 years for male and 23 years for females.

Discussion

Wass (1973) estimated maturity to occur at 3 years of age and produced k estimates of 0.4015 year^{-1} for males and 0.3745 year^{-1} for females by observing the growth rates of captive sharks. Wass also estimated maturity to occur at 10.2 and 13.1 years of age for males and females, respectively, using tooth replacement methodology and hypothesized the true value for the wild population to lie somewhere between the estimates from both methods used in his study. Using vertebral analyses, our estimated ages-at-maturity determined from the two-parameter von Bertalanffy models were 8 and 10 years of age for males and females, respectively, and produced k estimates of 0.12 and 0.10 year^{-1} for males and females, respectively.

Although all models fit the data well; statistically, the three-parameter von Bertalanffy model described the male size-at-age data better than the other three models. This model overestimated observed maximum size, but observed maximum size fell within the confidence intervals for this model. Size at birth was also overestimated by this model (47 cm PCL observed vs. 52 cm PCL predicted). The overestimate of size-at-birth could be due to the lack of newborns sampled during this study. The smallest male sampled was 46 cm PCL and only two females under 50 cm PCL were sampled. Despite the statistical ranking of the two-parameter von Bertalanffy model amongst the other models we feel the two-parameter von Bertalanffy growth model should be used

Table 2 Mean size at age for male and female sandbar sharks

	0	1	2	3	4	5	6	7	8	9	10	11	12	13	14	15	16	17	18	19	20	21	22	23
Females																								
PCL (cm)	53.5	63.9	72	79.6	74.5	88.5	90.9	103.7	99	105.3	117.4	124.3	124.3	124.3	121.7	131.1	139.0	135.0	137.0	133.0	142.0	137.0	141.0	146.0
SD	5.3	3.4	6.1	7.6	2.1	4.4	6.5	10.0	10.3	7.6	13.5	7.3	12.3	11.9	2.3	4.3	na	3.4	6	na	na	na	8.5	na
n	11	8	11	5	2	4	8	3	4	4	5	4	4	3	3	8	1	4	7	1	1	1	2	1
Males																								
PCL (cm)	50.9	63.4	70.3	76.5	81.3	83.5	89.3	93	99.6	100	114	120	118	117	132	119	131	128	125	128				
SD	3.12	3.52	5.05	9.47	4.86	4.8	6.89	na	8.26	na	5.77	na	6.8	10.4	na	na	na	1.41	na	na				
n	14	12	9	4	4	4	6	1	5	1	5	1	5	3	1	1	1	2	1	1				

All lengths are PCL (cm), SD = standard deviation, and na = not applicable

when describing growth of the male sandbar shark in Hawaii. This model provided a biologically realistic fit to the observed data. Predicted maximum asymptotic length agreed closely to observed data (132 cm PCL observed and 138.5 cm PCL predicted, Table 1) and it incorporated observed size at birth.

All models fit the female size-at-age data well. Statistically, the logistic and three-parameter von Bertalanffy models fit the data better than the other models. The asymptotic length estimate from the logistic model agreed with observed maximum size (147 cm PCL). The logistic model overestimated the size-at-birth (47 cm PCL observed vs. 56 cm predicted). The three-parameter von Bertalanffy model overestimated asymptotic length and size-at-birth. As with the male data, we feel the two-parameter von Bertalanffy growth model provided a more biologically realistic fit to the female data and should be employed when describing the growth of the female sandbar shark in Hawaii.

The Gompertz models fit the data well for both sexes, but underestimated the maximum asymptotic length. This inherently increased the rate at which asymptotic length was approached and, therefore, these models provided the highest growth coefficients. The logistic models for both males and females provided high growth coefficient values due to the overestimation of the size-at-birth and estimate of asymptotic maximum size. The combination of these factors effectively increased the rate at which asymptotic length was approached.

The three-parameter von Bertalanffy growth model overestimated both size-at-birth and asymptotic maximum length for both sexes. Both estimates were unrealistic and caused the estimated growth coefficients to be the lowest amongst all models.

Given the variability of growth rates within and between populations it is imperative to conduct rigorous examination of all possible methods to describe length-at-age data. As illustrated in this study, models fitted to size-at-age data can produce variable estimates of growth parameters. In this study, the growth coefficient estimates ranged from 0.09–0.19 year^{-1} for males to 0.08–0.17 year^{-1} for females between all models. Growth coefficients are often used in demographic analyses for stock assessment purposes. Researchers must consider statistical results and observed biological data when determining which model provides the best fit to the data. Often the two viewpoints do not agree, as in this study. Although the three-parameter von Bertalanffy model, which is often the only model used to describe the growth of fishes, provided the best statistical fit to these data, it produced unrealistic asymptotic lengths and sizes-at-birth. Thus, we suggest the use of the growth parameters estimated by the two-parameter von Bertalanffy model.

Age and growth of sandbar sharks in Hawaii differ from other populations that have been studied. Growth coefficients in Hawaii (K = 0.09–0.19 year^{-1} for males and 0.08–0.17 year^{-1} for females) are much higher than those reported for the northwest Atlantic Ocean (k = 0.057 year^{-1}, combined sexes, Sminkey and Musick 1995). Hawaiian sandbar sharks obtain smaller maximum sizes (132 cm PCL for males and 147 cm PCL for females, observed) and reach maturity earlier (8 years for males and 10 years for females) than those in the northwest Atlantic (172 cm PCL, observed and 15 years at maturity, sexes combined). Sandbar sharks in Taiwanese waters also reach larger maximum sizes (209 cm TL for males and 219 cm TL for females, observed—Joung et al. 2004) than those in Hawaii (179 cm TL for males and 196 cm TL for females, observed). Joung et al. (2004) estimated the growth coefficient for sandbar sharks in Taiwanese waters to be k = 0.17 year^{-1} for both sexes combined and the onset of maturity to occur at 8 years of age for both sexes. The youngest sharks sampled in the Taiwanese study were 4 years-of-age which led to estimates of length-at-birth (80.8–85.8 cm TL) that were much larger than observed (60–65 cm TL) for this population. Therefore, the estimated growth coefficients may have been overestimated by the use of the three-parameter von Bertalanffy model, which incorporated the t_0 parameter, and the age-at-maturity underestimated. It is likely that the life history parameters of the Taiwanese population are intermediate between the Hawaiian and northwest Atlantic populations.

Vertebrae from larger sharks were more difficult to read due to decreased band pair widths near the margin of the vertebrae. This contributed to the increased variability in age estimates between readers as shark size increased. This also contributed to discrepancies in mean size-at-age estimates for older ages of male and female sandbar sharks (Table 2). The low sample sizes for older male sharks also contributed to these discrepancies. Although there were difficulties in assigning age estimates to specimens, the oldest sharks estimated blind consensus was 22 years for females and 12 years for males.

Our ageing methodology was supported via OTC mark recapture, but tagging and OTC validation has only been shown for one at liberty shark under 78 cm PCL. A more robust tag recapture data set is needed to obtain empirical data on the growth rates of sandbar sharks in Hawaii and to investigate the long-term movements of sharks in Hawaii. During the period of the study seven tagged sharks were recaptured. Time at liberty ranged from 7 to 526 days. Growth of these sharks during time at liberty supported our growth models for sharks between 60 and 83 cm PCL. All recaptured sharks were under 100 cm PCL and thus do not offer any support of our models for maturing or mature sandbar sharks in Hawaii.

The age and growth estimates for this population of sandbar sharks in Hawaii supports the generalization that sharks are slow growing and have low reproductive output. Currently, a legal fishery does not exist for this population. Should a fishery open, caution in management of the fishery should be exercised. The life history parameters of this population, as with other populations of slow growing, late maturing and low fecundity fishes, render it extremely vulnerable to overfishing even at low levels of fishing effort (Musick 1999).

Acknowledgements We thank NOAA/NMFS for funding to the National Shark Research Consortium from which this study was supported. We thank the anonymous reviewers for their constructive comments. We thank John Carlson and Ken Goldman for their comments and for editing this volume. We thank the Hawaii Institute of Marine Biology for graciously allowing us the use of vessels and lab space. We would be remiss not to thank the students, staff, and volunteers that assisted in sample collection and preparation, Kim Holland, Toby Daly-Engel, Nick Whitney, Yannis Papastamatiou, RaeMarie Johnson, Kanesa Duncan, Dave Itano, Amanda Southwood, Amy Long, Todd Gedamke, Christina Conrath, and Demetria Christo. This paper is Contribution No. 2709 of the Virginia Institute of Marine Science, The College of William and Mary. This is Contribution No. 1209 of the Hawaii Institute of Marine Biology.

References

Akaike H (1973) Information theory as an extension of the maximum likelihood principle. In: Petrov BN, Csaki F (eds) Second international symposium on information theory. Akademiai Kiado, Budapest, pp 267–281

Bernard DR (1981) Multivariate Analysis as a means of comparing growth in fish. Can J Fish Aquat Sci 38:233–236

von Bertalanffy L (1938) A quantitative theory of organic growth (inquiries on growth laws. II). Hum Biol 10:181–213

Beverton RJH, Holt SJ (1957) On the dynamics of exploited fish populations. Ministry of Agriculture, Fisheries and Food Fishery Investigation Series II XIX, 533 pp

Bigelow HB, Schroeder WC (1948) Sharks. In: Tee-Van CMBJ, Hildebrand SF, Parr AE, Schroeder WC (eds) Fishes of the Western North Atlantic. Part 1. vol 1. Mem. Sears Foundation for Marine Research, Yale Univ., New Haven, CT

Branstetter S, Musick JA (1994) Age and growth estimates for the sand tiger in the northwestern Atlantic Ocean. Trans Am Fish Soc 123:242–254

Cailliet GM, Goldman KJ (2004) Age determination and validation in chondrichthyan fishes. In: Carrier J, Musick JA, Heithaus M (eds) The biology of sharks and their relatives. CRC Press, Boca Raton, FL, pp 399–447

Cailliet GM, Smith WD, Mollet HF, Goldman KJ (this issue) Age and growth studies of chondrichthyan fishes:an overview stressing terminology, verification, validation, and growth function fitting. Environ Biol Fish Special Volume: Age and growth of chondrichthyan fishes: new methods, techniques, and analyses

Carlson JK, Baremore IE (2005) Growth dynamics of the spinner shark (*Carcharhinus brevipinna*) off the United States southeast and Gulf of Mexico coasts: a comparison of methods. Fish Bull 103:280–291

Casey JG, Pratt HL Jr, Stillwell CE (1985) Age and growth of the sandbar shark (*Carcharhinus plumbeus*) from the Western North Atlantic. Can J Fish Aquat Sci 42:963–975

Castro JI (1993) The biology of the finetooth shark, Carcharhinus isodon. Environ Biol Fishes 36:219–232

Clark E, von Schmidt K (1965) Sharks of the Central Gulf coast of Florida. Bull Mar Sci 15:13–83

Compagno LJV (1984) FAO species catalogues. Vol. 4. Sharks of the World. An annotated and illustrated catalogue of shark species known to date. Part 2. Carcharhiniformes. FAO fisheries synopsis, (125)4, Pt. 2, pp 251–655

Driggers WB III, Oakley DA, Ulrich G, Carlson JK, Cullum BJ, Dean JM (2004) Reproductive biology of *Carcharhinus acronotus* in the coastal waters of South Carolina. J Fish Biol 64:1540–1551

Evans GT, Hoenig JM (1998) Testing and viewing symmetry in contingency tables, with application to readers of fish ages. Biometrics 54:620–629

Goldman KJ, Musick JA (2006) Growth and maturity of salmon sharks in the eastern and weatern North Pacific, with comments on back-calculation methods. Fish Bull 104:278–292

Hoenig JM (1995) Analysing differences between two age determination methods by tests of symmetry. Can J Fish Aquat Sci 52:364–368

Joung JS, Liao YY, Chen CT (2004) Age and growth of the sandbar shark, *Carcharhinus plumbeus*, in north-eastern Taiwan waters. Fish Res 70:83–96

Lawler EF (1976) The biology of the sandbar shark, *Carcharhinus plumbeus*, (Nardo, 1827) in the lower Chesapeake Bay and adjacent waters. Virginia Institute of Marine Science 1976:49

Musick JA (1999) Ecology and conservation of long-lived marine animals. In: Musick JA (ed) Life in the slow lane: ecology and conservation of long lived marine animals. Symposium 23. American Fisheries Society, Bethesda, Maryland, pp 1–10

Natanson LJ, Casey JG , Kohler NE (1995) Age and growth estimates of the dusky shark, *Carcharhinus obscurus*, in the western North Atlantic Ocean. Fish Bull 93:116–126

Quinn JT II, Deriso RB (1999) Quantitative fish dynamics. Oxford University Press, Oxford, England

Ricker WE (1975) Computation and interpretation of biological statistics of fish populations. Bull Fish Res Board Canada 191:1–382

Sminkey TR, Musick JA (1995) Age and growth of the sandbar shark, *Carcharhinus plumbeus*, before and after population depletion. Copeia 4:871–883

Sminkcy TR, Musick JA (1996) Dcmographic analysis of the sandbar shark, *Carcharhinus plumbeus*, in the western North Atlantic. Fish Bull 94:341–347

Springer S (1960) Natural history of the sandbar shark, Eulamia milberti. U S Fish Wildl Fish Bull 61:38

Wang YG, Milton DA (2000) On comparison of growth curves: How do we test whether growth rates differ? Fish Bull 98:874–880

Wass RC (1973) Size, growth and reproduction of the sandbar shark, Carcharhinus milberti, in Hawaii. Pac Sci 27:305–318

Environ Biol Fish (2006) 77:241–252
DOI 10.1007/s10641-006-9128-y

SI-ELASMOBRANCHS

A re-examination of the age and growth of sand tiger sharks, *Carcharias taurus*, in the western North Atlantic: the importance of ageing protocols and use of multiple back-calculation techniques

Kenneth J. Goldman · Steven Branstetter · John A. Musick

Received: 12 June 2006 / Accepted: 3 July 2006 / Published online: 6 September 2006
© Springer Science+Business Media B.V. 2006

Abstract Age and growth estimates for sand tiger sharks, *Carcharias taurus*, in the western North Atlantic were derived from 96 vertebral centra collected from sharks ranging from 94 to 277 cm total length (TL), and compared to previously published age and growth data. The oldest female and male sand tiger sharks aged in this study were 17 and 15 years of age, respectively. von Bertalanffy growth parameters derived from vertebral length-at-age data are $L_\infty = 295.8$ cm TL, $k = 0.11$ year^{-1}, and $t_0 = -4.2$ years for females, and $L_\infty = 249.5$ cm TL, $k = 0.16$ year^{-1}, and $t_0 = -3.4$ years for males. Sexual maturity is estimated to be 9–10 years for females and 6–7 years for males. Weight-to-length relationships determined for female and male sand tiger sharks in the western North Atlantic are; $W = 1.3 \times 10^{-4} \times L^{2.4}$ ($r^2 = 0.84$, $n = 55$) and $W = 9.0 \times 10^{-5} \times L^{2.5}$ ($r^2 = 0.84$, $n = 47$), respectively, and $7.9 \times 10^{-5} \times L^{2.5}$ ($r^2 = 0.84$) for the sexes combined. Our results show sand tigers possess a slower rate of growth than previously thought. This information is crucial for accurately assessing this population's ability to recover, and further justifies the need for this species to be fully protected.

Keywords Elasmobranch · Life history · Growth rate · Back-calculation · Reproduction

K. J. Goldman (✉)
Alaska Department of Fish and Game, 3298 Douglas Pl., Homer, AK 99603, USA
e-mail: ken_goldman@fishgame.state.ak.us

S. Branstetter
National Marine Fisheries Service, 9721 Executive Center Drive North, Koger Building, Suite 201, St Petersburg, FL 33712, USA

J. A. Musick
College of William and Mary, Virginia Institute of Marine Science, School of Marine Science, Route 1208, Greate Road, Gloucester Point, VA 23062, USA

Introduction

The sand tiger shark, *Carcharias taurus*, is a large coastal species that inhabits subtropical and temperate waters of the Atlantic, Indian, and western Pacific Oceans, as well as the Mediterranean Sea (Gilmore et al. 1983; Compagno 1984; Pollard et al. 1996). The sand tiger shark is fished or caught as bycatch in all areas it is found, but is of variable importance regionally (Pollard and Smith 2005). In the western North Atlantic Ocean, sand tiger sharks are exposed to several fisheries due to their highly migratory nature (Casey and Kohler 1990; Musick et al. 1993). As a result, the population has been depleted by an estimated 80–90% since the mid 1970s (Musick et al. 1993, 2000). To prevent further decline of sand tiger sharks in the western North Atlantic, the sand tiger shark was listed as a prohibited species in the amendment to the National Marine Fisheries Service (NMFS) Fishery Management Plan for Atlantic sharks in 1997 (NMFS 1999). They also have fully protected status in Australian waters and their capture in South African waters is being phased out (Pollard et al. 1996; Anonymous 2002). Although the western North Atlantic population may have stabilized since being listed as prohibited, recovery is not yet apparent (Musick et al. 2000).

There is still some uncertainty in the life-history parameters for sand tiger shark in the western North Atlantic Ocean. Whereas gestation and embryonic growth are well documented in sand tiger sharks, their reproductive periodicity has been a source of some contention. Gilmore (1993) stated that sand tiger sharks reproduce annually in the western North Atlantic Ocean and Gordon (1993) believed they may mate annually in Australian waters. Alternatively, Cliff (1989), and Branstetter and Musick (1994) presented evidence supporting a 2-year reproductive cycle for sand tiger sharks in South African waters and the western North Atlantic, respectively. However, successful captive reproduction over the past 10 years lends strong support to a 2-year reproductive cycle hypothesis (Henningsen et al. 2004). In terms of age and growth, Branstetter and Musick (1994) stated that sand tiger sharks may reach an age of 30–35 years and gave the following von Bertalanffy growth parameters for in the western North Atlantic: $L_\infty = 323.0$ cm PCL, $k = 0.14$ year^{-1}, and $t_0 = -2.6$ years for females; $L_\infty = 301.0$ cm PCL, $k = 0.17$ year^{-1}, and $t_0 = -2.3$ years for males; and $L_\infty = 321.0$ cm PCL, $k = 0.14$ year^{-1}, and $t_0 = -2.6$ for sexes combined. This was based on their estimation that this species forms two pairs of growth bands annually in the vertebral centra. These authors also stated a caveat about their uncertainty of that estimation. A re-calculation of the life-history parameters from Branstetter and Musick's (1994) mean back-calculated lengths-at-age was conducted assuming annual formation of a single pair of growth bands (Goldman 1998), which resulted in the following von Bertalanffy growth parameters: $L_\infty = 323.0$ cm PCL, $k = 0.07$ year^{-1}, and $t_0 = -5.1$ for females; and $L_\infty = 302.0$ cm PCL, $k = 0.08$ year^{-1}, and $t_0 = -4.5$ for males; and $L_\infty = 322.0$ cm PCL, $k = 0.07$ year^{-1}, and $t_0 = -5.2$ for sexes combined. The difference in these life-history parameter estimations, and their ramifications for management and conservation of this species (Musick et al. 1993, 2000), was the major reason for our re-examination of sand tiger age and growth. Herein, we report our findings, comment on the importance of ageing protocols and use of multiple back-calculation techniques for chondrichthyan fishes, and briefly discuss the reproductive periodicity of this species.

Materials and methods

Sand tiger shark vertebrae ($n = 96$) were obtained by the Virginia Institute of Marine Science (VIMS) ($n = 55$) and from the NMFS Narragansett, RI, laboratory ($n = 41$). (Some samples from each institute were tournament or sport caught.) Twenty-five samples (from the VIMS survey) previously used for analysis by Branstetter and Musick (1994) were re-examined for this study. Vertebral samples and weights of sand tiger sharks used in this study from the VIMS survey were taken between 1980 and 2001; those obtained from NMFS were taken between 1963 and 1991. Vertebral samples were not obtained from all animals that were measured and weighed (over this nearly 40-year period), hence the discrepancy in sample sizes.

Sand tiger sharks taken by VIMS were measured on a straight line from the tip of the snout to the tip of the tail while in a natural position, and precaudal, fork, and total length (TL) (PCL, FL, and TL) were recorded along with sex and weight (when possible). A 20–25 cm section of vertebrae was removed from the area midway between the first dorsal fin and the gills, and stored frozen. Samples provided by NMFS included at least one of the necessary measurements and the date and location of capture. Weights were obtained (by VIMS and NMFS) from 102 sand tiger sharks between 1963 and 1991, including 55 females (95–272 cm TL), and 47 males (100.7–259 cm TL). Data were fitted to the power equation, $W = aL^b$ (using SigmaPlot, SPSS Inc. 2000), where W = weight (kg) and L = length (cm TL). A likelihood ratio test was used to determine whether differences between female and male weight–length parameters were significant or if a single set of parameters better described the data (Kimura 1980; Cerrato 1990; Quinn and Deriso 1999; Haddon 2001) (SAS Institute Inc. 1999). We use TL measurements throughout this paper in order to make direct comparisons with previously published data on sand tiger shark growth parameters (Branstetter and Musick 1994). Linear regression equations based on measurements taken by VIMS and the NMFS lab, were developed for converting TL to FL and PCL.

Vertebral samples were thawed, cleaned of excess tissue, separated into individual centra, and stored in 70% ethyl alcohol for at least 24 h. Centra were sagittally sectioned immediately adjacent to the focus and then cut again approximately 1.5 mm off-center using an Isomet rotary diamond saw (Buehler, 41, Lake Bluff, IL, USA). The sections were pressed between two pieces of Plexiglas (to prevent warping), air-dried for 24 h under a ventilation hood, and then mounted onto microscope slides. After drying, sections were polished with wet fine grit sand paper (320, 400, and 600) to approximately 0.5 mm and air-dried. Sections were viewed using a binocular dissecting microscope with transmitted light.

Centrum radius (CR) and distance to each ring were measured to the nearest 0.001 mm as a straight line from the central focus to the outer margin of the corpus calcareum (Fig. 1) using a dissecting video microscope with the Optimus image analysis system (Media Cybernetics 1999). TL was plotted against CR to determine the proportional relationship between somatic and vertebral growth.

A banding pattern was readily distinguishable in sectioned centra, with wide translucent bands separated by distinct narrow opaque bands. This pattern occurred on both arms of the corpus calcareum and the band pairs extended across the intermedialia. A notch occurring on the outside edge of the corpus calcareum accompanied the distinct narrow bands (Fig. 1) providing an additional ageing feature, particularly in sections where the cut excluded the radials of the intermedialia. Each pair of wide–narrow bands was considered an annual growth cycle; the narrow bands, hereafter referred to as "rings", were counted (Fig. 1). An angle change in the intermedialia and a ring on the corpus calcareum were present approximately 6–7 mm from the focus of each centrum and considered to represent a birthmark. The "pre-birth rings" reported by Branstetter and Musick (1994) were present in most specimens, but were not counted nor measured.

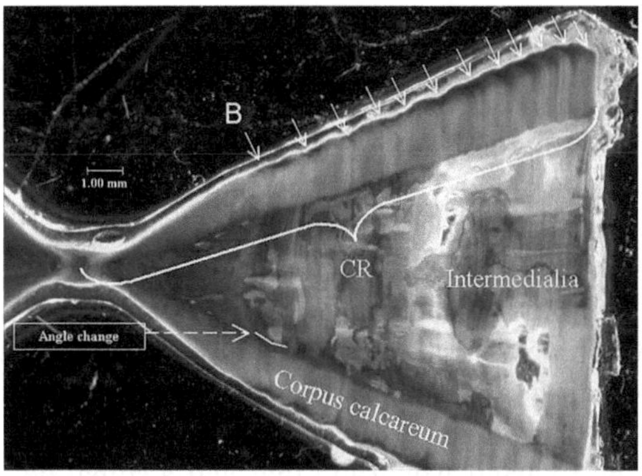

Fig. 1 Sagittal section of an 11-year-old sand tiger shark's vertebral centrum showing typical banding pattern. *CR* centrum radius. *Arrows* represent ages

Two readers independently aged all centra two times in blind, randomized trials. This allowed the calculation of within-reader precision, and between-reader precision twice. Percent agreement (PA = [no. agreed/no. read]100), and PA plus or minus 1 year (PA ± 1 year) were calculated for 10 cm length groups to test for precision (Cailliet and Goldman 2004; Goldman 2004; Goldman and Musick 2006; Cailliet et al., this issue). We used chi-square tests of symmetry to determine whether difference within and between readers were systematic (biased) or due to random error (Hoenig et al. 1995; Evans and Hoenig 1998; Campana 2001).

A relative marginal increment (RMI) analysis was used to verify the temporal periodicity of ring formation in the vertebrae. This is a standardized marginal increment analysis whereby the margin, or growth area of a centrum from the last narrow growth ring to the centrum edge, is divided by the width of the last fully formed growth increment (Branstetter and Musick 1994; Conrath et al. 2002; Cailliet et al., this issue). Resulting RMI values were compared to the month of capture. Age-zero animals were not included (they have no fully formed increments).

To assist in attempting to validate the periodicity of ring (annulus) formation, two male sand tiger sharks (152.5 and 157 cm TL) captured off of Cape May, NJ, USA, in 1998 for public display, were donated to this study, and injected with oxytetracycline (OTC) at a dose of 25 mg kg^{-1} body weight (Tanaka 1990; Gelsleichter et al. 1998). One was kept at Ripley's Aquarium, Myrtle Beach, SC, and the other at New England Aquarium, Boston, MA, USA. In 1999, an additional sand tiger (estimated at 164 cm TL) at the New England Aquarium that had a spinal deformity was offered for use in this study and fed OTC (that was injected into its food). Each individual was re-administered (injected or fed) OTC again approximately 1 year later, and two individuals were administered OTC a third time (Table 1). All three sharks were sacrificed using buffered MS222 in either December 2000 or January 2001.

The von Bertalanffy growth function was fit to the vertebral length-at-age data for sand tiger sharks with a nonlinear least squares regression algorithm ("nls" in S-Plus, Mathsoft Inc. 2000) to estimate parameters. The von Bertalanffy growth function is: $L_t = L_\infty \cdot [1 - \exp(-k(t - t_0))]$ where L_t = length at age "t," L_∞ = asymptotic or maximum length, k = the growth coefficient, and t_0 = age or time when length theoretically equals zero. Growth parameters were estimated for the sexes separately and combined. A likelihood ratio test was used to determine whether differences between female and male growth parameters were significant or if a single set of growth parameters better described the data (Kimura 1980; Quinn and Deriso 1999; Haddon 2001) (SAS Institute Inc. 1999).

Back-calculation is a method for describing the growth history of each individual sampled, and numerous variations in methodology exist (see Francis 1990 for a thorough review, and Goldman 2004 for description and application to elasmobranchs). Because our sample size was small and we did not have samples from all 12 months, lengths at previous ages were back-calculated from centra measurements for both sexes and fitted with the von Bertalanffy growth function. von Bertalanffy growth parameter estimates were then obtained from mean back-calculated length-at-age, and from a combination of back-calculated lengths-at-age and our sample data. The relationship between CR and TL for sand tiger sharks was investigated to determine the most appropriate method for back-calculating previous length-at-age. This is critical for obtaining accurate life-history parameter estimates from the von Bertalanffy growth function. To examine the statistical and biological accuracy of back-calculations relative to vertebral sample data (Goldman 2004; Cailliet and Goldman 2004; Goldman and Musick 2006), we compared several proportional back-calculation methods. Three different proportions methods were used and compared with our sample length-at-age data. First, we used the standard Dahl-Lea direct proportions method (Carlander 1969):

$$L_i = \left(\frac{L_c}{CR_c}\right) \cdot CR_i \qquad (1)$$

Table 1 Dates, aquarium, sex, lengths, and method of administering oxytetracycline (OTC) to three sand tiger sharks, *Carcharias taurus*

Shark #	Date	Aquarium	Sex	PCL	FL	TL	OTC method
OTC-1	15 October 1998	Ripley's	M	112	127	157	Injected
	15 October 1999			138	157	201	Injected
	18 January 2001			160	181	218.6	Euthanized
OTC-2	15 October 1998	New England	M	110	125	152.5	Injected
	4 November 1999			139	156	188	Injected
	7 October 2000			–	–	–	Fed
	13 December 2000			162	180	219	Euthanized
OTC-3	3 March 1999	New England	M	–	–	164	Fed
	21 November 1999			–	–	201	Fed
	13 October 2000			–	–		Fed
	13 December 2000			154	175	210	Euthanized

where L_i = length at ring "i", L_c = length at capture, CR_c = centrum radius at capture, and CR_i = centrum radius at ring "i". Next, we applied a modified version of the Dahl-Lea method that uses parameter estimates from the specific linear fit that described the TL-CR relationship (Francis 1990):

$$L_i = L_c \left[\frac{a + bCR_i}{a + bCR_c} \right] \quad (2)$$

where a and b are the linear fit parameter estimates.

Ricker (1992), Francis (1990), and Campana (1990) suggested that the point of origin of proportional back-calculations should be related to a biologically derived intercept (i.e., length at birth), so we also applied a "size-at-birth-modified" Fraser–Lee equation:

$$L_i = L_c + \left[\frac{(CR_i - CR_c)(L_c - L_{birth})}{(CR_c - CR_{birth})} \right] \quad (3)$$

where L_{birth} = length at birth and CR_{birth} = centrum radius at birth. (Based on Gilmore et al. 1983, and Branstetter and Musick 1994, 100 cm TL was used for L_{birth}).

Results

Length equations

Length measurements from 272 sand tiger sharks (137 female, 135 male) were obtained by VIMS and NMFS between 1963 and 2001. Females ranged from 95 to 277 cm TL, and males ranged from 98.4 to 248 cm TL. Fork length and PCL can be derived from TL by:

$$FL = 0.8471 \cdot TL - 0.592 \quad (r^2 = 0.99;\ n = 138)$$

$$PCL = 0.7736 \cdot TL - 5.05 \quad (r^2 = 0.97;\ n = 134)$$

Weight-to-length relationships for female and male sand tiger sharks in the western North Atlantic are; $W = 1.3 \times 10^{-4} \times L^{2.4}$ ($r^2 = 0.84$, $n = 55$) and $W = 9.0 \times 10^{-5} \times L^{2.5}$ ($r^2 = 0.84$, $n = 47$), respectively, and $7.9 \times 10^{-5} \times L^{2.5}$ ($r^2 = 0.84$) for the sexes combined. A likelihood ratio test showed that a single equation for the sexes combined better describe the data than separate equations for each sex individually ($\chi^2 = 5.3$; $df = 2$; $P = 0.07$).

Vertebral analysis

A linear regression gave a significant fit to the TL-CR data ($TL = 10.753 \times CR + 36.786$; $r^2 = 0.97$; $P < 0.0001$). However, it was important to compare the mean back-calculated results from Eq. 1 through Eq. 3 with our mean sample PCL data to see which method provided better biological accuracy for modeling growth (Goldman 2004; Goldman and Musick 2006).

Percent agreement was 68.0% for the first set of blind reads and 75.0% for the second set, and the within-reader PA was 75.0% for reader one (the lead author) and 72.9% for reader two. PA ± 1 - year was > 92% for all reader comparisons. When grouped by 10 cm length increments, agreement for sharks ≤ 200 cm TL was 100%, except for a single 1 year disagreement in each set of readings, and 100% ± 1 year for sharks ≤ 220 cm TL (Table 2). The chi-square tests of symmetry showed that differences between and within readers were due to random error rather than systematic error (chi-square test, $P > 0.05$ in all cases).

Although we were missing samples from some winter months, RMI analysis indicates that the first postnatal and all subsequent rings form annually between December and February. The smallest relative margins in our sample occurred in February, followed by a consistent increase in RMI with the largest relative margins occurring in October (Fig. 2).

The administration of OTC was successful in marking vertebrae in two of the three sand tiger sharks. The vertebrae from the animal with the spinal deformity did not mark, possibly due to the diseased nature of the spine or due to low weight estimates and subsequently low doses. The two sharks whose vertebrae were successfully marked by the OTC were both aged at 2 years for their date of capture in 1998. One animal had two OTC marks and the other had three (Table 1). The annual

Table 2 Percent agreement (PA) and PA ± 1 year, for both sets of readings for sand tiger sharks, *Carcharias taurus*, when placed into 10 cm length (TL) groups

Length group (cm)	Read	First set of readings				Second set of readings			
		Agree	Agree ± 1	PA	PA ± 1	Agree	Agree ± 1	PA	PA ± 1
90.1–100	7	7	7	100	100	7	7	100	100
100.1–110	7	7	7	100	100	7	7	100	100
110.1–120	7	7	7	100	100	7	7	100	100
120.1–130	7	7	7	100	100	7	7	100	100
130.1–140	1	1	1	100	100	1	1	100	100
140.1–150	9	9	9	100	100	9	9	100	100
150.1–160	2	1	2	50	100	2	2	100	100
160.1–170	5	5	5	100	100	4	5	80	100
170.1–180	2	2	2	100	100	2	2	100	100
180.1–190	0	0	0	–	–	0	0	–	–
190.1–200	4	4	4	100	100	4	4	100	100
200.1–210	5	2	5	40.0	100	2	4	40	80
210.1–220	7	3	7	42.9	100	3	7	43	100
220.1–230	7	2	5	28.6	71.4	3	6	43	86
230.1–240	9	2	7	22.2	77.8	4	8	44	89
240.1–250	7	4	7	57.1	100	4	7	57	100
250.1–260	3	0	2	0	66.7	2	3	67	100
260.1–270	4	2	4	50	100	3	3	75	75
270.1–280	3	0	1	0	33.3	1	3	33	100
n =	96	65	89			72	92		
Percent agree				67.7	92.7			75.0	95.8

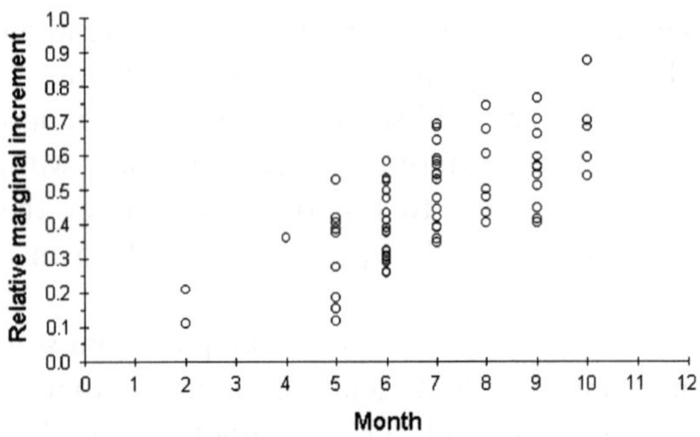

Fig. 2 Results of relative marginal increment analysis indicating annual ring formation likely occurs between December and February (n = 90)

OTC marks in both specimens were located near points where rings had formed (Fig. 3).

Vertebral length-at-age data from 48 female sand tiger sharks provided von Bertalanffy parameters of L_∞ = 295.8 cm TL, k = 0.11 year^{-1}, and t_0 = –4.2 years (Fig. 4). von Bertalanffy parameters from the linear-modified Dahl-Lea back-calculations (the most accurate back-calculation method; see below) gave slightly lower k coefficients and slightly higher L_∞ and t_0 values with or without sample data included (Table 3). Results from mean back-calculated data had the lowest standard error. Vertebral age data from 48 males provided Bertalanffy parameters of L_∞ = 249.5 cm TL, k = 0.16 year^{-1}, and t_0 = –3.4 years (Fig. 4). Back-calculated lengths-at-age for male sand tiger sharks (with or without sample data included) again provided slightly lower k coefficients, slightly higher L_∞, and t_0 parameters, and mean back-calculated data had the lowest standard error (Table 3). Vertebral age data for the sexes combined (n = 96) provided von Bertalanffy parameters of; L_∞ = 280.5 cm TL, k = 0.12 year^{-1}, and t_0 = –4.1 years. Parameters resulting from back-calculated data for the sexes combined produced results with similar trends as those for females and males (Table 3). A likelihood ratio test showed that separate von Bertalanffy growth models better describe the data for each sex than one model with the sexes combined (χ^2 = 22.8; df = 3; P = 0.000044).

The linear-modified Dahl-Lea method (Eq. 2) most accurately represented the mean sample length-at-age data. It produced mean back-calculated lengths-at-age within 9.1 cm of mean sample lengths-at-age for female sand tiger sharks, except

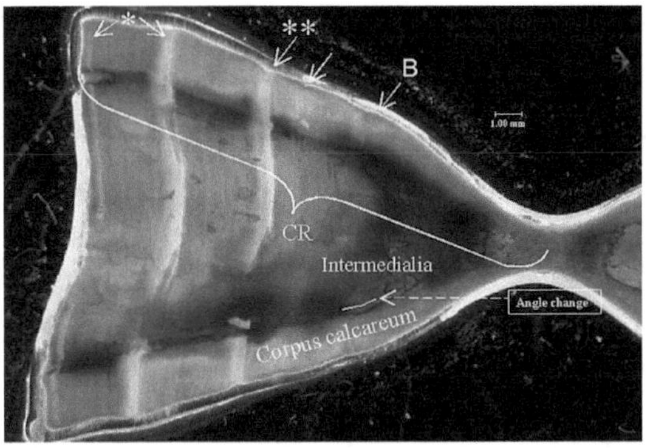

Fig. 3 Sagittally cut vertebral section of OTC sand tiger shark (OTC-2 (from Table 1). *Arrows without asterisks* represent wild growth. *Double asterisks* indicates initial OTC mark shortly after capture. *Arrows with single asterisks* indicate OTC marks and captive growth rings. All three OTC marks can clearly be seen

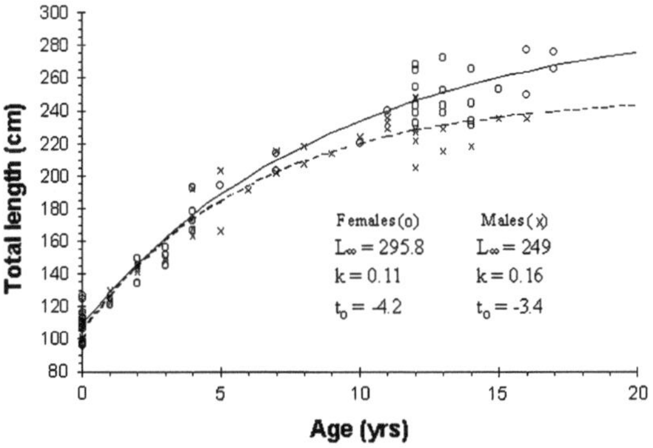

Fig. 4 von Bertalanffy growth curves fitted to female (n = 48) and male (n = 48) sample data for sand tiger sharks. Estimates for parameters of the von Bertalanffy growth function are summarized

for lengths of approximately 195 and 250 cm TL (Fig. 5a). When applied to males, Eq. 2 produced mean back-calculated lengths-at-age within 9.8 cm of mean sample lengths-at-age for sharks < 195 cm TL. At greater lengths, deviation from mean sample length-at-age ranged from 2.7 to 15.1 cm TL (Fig. 5b). Lee's phenomenon was present with the individual back-calculated data. There was a tendency for some age classes (usually older ones) to underestimate the mean sample length-at-age data after the first few back-calculated ages. This is not too surprising considering the time frame over which samples were obtained, and the potential for sampling variation (i.e., where in the spinal column the vertebrae were removed). However, Lee's phenomenon was not apparent in the mean back-calculation values, which were, overall, very similar to the mean sample length-at-age data.

Discussion

Branstetter and Musick (1994) suggested that sand tiger sharks form two rings per year in their vertebral centra whereas our results support a hypothesis that only one ring is formed annually. The difference in ageing protocols and the assumed time of band formation between the two studies is almost certainly the cause of the different results. We feel it is necessary to state the reasons that we did not follow the Branstetter and Musick (1994) ageing protocol in this study, then discuss our results and comment on their ramifications for recovery of the western North Atlantic population.

Branstetter and Musick (1994) "counted and measured rings in the intermedialia". When the intermedialia was damaged, they counted along the corpus calcareum. We found that using the intermedialia as the primary counting surface for ageing sand tiger sharks could lead to errors. In addition to the distinct banding pattern (i.e., band pairs making up rings) that extended completely across the intermedialia and corpus calcareum, and the presence of notches (at or near the rings) along the outside arm of the corpus calcareum, the intermedialia contained subtle or indistinct bands (similar to the fine feature "rings" discussed in Cailliet et al., this issue). These "indistinct bands" made using the intermedialia as the primary counting (and measuring) surface problematic for the following reasons: (1) they extended completely across the intermedialia, but did not appear on the corpus calcareum (nor were they associated with notches on the corpus calcareum); (2) they did not extend all the way across the intermedialia; (3) part-way across the intermedialia they blended into another indistinct band; (4) part-way across the intermedialia, they blended into a distinct band; (5) the number of indistinct bands (between distinct bands) was inconsistent ranging from 0–5 in number. Additionally, some indistinct bands were seen on the

Table 3 von Bertalanffy growth parameters of female, male, and sexes combined for sand tiger sharks in the western North Atlantic. Numbers in parentheses are standard errors

Females	L_∞	k	t_0
Sample data (n = 48)	295.8 (14.0)	0.11 (0.02)	–4.2 (0.5)
Back-calculations (n = 366[a])	302.4 (7.4)	0.09 (0.006)	–4.6 (0.22)
Mean back-calculation (n = 18[a])	308.3 (5.2)	0.09 (0.004)	–4.8 (0.21)
Back-calculations *W*/sample data (n = 414[a])	305.3 (6.9)	0.09 (0.006)	–4.7 (0.20)
Males			
Sample data (n = 48)	249.5 (7.2)	0.16 (0.02)	–3.4 (0.4)
Back-calculations (n = 278[a])	252.7 (6.1)	0.13 (0.009)	–4.1 (0.2)
Mean back-calculation (n = 16[a])	247.5 (3.6)	0.13 (0.009)	–4.2 (0.2)
Back-calculations *W*/sample data (n = 326[a])	256.1 (5.4)	0.12 (0.008)	–4.3 (0.2)
Combined			
Sample data (n = 96)	280.5 (9.4)	0.12 (0.01)	–4.1 (0.4)
Back-calculations (n = 644[a])	294.7 (6.5)	0.09 (0.005)	–4.8 (0.2)
Mean back-calculation (n = 34[a])	293.1 (16.0)	0.09 (0.01)	–5.0 (0.7)
Back-calculations *W*/sample data (n = 740[a])	295.2 (5.7)	0.09 (0.004)	–4.8 (0.2)

[a] Not independent

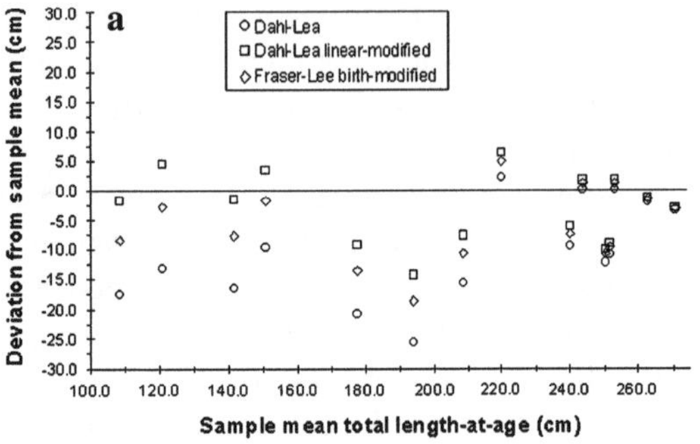

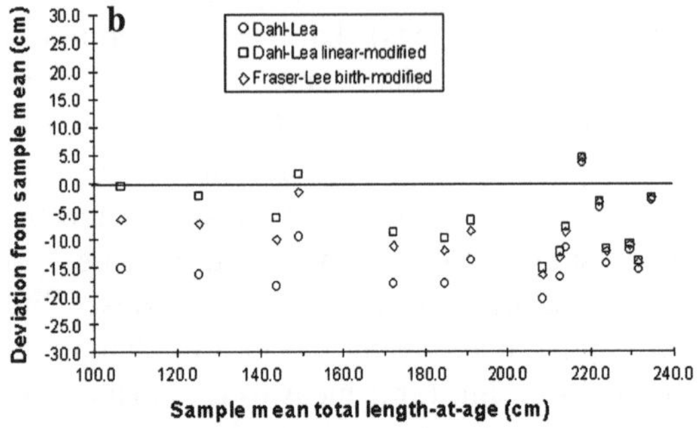

Fig. 5 Mean deviation, from mean sampled total length, of three proportional back-calculation methods for **a** female and for **b** male sand tiger sharks. Data points represent mean back-calculated lengths-at-age. A point on the x-axis would represent zero deviation from the sample mean length-at-age

corpus calcareum, but they did not match up with bands in the intermedialia and did not possess a corresponding notch along the outer arm of the corpus calcareum. These reasons were paramount in our decision to use the corpus calcareum as the primary counting and measuring surface, with the distinct rings in the intermedialia and notches as "confirmation" of an annulus.

The radials of the intermedialia of carcharhinid sharks are relatively hard, robust, and numerous, making it nearly solid. In contrast, the radials of the intermedialia in lamnoid sharks are less numerous, softer, and quite fragile, and the large amount of interstitial space often prevents its presence in sectioned vertebrae. When present, the outer edge of the intermedialia in lamniform vertebral sections can become warped in a concave manner. When this occurs, the rings near the outer edge of the intermedialia become "bunched up" and indistinguishable. The bands on the corpus calcareum can also become more tightly grouped at the outer edge, particularly in larger/older animals, however the rings have a tendency to remain distinguishable due to the stronger (more stout) nature of the structure. Based on these observations, we suggest that future studies on the age and growth of lamniform sharks use the corpus calcareum as the primary counting and measuring surface.

We found that vertebral growth significantly increased with somatic growth, which along with the reliability of the marginal increment analysis (Fig. 2) demonstrates vertebral growth patterns are a reliable indicator of age in sand tiger sharks. Precision was high between and within readers with limited differences (Table 2) that were attributable to random error. These results provided a high degree of confidence in the accuracy and precision of our age assessments (from sample data) used in the von Bertalanffy growth model, and hence in the resulting life-history parameter estimates. The similar von Bertalanffy growth parameter estimates generated from our sample data, back-calculated data, and the combination of the two indicate that our sample size was sufficiently large and encompassed the known size range of males, and all but the very upper end of the known size range of females. While these back-calculation results are, of course, dependent on the assumption that growth has not significantly changed over time, and are applicable only to sand tiger sharks, they demonstrate the importance of choosing the appropriate method in order to minimize error (Fig. 5a, b), which results in a greater ability to accurately model growth (Cailliet and Goldman 2004; Goldman 2004).

The RMI conducted by Branstetter and Musick (1994) indicated that specimens from the first part of the year (January–July) had an even number of bands while specimens from the second part of the year (August–December). We found no such relationship—animals from every month of the year from which we had more than a single sample possessed both even and odd numbers of rings. Our RMI analysis indicated that a single ring is formed annually sometime between December and February (Fig. 2). Additionally, the results from the two OTC sharks support the hypothesis that a single ring is formed annually. The annual OTC marks in both specimens were located near points where rings had formed (Fig. 3) even though they were kept under a constant photoperiod and ambient temperature.

Centrum banding patterns may be related to physiological changes induced by changes in environmental parameters such as temperature and photoperiod (Cailliet et al. 1986; Branstetter 1987), however, it has been shown not to be the case for some species such as the little skate, *Raja erinacea* (Natanson 1993), and the Pacific angel shark, *Squatina californica* (Natanson and Cailliet 1990; Cailliet et al. 1992). The north–south migration of sand tiger sharks has been proposed as a possible reason that two rings might be formed in the vertebral centra each year (Branstetter and Musick 1994) with the primary cue for the migration being either temperature or photoperiod. The primary cue for sand tiger migration has not been demonstrated, but the fact that captive sand tiger sharks appear to form a single ring in their vertebral centra each year makes those possibilities less likely. Vertebral growth is inevitably linked to food intake, and a lack of food for short periods of time can cause subtle bands to appear in vertebral centra of some species (J. Gelsleichter, personal communication). This may explain the "indiscrete bands" we saw in sand tiger vertebrae, and play a role in the annual formation of rings in the centra.

The life-history parameters presented by Goldman (1998) resulted in growth coefficients (k) that were half of those presented by Branstetter and Musick (1994). This is because Goldman (1998) simply doubled the age estimates from Branstetter and Musick (1994) by assuming that one ring was formed annually in the vertebral centra (e.g., a 5-year-old would then be considered to be 10 years of age), and not from the examination of any vertebral samples. In contrast, the estimations of sand tiger life-history parameters presented herein are the combined result of using an independently developed ageing protocol (applied to vertebral samples from 96 sand tiger sharks), a subsequent RMI analysis and the use of a direct validation method (OTC), all of which support a hypothesis that one ring is formed annually in vertebral centra. As such, we believe our results are a much better indicator of sand tiger shark growth parameters than either Branstetter and Musick (1994) or Goldman (1998).

Branstetter and Musick (1994) characterized sand tiger growth based on mean back-calculated lengths-at-age as follows; "growth was 25–30 cm per year for ages 0–1, 20–25 cm per year for ages 2–3, and 15–20 cm per year for ages 4–5. Subadults and adults had a growth rate of 10–15 cm per year for ages 6–7, and growth declined to 5–10 cm per year for ages 8 and greater". They also characterized growth in females and males to be nearly identical throughout life. Our results indicate that growth is similar up to age 5 at which point-in-time females begin outgrowing males at a significant rate, and that two separate growth curves should be used to describe the rate at which each sex attains their maximum size (Fig. 4). Mean back-calculated lengths-at-age from this study indicate that growth averaged 14.5–18.5 cm per year for ages 0–1, 13–14.5 cm per year for ages 2–3, and 11 cm per year for age 4. For ages 5 and 6, growth averaged 10–11.2 cm for females and 9.2–9.5 for males; and for age 7 females averaged 9.3 cm per year whereas males averaged 6.8 cm per year. For sand tiger sharks (8 years of age, the growth rates declined to an average of 6.4 cm per year for females and 4.6 cm per year for males.

Female sand tiger sharks become sexually mature at a length of 220–230 cm TL and males mature at 190–195 cm TL (Gilmore et al. 1983). Previous estimates of age at maturity stated that the equivalent ages to those lengths are 6 years for females and about 4 years for males (Branstetter and Musick 1994). Our length-at-age results place age at sexual maturity at 9–10 years for female sand tiger sharks, and at 6–7 years for males. Reproduction in captivity (from copulation to parturition) has been documented in South Africa and Australia. A female sand tiger shark born at Underwater World aquarium, Australia, in 1992 became pregnant in 2000 (resulting in two preterm stillborn pups ~70–80 cm TL) (Henningsen et al. 2004). Additionally, a male sand tiger born at the New York aquarium in 1994 was exhibiting pre-copulatory behavior in 2001–2002 (H. Walters, personal communication). These data, although captive, provide corroboration to our estimates of age at sexual maturity.

Gilmore (1990) stated that all female sand tiger sharks he examined from Florida to North Carolina between March and January were impregnated and that no resting stage took place in the reproductive cycle of this species. However, Branstetter and Musick (1994) presented strong evidence that a resting stage did occur and that the reproductive cycle was at least 2 years. They presented records from the VIMS longline survey from 29 mature female sand tiger sharks caught off Virginia (during the 1980s and early 1990s) that were all noted as either postpartum or in a resting state with small ovarian egg follicles. Since then, we have records (from the VIMS survey) for an additional 17 mature females that were in the same postpartum or resting stage condition. The mother of the aforementioned female born at Underwater World, Australia was captured pregnant in 1992, and while in captivity has given birth two more times; in 1997 and 1999 (Henningsen et al. 2004). A gestation period of 9–12 months leaves little to no energetic "turn-around" time for a female to build up her reserves and go through another reproductive cycle, unless there is a resting year. If a female carried pups through a 12-month gestation period, she would immediately have to mate again in order to reproduce the following year. Any time delay in mating would "throw off" her timing for parturition, and at some point-in-time she would require a resting period. The current body of evidence strongly supports a 2-year reproductive cycle for female sand tiger sharks. However, preliminary evidence from captive male sand tiger sharks indicates that they may mate annually (A. Henningsen, personal communication).

Maximum observed age for female and male sand tiger sharks in this study was 17 and 15, respectively. These ages are close to the maximum documented ages of sand tiger sharks in captivity. Govender et al. (1991) reported a male that had been in captivity at an aquarium in Durban, South Africa for 16 years, and there is currently a 20-year-old female (as of 2002) at the National Aquarium in Baltimore, MD, USA (A. Henningsen, personal communication). Branstetter and Musick (1994) estimated longevity at 30–35 years by extrapolating their von Bertalanffy curves generated from back-calculated data. Applying the same technique to our data, longevity may be as high as 40 years for females and 30 years for males. However, estimating longevity using the time required to reach 95% of its asymptotic length (i.e., $5 \times \ln 2/k$, see Fabens 1965; Cailliet et al. 1992) provided longevity estimates of 32–38 for females and 22–27 for males.

Branstetter and Musick (1994) provided a weight–length equation for sand tiger sharks of; $W = 1.62 \times 10^{-6} \times L^{3.15}$. This equation was later found to have an error in it, which has not been published. The corrected weight–length equation is: $W = 1.62 \times 10^{-6} \times L^{3.24}$. Our data for length conversions and maximum size of sand tiger sharks are similar to those from the corrected weight–length equation. The increased sample size (particularly for smaller individuals) enhances the accuracy and precision of these estimates. More importantly, we were able to determine that a single length–weight curve combined is adequate for describing that relationship for both sexes.

Our estimates of sand tiger shark life-history parameters show that they are much slower growing to adulthood and maximum length than previously assumed. Since our mean back-calculated lengths-at-age gave the smallest standard error in von Bertalanffy estimates (Table 3), they may better represent the life-history parameters of this species and should be considered more valid when used as inputs for determining vital rates. Considering the large population depletion suffered by sand tiger sharks in the western North Atlantic over the past 20 years, this information is crucial for accurately assessing the ability of the population to recover, and further justifies the need for this species to be fully protected. The life-history parameters presented here also allow for a re-adjustment of previously predicted vital rate estimates that can aid managers in taking appropriate steps for sand tiger shark protection and conservation.

Acknowledgments We are thankful to many people for their contributions to this research. We thank Christina Conrath for her time and effort in ageing sharks as our second reader, and Robert Latour, John Hoenig, Todd Gedamke, John Walters, and Dan Hepworth for assistance with statistical questions and providing SAS code for us to use. John Demerest provided an enormous amount of volunteer time to assist in preparing vertebral samples for microscope viewing, and we are grateful to him for his time and effort. We also thank R. Dean Grubbs, Richard Krauss, Jason Romine, and Captain Durand Ward (R/V Bay Eagle) for their invaluable assistance in the field. We thank Ian Gordon, Hans Walters, and particularly Alan Henningsen for their assistance with questions regarding sand tiger shark catches in their areas and data about sand tiger shark captive mating and associated behaviors. We owe many thanks to Joe Choromanski, Tim Handsel, and the staff at the Ripley's Aquarium in Myrtle Beach, NC, USA, and to Holly Martel Bourbon, Steve Bailey, and the staff at the New England Aquarium, Boston, MA, USA, for housing and caring for sand tiger sharks used to attempt validating the age–growth relationship for the species. Special thanks go to Dr Lisa Natanson for providing us with over 40 sand tiger shark vertebral samples. Partial support for travel to this symposium was provided by California Sea Grant, the American Elasmobranch Society, and the Florida Aquarium, Tampa, FL, USA.

References

Anonymous (2002) Preliminary draft recovery plan—grey nurse shark. New South Wales Fisheries, Australia

Branstetter S (1987) Age and growth validation of newborn sharks held in laboratory aquaria, with comments on the life history of the Atlantic sharpnose shark, *Rhizoprionodon terraenovae*. Copeia 1987:291–300

Branstetter S, Musick JA (1994) Age and growth estimates for the sand tiger in the Northwestern Atlantic Ocean. Trans Am Fish Soc 123:242–254

Cailliet GM, Goldman KJ (2004) Age determination and validation in chondrichthyan fishes. In: Carrier J, Musick JA, Heithaus M (eds) The biology of sharks and their relatives. CRC, Boca Raton, pp. 399–447

Cailliet GM, Radke RL, Weldon BA (1986) Elasmobranch age determination and verification: a review. In: Uyeno T, Arai R, Taniuchi T, Matsuura K (eds) Indo-Pacific fish biology: Proceedings of the 2nd international conference on Indo-Pacific fishes. Ichthyological Society of Japan, Tokyo, pp. 345–359

Cailliet GM, Mollet HF, Pittenger GG, Bedford D, Natanson LJ (1992) Growth and demography of the Pacific angel shark (*Squatina californica*), based on tag returns off California. Aust J Mar Freshw Res 43:1313–1330

Cailliet GM, Smith WD, Mollet HF, Goldman KJ (this issue) Age, growth studies of chondrichthyan fishes: an overview stressing terminology, verification, validation, and growth function fitting. In: Goldman KJ, Carlson JK (eds) Special volume from symposium of the American Elasmobranch Society, July 2005. Environ Biol Fishes, pp. 000–000

Campana SE (1990) How reliable are growth back-calculations based on otoliths? Can J Fish Aquat Sci 47:2219–2227

Campana SE (2001) Accuracy, precision and quality control in age determination, including a review of the use and abuse of age validation methods. J Fish Biol 59:197–242

Carlander KD (1969) Handbook of freshwater fishery biology, vol. 1. Iowa University Press, Ames

Casey JG, Kohler NE (1990) Long distance movements of Atlantic sharks from the NMFS cooperative shark tagging program. In: Gruber SH (ed) Discovering sharks. American Littoral Society, Special Publication 14, Highlands, pp. 87–91

Cerrato RM (1990) Interpretable statistical tests for growth comparisons using parameters in the von Bertalanffy equation. Can J Fish Aquat Sci 47:1416–1426

Cliff G (1989) Breeding migration of the sand tiger shark *Carcharias taurus* in southern African waters. In: Abstracts of the 5th annual meeting of the American Elasmobranch Society, San Francisco, June 1989

Compagno LJV (1984) FAO species catalogue. Vol. 4. Sharks of the World. An annotated and illustrated catalogue of shark species known to date. Part 1. Hexanchiformes to Lamniformes. FAO Fish Synopsis 125 4(Pt 1):249

Conrath CL, Gelsleichter J, Musick JA (2002) Age and growth of the smooth dogfish (*Mustelus canis*) in the northwest Atlantic Ocean. Fish Bull 100:674–682

Evans GT, Hoenig JM (1998) Testing and viewing symmetry in contingency tables, with application to readers of fish ages. Biometrics 54:620–629

Fabens AJ (1965) Properties and fitting of the von Bertalanffy growth curve. Growth 29:265–289

Francis RICC (1990) Back-calculation of fish length: a critical review. J Fish Biol 36:883–902

Gelsleichter J, Cortés E, Manire CA, Hueter RE, Musick JA (1998) Evaluation of toxicity of oxytetracycline on growth of captive nurse sharks, *Ginglymostoma cirratum*. Fish Bull 96:624–627

Gilmore RG (1990) The reproductive biology of lamnoid sharks. In: Gruber SH (ed) Discovering sharks. American Littoral Society, Special Publication 14, Highlands, pp. 64–67

Gilmore RG (1993) Reproductive biology of lamnoid sharks. Environ Biol Fishes 38:95–114

Gilmore RG, Dodrill JQ, Linley PA (1983) Reproduction and embryonic development of the sand tiger shark, *Odontaspis taurus* (Rafinesque). Fish Bull 81:201–225

Goldman KJ (1998) Demographic analysis for sand tiger sharks and dusky sharks. Part 3. In: Musick JA, Gelsleichter J, Grubbs RD, Goldman KJ (eds) VIMS Shark Ecology Program. Appendix SB-IV-13. National Oceanic and Atmospheric Administration/National Marine Fisheries Service Shark Evaluation Workshop, Panama City, 3 pp., June 1998

Goldman KJ (2004) Age and growth of elasmobranch fishes. In: Musick JA, Bonfil R (eds) Elasmobranch Fisheries Management techniques. Asia Pacific Economic Cooperation, Singapore, pp. 97–132, 370 pp. Electronic version available at http://www.flmnh.ufl.edu/fish/organizations/ssg/EFMT2004.htm

Goldman KJ, Musick JA (2006) Growth and maturity of salmon sharks in the eastern and western North Pacific, and comments on back-calculation methods. Fish Bull 104:278–292

Gordon I (1993) Pre-copulatory behaviour of captive sand tiger sharks, *Carcharias taurus*. Environ Biol Fishes 38:159–164

Govender A, Kistnasamy N, Van Der Elst RP (1991) Growth of spotted ragged-tooth sharks *Carcharias taurus* (Rafinesque) in captivity. S Afr J Mar Sci 11:15–19

Haddon M (2001) Modeling and quantitative methods in fisheries. Chapman & Hall/CRC, Boca Raton

Henningsen AD, Smale M, Garner R, Gordon I, Marin-Osorno R, Kinnunen N (2004) Captive breeding and sexual conflict. In: Elasmobranch husbandry manual. Ohio Biological Survey, Columbus, Ohio, pp. 239–250

Hoenig JM, Morgan MJ, Brown CA (1995) Analyzing differences between two age determination methods by tests of symmetry. Can J Fish Aquat Sci 52:364–368

Kimura DK (1980) Likelihood methods for the von Bertalanffy growth curve. Fish Bull 77:765–773

Mathsoft Inc (2000) S-Plus 2000 Professional Release 1. Mathsoft Inc., Seattle

Media Cybernetics (1999) Optimus, version 6.5. Media Cybernetics, Silver Spring

Musick JA, Branstetter S, Colvocoresses JA (1993) Trends in shark abundance from 1974–1991 for the Chesapeake Bight region of the U.S. mid-Atlantic coast. National Oceanic and Atmospheric Administration, Technical Report. National Marine Fisheries Service 115:1–18

Musick JA, Harbin MM, Berkeley SA, Burgess GH, Eklund AM, Findley L, Gilmore RG, Golden JT, Ha DS, Huntsman GR, McGovern JC, Parker SJ, Poss SG, Sala E, Schmidt TW, Sedberry GR, Weeks H, Wright SG (2000) Marine, estuarine, and diadromous fish stocks at risk of extinction in North America (exclusive of Pacific salmonids). Fisheries 25(11):6–30

Natanson LJ (1993) Effect of temperature on band deposition in the little skate, *Raja erinacea*. Copeia 1993:199–206

Natanson LJ, Cailliet GM (1990) Vertebral growth zone deposition in angel sharks. Copeia 1990:1133–1145

NMFS (National Marine Fisheries Service) (1999) Final fishery management plan for Atlantic tunas, swordfish and sharks, vols. I–III. U.S. Department of Commerce National Oceanic and Atmospheric Administration, National Marine Fisheries Service, Silver Spring, April 1999

Pollard DA, Smith AK (2005) IUCN Red List assessment of Alopias vulpinus. In: Fowler SL, Cavanagh RD, Camhi M, Burgess GH, Cailliet GM, Fordham SV, Simpfendorfer CA, Musick JA (eds) Sharks, rays and chimaeras: the status of the chondrichthyan fishes. IUCN SSC Shark Specialist Group. IUCN, Gland, Switzerland and Cambridge, UK, pp. 250–252

Pollard DA, Lincoln Smith MP, Smith AK (1996) The biology and conservation status of the grey nurse shark (*Carcharias taurus* Rafinesque 1810) in New South Wales, Australia. Mar Freshw Ecosyst 6:1–20

Quinn TJ, Deriso RB (1999) Quantitative fish dynamics. Oxford University Press, New York

Ricker WE (1992) Back-calculation of fish lengths based on proportionality between scale and length increments. Can J Fish Aquat Sci 49:1018–1026

SAS Institute Inc. (1999) Version 8.0. Statistical Analysis Systems Institute Inc., Cary

SigmaPlot (2000) Version 6.0. SPSS Inc., Chicago, IL

Tanaka S (1990) Age and growth studies on the calcified structures of newborn sharks in laboratory aquaria using tetracycline. In: Pratt HL Jr, Gruber SH, Taniuchi T (eds) Elasmobranchs as living resources: advances in the biology, ecology, systematics, and the status of the fisheries. National Oceanic and Atmospheric Administration, Technical Report, National Marine Fisheries Service 90, pp. 189–202

Environ Biol Fish (2006) 77:253–264
DOI 10.1007/s10641-006-9130-4

ELASMOBRANCHS

Comparing external and internal dorsal-spine bands to interpret the age and growth of the giant lantern shark, *Etmopterus baxteri* (Squaliformes: Etmopteridae)

Sarah B. Irvine · John D. Stevens · Laurie J. B. Laurenson

Received: 27 June 2006 / Accepted: 12 July 2006 / Published online: 8 November 2006
© Springer Science+Business Media B.V. 2006

Abstract The giant lantern shark, *Etmopterus baxteri*, is taken as bycatch of commercial fisheries that operate in deepwater off southeastern Australia. Bands on the second dorsal spine were used to obtain age estimates. The number of bands on the external surface of the spine and within the inner dentine layer increased with animal length. Most spines had more bands on the external surface, and the rate of band formation was significantly different between the external surface and the inner dentine layer. Females had a maximum of 57 external bands and 26 internal bands, while males had up to 48 external bands and 22 internal bands. Age estimates from external bands suggest maturity (A_{50}) at 20 years for males and 30 years for females. Internal band age estimates suggest maturity at 10.5 years for males and 11.5 years for females. Although there is a large discrepancy between these two preliminary (i.e., unvalidated) age estimates, they both suggest that *E. baxteri* is a long-lived and late maturing species that is likely to be susceptible to over fishing.

S. B. Irvine (✉) · L. J. B. Laurenson
School of Ecology and Environment, Deakin University, PO Box 423, Warrnambool, VIC 3280, Australia
e-mail: sbirvine@aapt.net.au

S. B. Irvine · J. D. Stevens
CSIRO Marine and Atmospheric Research, GPO Box 1538, Hobart, TAS 7001, Australia

Keywords Dorsal spine · Deepwater dogfish · Francis model

Introduction

Elasmobranchs are typically aged using annuli on the vertebral centra. However, the centra of most dogfish are poorly calcified and do not produce reliable band counts (Wood et al. 1979), creating the necessity to find an alternate ageing structure. Fortunately, many dogfishes possess dentine spines (one anterior to each dorsal fin) that may be used to investigate age. Kaganovskaia (1933) was the first to estimate age using the dorsal-fin spines of the spiny dogfish, *Squalus acanthias*, and this species is by far the most common shark to be aged using the dorsal spines (see Henderson et al. 2002 and references within). The dorsal-fin spine structure has been described by Maisey (1979), Beamish and McFarlane (1985) and Tanaka (1990). Further, Clarke and Irvine (2006) clarify the terminology used to describe the main structural components of deepwater dogfish spines.

Studies on deepwater elasmobranchs have frequently used dorsal fin spines for ageing. Tanaka (1990) used the internal dentine bands found in transverse spine sections to age *Centrophorus acus* from Japanese waters. Subsequent deepwater dogfish studies have described the age and growth to varying degrees using these internal dentine bands (Guallart Furio 1998; Machado and Figueiredo 2000; Sion et al. 2002; Clarke et al. 2002a, b; Francis and Maolagáin 2004). However, Irvine et al. (2006) reported an alternative method of ageing the golden dogfish *Centroselachus crepidater*, using bands on the external surface of the spine. These bands were first reported on *S. acanthias* spines (Holden and Meadows 1962), and are caused by the lack of synchrony in the deposition of dentine at the spine base and upward spine growth. Irvine et al. (2006) also counted the internal dentine bands and found they did not form at the same rate as external bands.

The use of typical age validation techniques is impractical for most deepwater fish studies (Cailliet et al. 2001). However, Irvine et al. (2006) verified the annual formation of external bands in *C. crepidater* using preliminary radiometric age estimates from Fenton (2001). As internal bands in *C. crepidater* spines were either consistently difficult to read or did not form annually, Irvine et al. (2006) suggested that internal bands may underestimate fish age. The aim of this study was to estimate the age and growth of the giant lantern shark, *Etmopterus baxteri.* Age estimates from both external and internal dorsal-fin spine bands were used to investigate growth, longevity and age at maturity. The number of external bands was compared with the number of internal bands to assess if they were formed simultaneously.

Materials and methods

Specimen collection

Giant lantern sharks were opportunistically collected from the bycatch of commercial trawl fisheries operating in southeastern Australia between November 2000 and July 2002 (Fig. 1). Trawls typically had a cod-end mesh size of 90–110 mm and targeted orange roughy, *Hoplostethus atlanticus*, in 600–1200 m. Most *E. baxteri* (71%) were caught at 830–1200 m over seamounts on the Cascade Plateau to the southeast of Tasmania.

Biological examination

Each dogfish was sexed, and the total length (TL) and fork length (FL) were measured as a straight line (±1 cm) by allowing the caudal fin to take a natural position. All animal lengths hereafter are TL. For those dogfishes with a damaged caudal fin, FL was converted to TL using the linear relationship: $TL = 1.07FL + 3.70$, ($n = 874$, $r^2 = 0.97$). Dogfishes were weighed (±10 g) on a top-loading digital scale, and the relationship between weight and length was examined and compared between each sex using an analysis of covariance (ANCOVA).

Age and growth analysis

Vertebrae

The vertebral centra of *E. baxteri* are small and very deep-coned, and appeared to be adequately calcified for age determination. A block of 5–10 vertebrae was removed from the vertebral column below the origin of the first dorsal-fin. A subsample of vertebrae representing both sexes and various size classes were cleaned by soaking in 4% sodium hypochlorite (house-hold bleach) solution for 5–10 min. Vertebrae were then thoroughly washed in tap water and examined for the presence of bands on the centrum face using a magnification lamp or binocular dissecting microscope at 10–60 times magnification. A few discontinuous and irregularly spaced concentric bands were visible, although it was decided that whole, clean vertebrae were not suitable for age determination.

Various treatments for enhancing bands on the centrum face were performed. Whole centra were stained with silver nitrate (Stevens 1975); alizarin red S (La Marca 1966); crystal violet (Schwartz 1983); cobalt nitrate (Hoenig 1979); ninhydrin (Davenport and Stevens 1988) and graphite powder (Ferreira and Vooren 1991) or decalcified in nitric acid (Correia and Figueiredo 1997). Calcium affinity stains ($AgNO_3$ and $CoNO_3$) accumulated superficially on the vertebra but did not penetrate into a readable concentric pattern. All other stains and treatments had no effect and it was therefore decided that

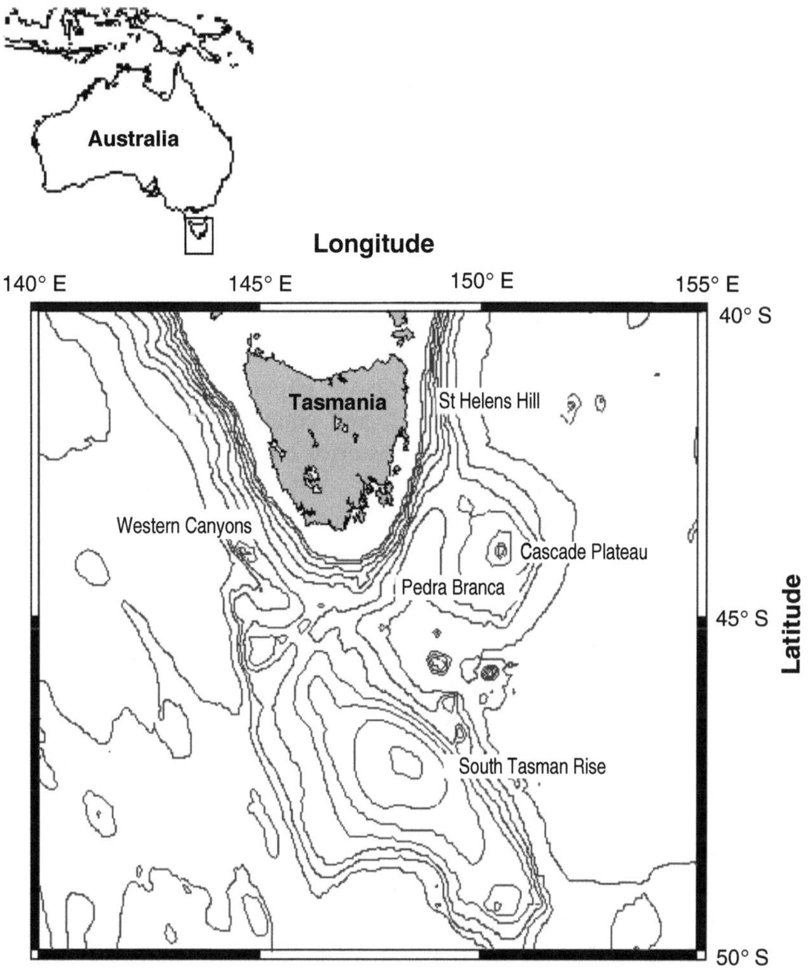

Fig. 1 Topographic map of southeastern Australia indicating the fishing locations where *E. baxteri* specimens were collected as bycatch

chemically treated whole centra were not suitable for age determination.

Both stained and un-stained whole vertebrae were air-dried for up to 2 h, placed upright in Wonderflex® silicon cookware 'brownie' moulds and embedded in epoxy resin (Renlam M-1 AU resin and HY951 hardener at 9:1 w/w). Each centrum was individually sectioned using a lapidary saw (RPM = 1250) fitted with one 0.6 mm wide diamond-tipped blade. Sectioning twice through the focus along the sagittal plane resulted in "bow-tie" sections with a thickness of 500 ± 100 μm. Each section was individually examined using a binocular dissection microscope equipped with a micrometer. Immersion oil was used to disguise any saw-blade marks made during sectioning.

No clear banding pattern was found on the centrum face, corpus calcareum or intermedialia of sectioned vertebrae. Thus, the vertebral centra of *E. baxteri* could not be used for age determination.

Dorsal-fin spines

Spine preparation. Spines were collected, measured and prepared following Irvine et al. (2006). As the second spine is larger and is less often damaged than the first spine, it was chosen for age estimation. Excess muscle and connective tissue was removed using a sharp scalpel and by immersing in hot water for short periods. Spine morphometrics, including total spine length (TSL), external spine length (ESL) and external spine width (ESW) were measured using digital callipers (±0.01 mm) following Clarke and Irvine (2006). Spine growth was investigated by examining the relationship between spine length and total length.

Growth band enhancement. External base bands were enhanced following the techniques outlined by Irvine et al. (2006), which included staining with an alizarin red derivative for 3–5 days. All spines were kept wet and examined either under a low power (6×) dissecting microscope, with a magnifying lamp, by the naked eye, or a combination of all three depending on the particular spine.

One external band describes either a ridge of dentine or a white band that resulted after rubbing a stained spine with abrasive paper. Counting started at the spine base, and counts were occasionally made along the posterior edge due to increased band clarity.

Once external band examination was completed, spines were embedded and sectioned following Irvine et al. (2006). Between 5 and 15 transverse sections were taken from the tip of each spine and section thickness was generally 250–350 μm. The internal spine structure consists of three dentine layers. Although bands were present in each dentine layer, bands were more widely spaced and easier to interpret in the inner layer. The inner layer was widest below the pulp cavity and most sections were taken immediately below the tip of the pulp cavity. Spine sections were mounted onto glass slides using epoxy resin. Slide sections were examined using an F-View Soft Imaging System on a B(51A Olympus compound microscope using the differential interference contrast. Within the internal dentine layers, 'one band' refers to a dark (opaque) and light (translucent) concentric ring. Counting started at the lumen (spine centre) and continued outwards until the first trunk primordium (the junction between the inner and middle layers).

Reading precision and accuracy. Three non-consecutive band counts were made for each spine/section without prior knowledge of the animal's length, sex or the previous band counts. A subjective measure of band readability was used with a sliding scale from 1 (samples with unambiguous bands and excellent readability) to 5 (an unreadable sample). Samples with a readability score of 1–3 were used in age analysis. Approximately 20% of the samples were re-examined 1–3 months after the first examination to imitate the second (independent) reader as suggested by Cailliet and Goldman (2004).

Precision was calculated using the coefficient of variance (CV) across all fish ages (Change 1982). This gave an assessment of the ease of ageing dogfish spines and tested the within reader reproducibility of age determinations (Campana 2001). The upper limit for the CV was set at 20% for each spine section (adapted from the index of average percent error (IAPE) for vertebrae analysis (Beamish and Fournier 1981). Samples were not included in the growth analysis if the CV was >20%. The mean age for each of the three counts defined the age estimate for each shark (Casey et al. 1985).

Growth analysis

The von Bertalanffy (1938) growth model (VBGM) is the most commonly used model to describe fish growth in length or weight (Campana 2001). However, it does not always provide a particularly good fit and there has been a wide array of criticisms (Haddon 2001). The VBGM curve is strongly determined by values of L_∞ and t_0, which are at either end of the curve, generally where there is the least amount of data (Sainsbury 1980). The stochastic model from Francis (1988) suggests new biologically significant parameters for the VBGM, and Moulton et al. (1992) and Irvine et al. (2006) used this model for shark age and growth analysis:

$$L = l_\phi + (l_\psi - l_\phi)(1 - r^{2(T-\phi)/(\psi-\phi)})/(1 - r^2)$$

where the three VBGM parameters (L_∞, t_0 and k) are replaced with l_ϕ (mean length at reference age ϕ), l_ψ (mean length at reference age ψ), l_χ (mean length at reference age $(\phi + \psi)/2$), T is age and $r = (l_\psi - l_\chi)/(l_\chi - l_\phi)$.

Assuming bands are formed annually, the Francis model was fitted to the observed length-at-age data for each sex. In order to compare growth between each sex, the same reference ages ($\phi = 5$, $\psi = 40$ external bands and $\phi = 3$, $\psi = 18$ internal bands) were chosen to avoid unnecessary extrapolations when comparing growth between sexes.

In order to present the VBGM equations for each sex, the Francis parameters were related to the conventional VBGM parameters using:

$$L_\infty = l_\phi + (l_\psi - l_\phi)/(1 - r^2)$$

$$k = -(2\log_e r)/(\psi - \phi)$$

$$t_0 = \phi + (1/k)\log_e((L_\infty - l_\phi)/L_\infty)$$

Confidence intervals (95%) around the Francis parameters gave a direct measure of the heterogeneity in length-at-age and were calculated using the non-linear regression function in Systat 8.0 (Systat Software Inc., Point Richmond, California USA).

The Francis parameters are biologically significant and can therefore be directly compared between sexes. However, Kimura (1980) suggested the use of χ^2 tests on each likelihood ratio to compare the growth curves. This method, advocated by Moulton et al. (1992) and Haddon (2001) is a more reliable means of finding a difference in growth. Chi-square tests on Francis parameters (by sex and internal/external bands) were performed in Microsoft Excel similar to that outlined for the VBGM by Haddon (2001).

Correlation between inner and outer counts

In order to determine if external bands were formed at the same rate as inner dentine bands, the number of external bands was compared to the number of internal bands in each spine. The rate of band deposition in each growth zone was assessed via the relationship between the number of external band and the mean number of internal bands using the non-linear (mode/loss) regression function in Systat 8.0.

Longevity. Band counts provided an initial estimate of theoretical longevity, although these values may be underestimated if the population has been commercially fished. Longevity was investigated using the equations of Taylor (1958), where the age at which 95% of L_∞ is reached = $t_0 + (\log_0(1 - 0.95)/k)$, Fabens (1965) where the age at which 99% of L_∞ is reached = $7((\ln 2)/k)$ and Ricker (1979) where the age at which 95% of L_∞ is reached = $5((\ln 2)/k)$ where t_0 and k are VBGM parameters derived from Francis growth parameters.

Maturity. Sexual maturity in males was determined by clasper condition and macro-examination of the testes. Males were considered mature when claspers had elongated and were fully calcified, and when the testes showed signs of lobulation. The reproductive status of females was based on the condition of the ovaries and uteri adapted from Wetherbee (1996) and Stehmann (2002). Females were assumed in a mature condition when distinct oocytes were present in the ovaries and/or the uteri had expanded away from the central axis of the body cavity.

Age at maturity was estimated from the relationship between the proportions of immature versus mature specimens within 2-year age classes. Each interval was indicated by its lower value. A logistic curve was fitted to the proportion P_A of sexually mature individuals by age (A) using:

$$P_A = \frac{1}{1 + \exp\left[-\ln(19)\frac{(A - A_{50})}{(A_{95} - A_{50})}\right]}$$

where ln is the natural logarithm, and A_{50} and A_{95} correspond to the age at which a proportion of 0.5 (or 50%) and 0.95 (or 95%) are mature. A non-linear regression function was fitted using Systat 8.0 where the sum of squares was minimised and the standard error of A_{50} was calculated.

Results

The relationship between weight (Wt) and total length (TL) differed significantly with sex (ANCOVA; $p < 0.05$):

$$\text{Male}: \text{Wt} = 0.0107\,\text{TL}^{2.7943}(r^2 = 0.76, n = 74, \text{length range } 24\text{–}74\,\text{cm})$$

$$\text{Female}: \text{Wt} = 0.0012\,\text{TL}^{3.341}(r^2 = 0.88, n = 137, \text{length range } 30\text{–}80\,\text{cm}).$$

Age and growth

A total of 424 *E. baxteri* were examined for age and growth analysis, and the size ranges were 30–80 cm for females and 24–74 cm for males. The

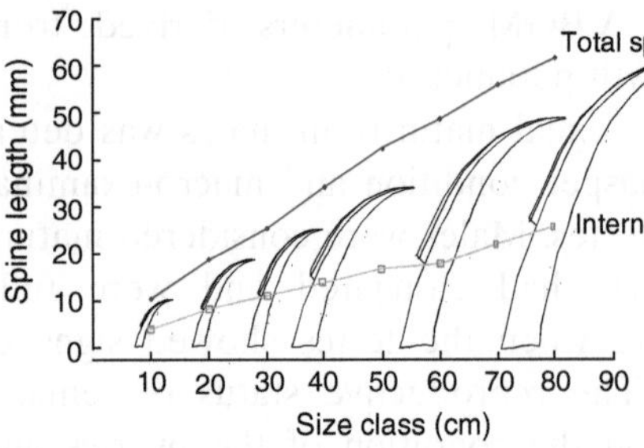

Fig. 2 Relationship between animal length (TL) and the mean total spine length (TSL) and internal spine length (ISL) for *E. baxteri*

spines of the four larger females (> 80 cm) were broken and were unable to be used. The spines of eleven embryos (12–22 cm) were also examined.

Spine structure and growth

The first and second dorsal-fin spines of *E. baxteri* are structurally similar, although the second is more than twice the length of the first. The first spine is relatively straight, while the second spine is strongly curved and is higher than the second dorsal fin. The spines are ivory in colour with an enamel anterior dentine portion that is often brown. Spines were visible in embryos > 6 cm TL, full-term embryos (~20 cm TL) had fully developed spines measuring 18 mm TSL. Of the 424 spines cleaned and measured, 378 (89%) had no damage, 26 had a worn/blunt tip, 19 had a broken tip, and three were broken and therefore discarded. The relationship between TL and TSL for non-damaged spines was linear [TL = 1.15TSL + 4.36 (r^2 = 0.82; n = 362)] and did not differ between sexes (ANCOVA; p = 0.399; F = 0.711). Pre-natal spine growth (embryos 10 cm TL to birth at 20 cm TL) and post-natal spine growth was similar, with the most rapid growth between 30 and 50 cm TL (Fig. 2). More than 59% of the spine length was external (outside the body), and this proportion increased with animal size.

External bands

A countable banding pattern (Fig. 3) was found on the surface of 294 (70%) spines. Over-staining was the main cause of poor readability and most spines had a readability score of three. The average reading precision (CV) of readable counts was 1.45%. Ten ages had a CV above 20% and were excluded from all further age analysis.

No bands were found close to the spine tip, and the number of external bands (No.) increased with spine size (TSL) [Male TSL = $54(1 - e^{(-0.05(\text{No.}+7.8))})$ r^2 = 0.80, n = 74; Female = $69(1 - e^{(-0.03(\text{No.}+7.3))})$ r^2 = 0.83, n = 137].

The greatest number of external bands observed was 57 (76 cm TL) for females and 48 (53 cm TL) for males. The largest female with a non-broken spine (80 cm TL) had 46 bands, and the largest male (72 cm TL) had 36 bands. Near-term embryos had no external bands, and the birth-band subtraction was not required. The base of larger (older) spines was fragile and was sometimes only partly formed.

Internal bands

Cross-sections of 420 spines were examined, although only 193 (46%) had a readable banding pattern in the inner dentine layer (Fig. 4). The

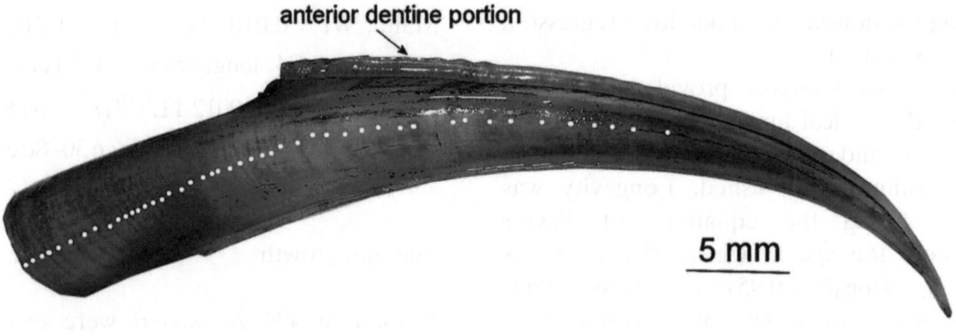

Fig. 3 Second dorsal fin spine of 64 cm TL male *E. baxteri* stained with an alizarin red derivative to enhance the external banding pattern (46 bands). Note enamel bands/ridges on the anterior dentine portion

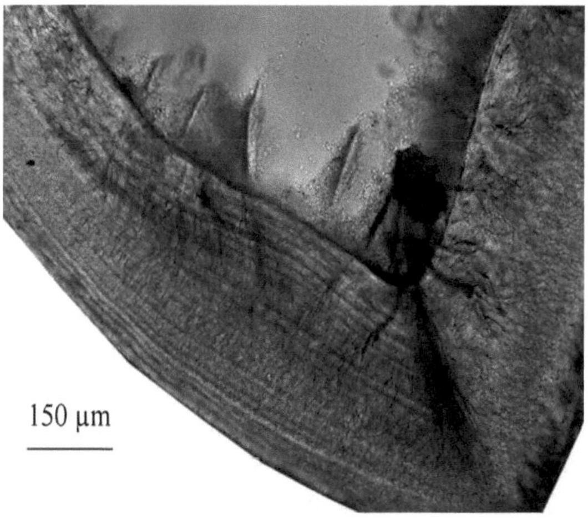

Fig. 4 Transverse-section photomicrograph (magnification 100×) of the second dorsal fin spine of a 69 cm female *E. baxteri* with 22 bands

optimal sectioning position (where band clarity was the best) was 10–15 mm from the spine tip.

The number of internal bands increased with animal size and the greatest number of bands observed was 22 (62 cm TL) for males and 26 (68 cm TL) for females. The spines of the largest male with readable bands (72 cm TL) had 15 internal bands, while the largest female (78 cm TL) has 19 internal bands. The CV of these samples was 3.1%. Only nine spines were excluded due to poor reading precision (CV > 20%). Near-term embryos had a poor inner spine structure and no bands were found in the dentine layers of embryos, making birth-band subtractions unnecessary.

Table 1 Francis growth parameter estimates for male and female *E. baxteri* using external bands and internal bands (including standard error and 95% confidence limits)

Parameters	Male	Female
External bands		
l_5 (cm)		
Estimate	24.86	23.11
SE	2.70	3.09
95% CI	6.12	5.38
$l_{22.5}$ (cm)		
Estimate	53.77	53.96
SE	0.66	0.76
95% CI	1.32	1.51
l_{40} (cm)		
Estimate	60.648	68.35
SE	0.56	0.42
95% CI	1.11	0.83
Sample size (n)	83	158
Coefficient of determination (r^2)	0.78	0.78
Internal bands		
l_3 (cm)		
Estimate	33.24	28.50
SE	1.65	3.01
95% CI	3.30	5.99
$l_{10.5}$ (cm)		
Estimate	53.64	57.35
SE	0.61	0.93
95% CI	1.22	1.83
l_{18} (cm)		
Estimate	59.64	69.43
SE	0.84	0.88
95% CI	1.67	1.75
Sample size (n)	54	99
Coefficient of determination (r^2)	0.79	0.74

r^2 coefficient of determination; l_5 is the mean TL at age 5; $l_{22.5}$ is the mean TL at age 22.5; l_{40} is the mean TL at age 40; l_3 is the mean TL at age 3; $l_{10.5}$ is the mean TL at age 10.5; l_{18} is the mean TL at age 18

Growth

Assuming external bands were formed annually, the Francis model fitted the observed length-at-age data well for both sexes (male r^2 = 0.78; female r^2 = 0.78) (Fig. 5a). Males and females exhibited similar growth (29–30 cm) between the

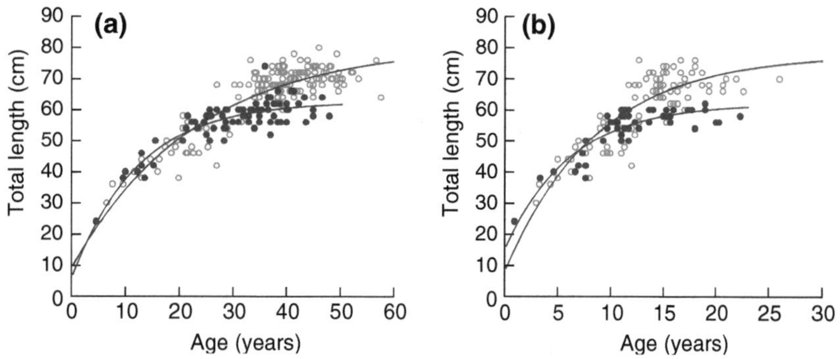

Fig. 5 Growth models for male (•) and female (○) *E. baxteri* using (**a**) external spine age estimates; (**b**) internal spine age estimates

ages of 5 and 22.5 years. Over the next 17.5 years (between 22.5 and 40 years), males grew an average of 7 cm compared with 14 cm for females (Table 1). Kimura's likelihood ratio test calculated a difference in the growth curves for male and female *E. baxteri* (coincident curve: $\chi^2 = 70.96$, $p < 0.001$). The curve intercepted the y-axis at 7.13 cm for males and 11.87 cm for females. The von Bertalanffy equations for external bands were: male $L_t = 60.59(1 - \exp(-0.082(t + 1.43)))$; female $L_t = 68.13(1 - \exp(-0.040(t + 4.51)))$.

Assuming internal bands were formed annually, the Francis model (Fig. 5b) fitted the male length-at-age data slightly better ($r^2 = 0.79$), than the female data ($r^2 = 0.73$). The Francis parameters indicated that males and females grow at a similar rate (29–30 cm) between ages 3 and 18. Over the next 7.5 years (10.5–18 bands) males only grew 6 cm on average, compared with 12 cm for females (Table 1).

Kimura's likelihood ratio test calculated a difference in the growth curves for male and female *E. baxteri* (coincident curve: $\chi^2 = 61.12$, $p < 0.001$). The curve intercepted the y-axis at 18.69 cm for males and 9.27 cm for females. The von Bertalanffy equations for internal bands were: male $L_t = 59.55(1 - \exp(-0.163(t + 2.00)))$; female $L_t = 69.25(1 - \exp(-0.116(t + 1.56)))$.

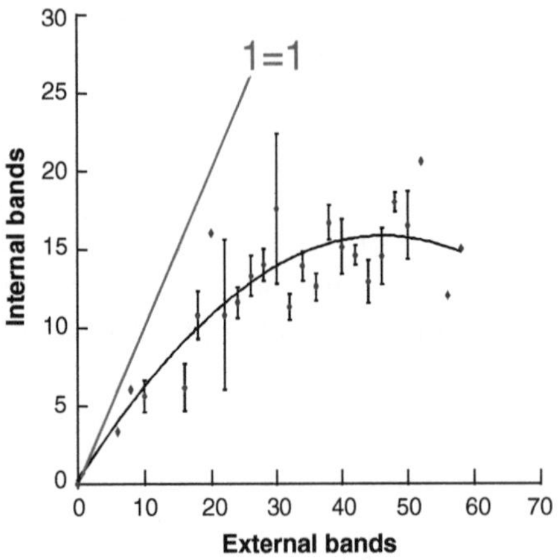

Fig. 6 Relationship between the number of external bands (EB) (age class) and the mean number of internal bands (IB) for *E. baxteri* from southeastern Australia. Error bands represent one SE and the polynomial relationship is $IB = -0.0074(EB^2) + 0.679(EB) + 0.1745$; $r^2 = 0.785$, $n = 101$. The linear line indicates the exponential relationship if one EB equals one IB

Correlation between external and internal growth bands

A curvilinear relationship was found between the number of external bands (EB) and internal bands (IB) [$IB = -0.0074(EB^2) + 0.679(EB) + 0.1745$, $r^2 = 0.785$, $n = 101$; Fig. 6]. Most spines had more external bands than internal bands. The polynomial relationship showed that the first two bands were formed at the same rate. Spines with 2–25 external bands had half as many internal bands, and the proportion of internal bands rapidly declined in spines with more than 25 external bands. After 30 external bands, no further internal bands were deposited. The growth curves from external bands were significantly different to the growth curves from internal bands. Francis parameters differed greatly for males and Kimura's likelihood ratio test calculated a significant difference in the growth curves from external and internal length-at-age data for male and female *E. baxteri* (male coincident curve: $\chi^2 = 118.87$, $p < 0.001$; female coincident curve: $\chi^2 = 61.29$, $p < 0.001$).

Longevity

Internal bands suggest longevity of 26 years for females and 22 years for males. The maximum number of external bands suggests female *E. baxteri* live longer than males; the oldest male examined was aged at 48 years and the oldest female was aged at 57 years. The corrected longevity estimates for a commercially fished population were much higher than the maximum band counts (Table 2).

Table 2 Longevity estimates for *Etmopterus baxteri* using age and growth data from external and internal bands

Method	External bands		Internal bands	
	Female	Male	Female	Male
Maximum band count	57	48	26	22
Taylor (1958)	126	72	27	27
Fabens (1965)	121	59	42	30
Ricker (1979)	87	42	30	21

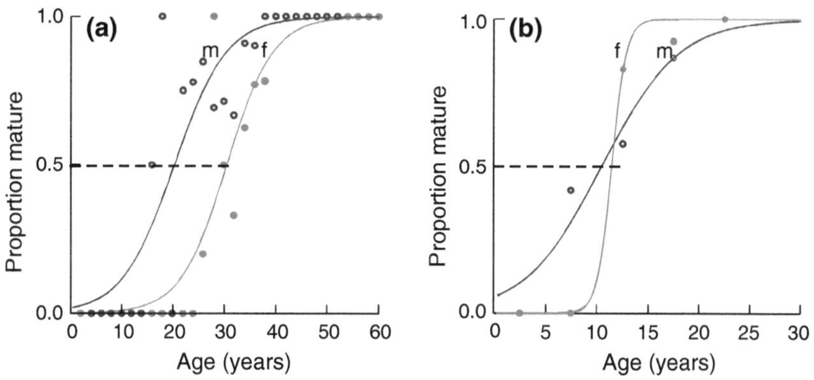

Fig. 7 Age at maturity ogives for (○) male and (•) female *E. baxteri* from southeastern Australia. (**a**) External band age estimates, male $= 1/(1 + e^{(-\ln(19)((AGE-20.3)/(35.3-20.3)))})$, $r^2 = 0.79$, $n = 111$; female $= 1/(1 + e^{(-\ln(19)((AGE-30.5)/(43.7-30.5)))})$, $r^2 = 0.91$, $n = 181$; (**b**) internal band age estimates, male $= 1/(1 + e^{(-\ln(19)((AGE-10.5)/(21.4-10.5)))})$, $r^2 = 0.97$, $n = 57$; female $= 1/(1 + e^{(-\ln(19)((AGE-11.4)/(13.44-11.42)))})$, $r^2 = 0.96$, $n = 100$

Age at maturity

External band maturity-at-age data suggest that male maturity occurs over a 21-year period, the youngest mature male was 16, and the oldest immature male was 37 years. Male A_{50} was estimated to occur at about 20 years (SE = 1.6) (Fig. 7a); this is 53 % of the maximum age (A_{max}). Female maturity also occurred over a broad age period (27–41 years), and A_{50} was estimated to occur at 30 years (SE = 1.1) (Fig. 7a); this is 42% of A_{max}. The youngest female in the early stages of pregnancy (fertilised eggs in utero) was 35 years, whereas the youngest near-term pregnant female was 39 years.

Internal band maturity-at-age estimates suggest that males mature at 47% of A_{max}, with an A_{50} of 10.5 years (range 8–20 years). Females matured between 7 and 16 years (A_{50} = 11.5 years); this is 44% of A_{max} (Fig. 7b).

Discussion

Age and growth

The vertebral centra of *E. baxteri* do not have a concentric banding pattern suitable for age determination. However, Gennari et al. (2002) used vertebral bands to estimate the age of *E. spinax* from the Mediterranean. Although these age estimates were not validated, the presence of bands in *E. spinax* vertebrae is very interesting and may be attributed to the warmer temperatures of the Mediterranean deepsea (~13°C below the thermocline) (Salat and Font 1987).

The precision of the age data as indicated by the CV values showed that external bands were easier to interpret than internal bands. Both internal and external bands increased in number with animal length, although each spine had more external bands. The relationship between internal and external counts suggests that only the first two bands were formed at the same time. A similar relationship was found between external and internal bands in *Centroselachus crepidater* dorsal spines (Irvine et al. 2006), and it was suggested that adult spines stop growing when the space above the pulp cavity has been filled by dentine. The present study also supports the possibility that internal dentine stops forming or that the internal bands become undistinguishable when tightly packed.

Verification and validation of the periodicity of band formation is essential, although most traditional techniques are unsuitable for deepwater species. Marginal increment analysis and edge analysis on internal dentine bands of *C. crepidater* proved unsuccessful (Irvine et al. 2006). Gullart-Furio (1998) outlined the problems associated with using these analyses on internal bands in dorsal spines, including the requirement of obtaining sections from the same location. These analyses are unsuitable on external bands as the banding pattern often consists of ridges of dentine rather than dark/light growth rings. Independent ageing techniques including bomb dating or radiometric isotope analysis may offer the best means of verifying the age estimates from band counts. Campana et al. (this issue) showed that *S. acanthias* spines could be aged using bomb carbon (^{14}C) dating. However, ^{14}C dating may be problematic in deepwater species because of a potential depleted bomb signal (i.e., phase lag) in areas where there are low deepwater mixing rates (Keer et al. 2004). The results of Fenton (2001) and Irvine et al. (2006) imply that further research is required into the usefulness of radiometric isotope analysis on both dogfish vertebrae and spines.

Age at maturity

Female age at maturity using external bands (30 years) differed from the age of the youngest pregnant female (35 years). However, no difference was found between the size at maturity (L_{50}) and the size of the smallest pregnant female (both 62 cm TL) (Irvine 2004), suggesting that females may not grow during reproductive activity.

Etmopterus baxteri exhibits slower growth and higher longevity than many other dogfishes, including a sympatric species (Irvine et al. 2006). However, *Centrophorus squamosus* from the northeastern Atlantic lives to 70 years (Clarke et al. 2002a), and is the oldest deepwater dogfish aged to date.

Implications

The resilience of a species to fishing pressure depends on vulnerability to the fishing gear and biological productivity (Stevens et al. 2000). The high longevity and late age at maturity of *E. baxteri* from southeastern Australia are indicators of a low productivity. Non-selective trawl fishing off southeastern Australia collected a wide age range of *E. baxteri*, although most females collected had only recently matured (24–34 years). The removal of this important population component is likely to be unsustainable. Although robust management arrangements have recently been announced for some deepwater fisheries, the accurate identification and recording of discarded species is crucial to ensure that the level of bycatch is monitored. Understanding the movement patterns of deepwater dogfishes is also of great importance for identifying sensitive habitats, and effective management and conservation may require the collaboration of different States or nations.

Acknowledgements We gratefully acknowledge the skippers and crew of the many orange roughy trawlers who collected dogfish samples. Thanks to C. Hunter for statistical help, and to those who provided technical assistance, especially R. Daley. Financial support for SBI to travel to this symposium was provided by California-Sea Grant and NOAA Fisheries Service-Southeast Fisheries Science Centre.

References

Beamish RJ, Fournier DA (1981) A method for comparing the precision of a set of age determinations. Can J Fish Aquat Sci 38(8):982–983

Beamish RJ, McFarlane GA (1985) Annulus development on the second dorsal spine of the spiny dogfish (*Squalus acanthias*) and its validity for age determination. Can J Fish Aquat Sci 42(11):1799–1805

Cailliet GM, Andrews AH, Burton EJ, Watters DL, Kline DE, Ferry-Graham LA (2001) Age determination and validation studies of marine fishes: do deep-dwellers live longer? Exp Gerontol 46:739–764

Cailliet GM, Goldman KJ (2004) Age determination and validation in chondrichthyan fishes. In: Carrier JC, Musick JA, Heithaus MR (eds) Biology of sharks and their relatives. CRC Press, USA, pp 399–447

Campana SE (2001) Review paper: accuracy, precision and quality control in age determination, including a review of the use and abuse of age validation methods. J Fish Biol 59:197–242

Campana SE, Jones C, McFarlance GA, Myklevoll S (2006) Bomb dating and age validation using the spines of spiny dogfish (*Squalus acanthias*). Environ Biol Fishes

Casey JG, Pratt HLJ, Stillwell CE (1985) Age and growth of the sandbar shark (*Carcharhinus plumbeus*) from the western North Atlantic. Can J Fish Aquat Sci 42(5):963–975

Chang WYB (1982) A statistical method for evaluating the reproducibility of age determination. Can J Fish Aquat Sci 39:1208–1210

Clarke MW, Connolly PL, Bracken JJ (2002a) Age estimation of the exploited deepwater shark *Centrophorus squamosus* from the continental slopes of the Rockall Trough and Porcupine Bank. J Fish Biol 60(3):501–514

Clarke MW, Connolly PL, Bracken JJ (2002b) Catch, discarding, age estimation, growth and maturity of the squalid shark *Deania calceus* west and north of Ireland. Fish Res 56(2):139–153

Clarke MW, Irvine SB (2006) Terminology for the ageing of chondrichthyan fish using dorsal-fin spines. Environ Biol Fishes

Correia JP, Figueiredo IM (1997) A modified decalcification technique for enhancing growth bands in deep-coned vertebrae of elasmobranchs. Environ Biol Fishes 50(2):225–230

Davenport S, Stevens JD (1988) Age and growth of two commercially important sharks (*Carcharhinus tilstoni* and *C. sorrah*) from northern Australia. Aust J Mar Freshw Res 39(4):417–433

Fabens AJ (1965) Properties and fitting of the von Bertalanffy growth curve. Growth 29:265–289

Fenton GE (2001) Radiometric ageing of sharks. Fisheries Research and Development Corporation, Canberra, Australia. FRDC Final Report 1994/021

Ferreira BD, Vooren CM (1991) Age, growth and structure of vertebra in the school shark, *Galeorhinus galeus* (Linnaeus, 1758) from southern Brazil. Fish Bull 89:19–31

Francis RICC (1988) Are growth parameters estimated from tagging and age-length data comparable? Can J Fish Aquat Sci 45:936–942

Francis MP, Maolagáin ÓC (2004) Feasibility of ageing shovelnose dogfish (*Deania calcea*), leafscale gulper shark (*Centrophorus squamosus*), and seal shark (*Dalatias licha*). Final research report for Ministry of Fisheries research project ENV2001/05, Objective 2, 31 pp

Gennari E, Scacco U, Vacchi M (2002) A simple technique for a preliminary vertebral ageing study in the velvet belly, *Etmopterus spinax* (Squalidae) in Central Mediterranean Sea. 6th European Elasmobranch Association Meeting, 6–8 September. National Museum and Galleries of Wales, Cardiff Wales, UK

Guallart Furio J (1998) Contribution to the knowledge of the biology and taxonomy of the deepsea shark, *Centrophorus granulosus* (Bloch and Schneider, 1801) (Elasmobranchii, Squalidae) in the Balearic Sea (western Mediterranean). PhD Thesis (in Spanish). Univertat de Valencia

Haddon M (2001) Modelling and quantitative methods in fisheries. Chapman & Hall/CRC, USA

Henderson AC, Flannery K, Dunne J (2002) Growth and reproduction in spiny dogfish *Squalus acanthias* L. (Elasmobranchii: Squalidae), from the west coast of Ireland. Sarsia 87:350–361

Hoenig JM (1979) The vertebral centra of sharks and their age determination. M.Sc. Thesis. University of Rhode Island, USA

Holden MJ, Meadows PS (1962) The structure of the spine of the spur dogfish (*Squalus acanthias*) and its use for age determination. J Mar Biol Assoc UK 42:179–197

Irvine SB (2004) Age, growth and reproduction of deepwater dogfishes from southeastern Australia. PhD thesis, Deakin University Warrnambool, Victoria, Australia

Irvine SB, Stevens JD, Laurenson. LJB (2006) Surface bands on deepwater squalid dorsal-fin spines: an alternative method for ageing the golden dogfish *Centroselachus crepidater*. Can J Fish Aquat Sci 63:617–627

Kaganovskaia SM (1933) A method for determining the age and composition of the catches of *Squalus acanthias*. Bull Far-East Br Acad Science USSR 1(3): 5–6 (Translated from Russian by WE Ricker). Fisheries Research Board of Canada, Translation Series No. 281, 1(3): 5–6

Keer LA, Andrews AH, Frantz BR, Coale KH, Brown TA, Cailliet GM (2004) Radiocarbon in otoliths of yelloweye rockfish (*Sebastes ruberrimus*): a reference time series for the coastal waters of southeast Alaska. Can J Fish Aquat Sci 61:443–451

Kimura DK (1980) Likelihood methods for the von Bertalanffy growth curve. Fish Bull 77(4):765–776

La Marca MJ (1966) A simple technique for demonstrating calcified annuli in the vertebrae of large elasmobranchs. Copeia Ichthological Notes 2:351 352

Machado PB, Figueiredo I (2000) A technique for ageing the birdbeak dogfish (*Deania calcea* Lowe, 1839) from dorsal spines. Fish Res 45(1):93–98

Maisey JG (1979) Finspine morphogenesis in squalid and heterodontid sharks. Zool J Linnean Soc 66:161–183

Moulton PL, Walker TI, Saddlier SR (1992) Age and growth studies of gummy shark, *Mustelus antarcticus* Günther, and school shark *Galeorhinus galeus* (Linnaeus) from southern Australian waters. Aust J Mar Freshw Res 43:1241–1267

Ricker WE (1979) Growth rates and models In: Hoar WS, Randall DJ, Brett JR (eds) Fish physiology, vol. 8. Academic Press, New York, USA, pp 678–743

Sainsbury KJ (1980) Effect of individual variability on the von Bertalanffy growth equation. Can J Fish Aquat Sci 37:241–247

Salat J, Font J (1987) Water mass structure near and offshore the Catalan coast during winters of 1982 and 1983. Ann Geophysicate 5:49–54

Schwartz FJ (1983) Shark ageing methods and age estimation of Scalloped Hammerhead, *Sphyrna lewini*, and Dusky, *Carcharhinus obscurus*, Sharks based on vertebral ring counts, pp 167–174. In: Prince ED, Pulos LM (eds) Proceedings of the international workshop on age determination of ocean pelagic fishes, tuna, billfishes and sharks. NOAA Technical Report NMFS 8

Sion L, D'Onghia G, Carlucci R (2002) A simple technique for ageing the velvet belly, *Etmopterus spinax* (Squalidae). In: Vacchi M, La Mesa G, Serena F, Séret B (eds) Proceedings of the 4th European Elasmobranch Association Meeting, Livorno (Italy) 2000, pp 135–139

Stehmann MFW (2002) Proposal of a maturity stage scale for oviparous and viviparous cartilaginous fishes (Pisces: Chondrichthyes). Arch Fish Mar Res 50(1):23–48

Stevens JD (1975) Vertebral rings as a means of age determination in the blue shark (*Prionace glauca* L.). J Mar Biol Assoc UK 55(3):657–665

Stevens JD, Bonfil R, Dulvy NK, Walker PA (2000) The effects of fishing on sharks, rays, and chimaeras (chondrichthyans), and the implications for marine ecosystems. ICES J Mar Sci 57(3):476–494

Tanaka S (1990) The structure of the dorsal spine of the deep sea squaloid shark *Centrophorus acus* and its

utility for age determination. Nippon Suisan Gakkaishi Bull Jpn Soc Sci Fish 56(6):903–909
Taylor CC (1958) Cod growth and temperature. Journal du Conseil International pour l'exploration de la Mer 23:366–370
von Bertalanffy L (1938) A quantitative theory of organic growth. Human Biol 10:181–213
Wetherbee BM (1996) Distribution and reproduction of the southern lantern shark from New Zealand. J Fish Biol 49(6):1186–1196
Wood CC, Ketchen KS, Beamish RJ (1979) Population dynamics of spiny dogfish (*Squalus acanthias*) in British Columbia waters. J Fish Res Board (Canada) 36(6):647–656

Environ Biol Fish (2006) 77:265–272
DOI 10.1007/s10641-006-9114-4

ORIGINAL PAPER

The potential use of caudal thorns as a non-invasive ageing structure in the thorny skate (*Amblyraja radiata* Donovan, 1808)

Michael J. Gallagher · Marianne J. Green · Conor P. Nolan

Received: 16 June 2006 / Accepted: 6 July 2006 / Published online: 3 August 2006
© Springer Science+Business Media B.V. 2006

Abstract The thorny skate, *Amblyraja radiata*, is the most widely distributed and abundant of all skate species worldwide, found on both sides of the north Atlantic Ocean. Large inter-regional size differences exist for this species and the few age and growth studies undertaken have revealed marked differences in life history traits for geographically distinct stocks. To facilitate the progression of further age and growth studies for this commercially important species, the effectiveness of caudal thorns as a rapid ageing tool was assessed. Twenty-eight male and 24 female thorny skates were collected off Greenland, covering the full size range of the species. Replicate age readings of crystal violet stained vertebral sagittal sections and whole silver nitrate stained caudal thorns revealed mean intra-reader age reading precision was higher for caudal thorns (Covariance (CV): reader 1 = 9.07, reader 2 = 9.73) than vertebrae (CV: reader 1 = 14.91, reader 2 = 14.27). Age bias plots revealed minimal inter-structure bias, apart from a higher average thorn age reading of 0.76 years from age classes 5–11 years for reader 1. Minor inter-reader bias was evident for vertebrae only; averaging 0.90 years higher for reader 1 from age classes 11 to 15 years. Preliminary evidence suggests caudal thorns could prove an effective non-invasive ageing tool for thorny skates.

Keywords Skate · Age · Caudal thorns · Vertebrae · Bias

M. J. Gallagher (✉) · C. P. Nolan
Irish Sea Fisheries Board, Killybegs,
Co. Donegal, Ireland
e-mail: gallagher@bim.ie

M. J. Green
Killybegs Fishermen's Organisation, Killybegs,
Co. Donegal, Ireland

Introduction

Amblyraja radiata (Donovan, 1808), commonly known as the thorny skate or starry ray, has an almost ubiquitous boreal distribution, spanning the northern Atlantic Ocean (Bigelow and Schroeder 1954). In the eastern North Atlantic Ocean, it is distributed from east Greenland, Iceland, the Barents Sea, Baltic Sea and as far south as the mid-North Sea and southern English Channel (Stehmann and Burkel 1984). In the western North Atlantic Ocean, it can be found from off the west Greenland coast, Hudson Bay, Canada to South Carolina (Bigelow and Schroeder 1954; Compagno et al. 1989). This species accounts for over 80% of the estimated biomass of all skates in the North Sea, east coast of Greenland (Rätz 1992, 1999), Barents Sea (Dolgov 2004), and Grand Banks (Kulka and Mowbray 1999).

Despite widespread abundance, marked contrasts in levels of commercial interest are evident for the thorny skate, particularly between the Northwest and Northeast Atlantic. In the North Sea (NE), it is discarded due to its small size (maximum size 54 cm total length [TL]) (Walker 1999). However, in the Grand Banks directed skate fishery (NW), the large growing thorny skates (maximum size 108 cm [TL]) is the most important component of directed and by-catch skate fisheries landing approximately 12,000 t annually (Kulka and Miri 2003).

Inter-regional heterogeneity in life history traits for thorny skates also appear to exist possibly due to a combination of both differing environmental conditions (Templeman 1987; Del Río 2002) and compensatory responses to intense exploitation (Walker and Hislop 1998). Therefore, robust area-specific age and growth studies are a basic requisite to enable meaningful assessments and subsequent management strategies to be implemented (Anderson 1990; Frisk et al. 2001). Despite being the most studied skate species, only a relatively small number of age and growth studies have been undertaken (Vinther 1989; Walker 1999; Sulikowski et al. 2005). In part, the paucity of age and growth studies can be attributed to the use of vertebrae for age assessment of skates, which is labour intensive and necessitates dissection for removal, cleaning, separation, and often the subsequent application of various enhancement techniques to improve band resolution (Cailliet 1990, Cailliet and Goldman 2004). Additionally, interpretation problems can often arise with reduced band resolution both at the focus and at the periphery of these structures, particularly for larger specimens (Cailliet et al. 1990; Francis and O'Maolagain 2005). These problems have also been encountered in a northeast Atlantic Ocean population of thorny skates using vertebrae as the ageing structure, which has lead to, in part, uncertainties in their derived growth data (Walker 1999). Gallagher and Nolan (1999) presented an alternative ageing technique using caudal thorns, which proved both rapid and reliable. Close association between vertebral and thorn band estimates was evident for four commercial skate species from the Falkland Islands and derived growth rate data (Gallagher 2000) were substantiated in a subsequent study utilizing caudal thorns (Hendersen et al. 2004).

The thorny skate, as its name implies, has a spinulose covering, distributed both dorsally and ventrally that vary in abundance depending on size, sex and maturity status (Bigelow and Schroeder 1954; Stehmann and Burkel 1984; Templeman 1987). Of importance however, is a distinct row of approximately 10 median caudal thorns that are consistently present from pre-hatching through adulthood (Berestovskii 1994; Stehmann and Merrett 2001). If these caudal thorns prove effective as a non-invasive, rapid ageing tool it would undoubtedly enable the proliferation in necessary age and growth studies for this ubiquitous species. This study set out to ascertain the effectiveness of these structures in this regard for the thorny skates by comparing thorn and vertebral band data from multiple readers.

Materials and methods

Sample collection

In October 2004, 52 thorny skates were collected onboard the 'Walther Herwig III', a German research vessel undertaking an annual groundfish survey off Greenland. Samples collected using otter trawl gear to a maximum depth of 400 m were labelled, blast frozen onboard, and subsequently transported to Ireland for further analysis.

Sample processing

In the laboratory each skate was sexed, TL (mm) (straight line distance from the tip of the rostrum to the end of the tail) and disc width (mm) (straight line distance between the tips of the widest portion of the pectoral fins) were measured. A vertebral segment containing approximately 10 vertebrae was extracted from each specimen from the abdominal cavity region. An incision was made under the thorn base tissue of the anterior caudal thorns towards the tail region, which enabled the removal of 7–10 caudal thorns. Excess tissue was removed from both vertebrae and thorns by immersion in hot water (circa. 70–85°C) followed by a light scraping with a fine scalpel. Both sets of structures were air-dried and then immersed in a 5% solution of trypsin (hog pancreatin) at 35–40°C for 24 h to remove neural and haemal arches and connective tissues from vertebrae and remove remaining tissue from caudal thorns. Following this process, centra and thorns were washed for 30 min in running water and stored in 70% industrial methylated spirits. For each sample, a single vertebra and thorn was selected. Vertebral and caudal thorn processing followed the protocol of Gallagher and Nolan (1999). Sagittal vertebral sections were stained with crystal violet and whole caudal thorns were stained with silver nitrate.

Age reading

For age reading purposes, a growth increment or 'band' on vertebral sagittal sections was defined as a translucent unstained band appearing at relatively regular intervals along the corpus calcareum, and separated by an opaque band (Fig. 1). As the intermedialia was often over-stained or ground away, the majority of band counts were derived from counts of the heavily stained bands on the corpus calcareum. The birth or hatching mark (age zero) was defined as the first mark after the change in the angle of the corpus calcareum. All age readings were made from the focus towards the distal edge on a single axis exhibiting greatest resolution. Band counts were corroborated with readings made along the remaining three axes. For whole caudal thorns, bands were defined as concentric surface ridges separated by broader bands, distal to the base of the birth-mark ridge (Fig. 2). Caudal thorn band counts were made from the tip towards the distal margins. The location of the birthmark (Fig. 3) was defined by the examination of a recently hatched juvenile of 11.3 cm (TL). Three non-consecutive band counts were made independently by two readers for each vertebral section and caudal thorn without prior knowledge of the specimen's details.

A dissecting microscope (magnification 6-40×) was used to view both sets of structures. A fibre-optic source facilitated the illumination of vertebral sections with transmitted light, and whole thorns were viewed under both reflected and transmitted light. Each vertebral section was measured from focus to distal margin, and the base of each thorn was measured from anterior to

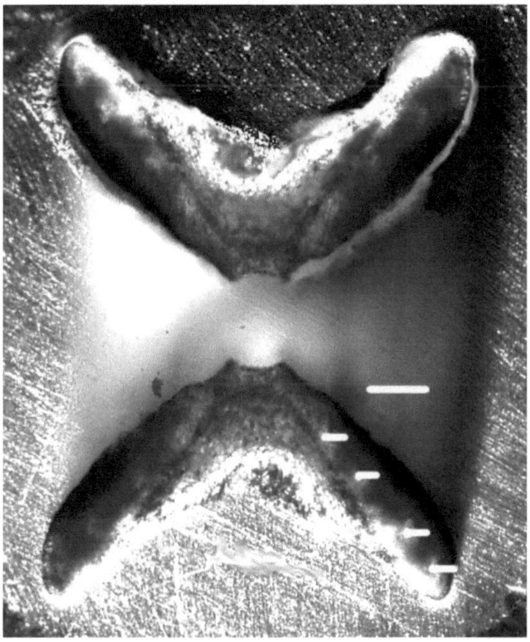

Fig. 1 Sagittal section of crystal violet stained vertebral centra from a thorny skate *A. radiata* measuring 356 mm TL estimated to be 4 years (scale bar = 0.8 mm)

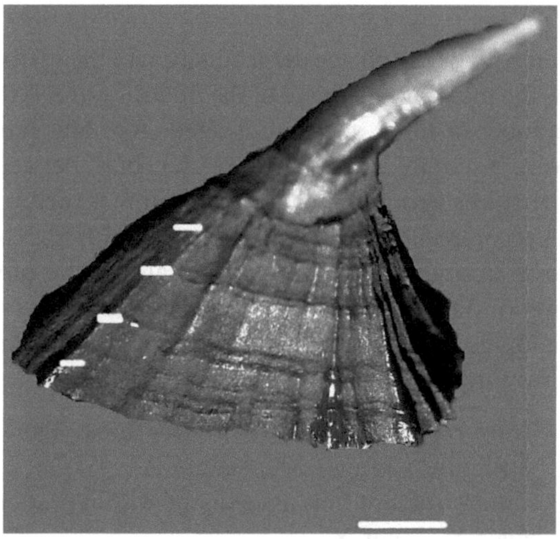

Fig. 2 A whole silver nitrate stained caudal thorn from a thorny skate *A. radiata* measuring 316 mm TL estimated to be 4 years (scale bar = 1.1 mm)

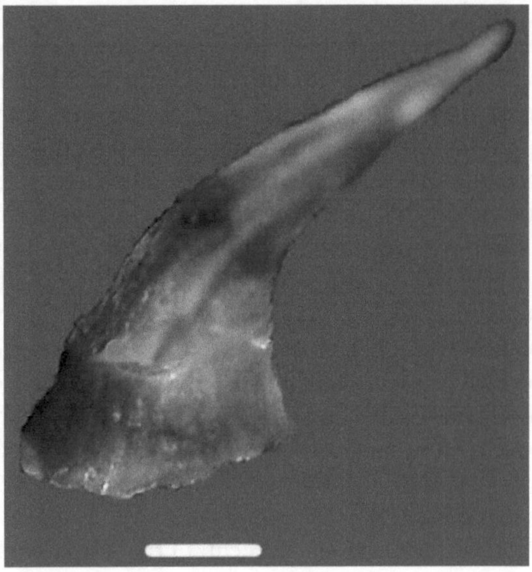

Fig. 3 A whole un-stained caudal thorn from a recently hatched thorny skate *A. radiata* measuring 113 mm TL (scale bar = 0.9 mm)

posterior using a graticule on the microscope, to relate structural to somatic growth.

Precision and bias

Coefficient-of-variation (CV) (Chang 1982) was used to assess precision, described as

$$CV = 100x\frac{\sqrt{\sum_{i=1}^{R}\frac{(X_{ij}-X_j)^2}{R-1}}}{X_j}$$

where X_{ij} is the ith age determination of the jth fish, X_j the mean age estimate of the jth fish, and R is the number of times each fish is aged. Age bias plots were used to assess intra reader bias between structures and inter-reader bias within structures, as recommended by Campana et al. (1995). This form of graphical representation allows the plotting of one set of readings versus a second set of readings with reference to an equivalence line. Two forms of bias were assessed;

- intra-reader bias between structures: ager X structure 1 = ager X structure 2
- inter-structure bias assessment within readers: ager X = ager Y

The age readings of ager Y, or ager X structure 2 are presented as the mean and 95% confidence interval corresponding to each of the age categories reported by ager X or ager X structure 1 (Campana et al. 1995).

Results

Structure growth and band morphologies

A strong linear relationship was evident ($r^2 = 0.93$) for vertebral radius and TL (mm) (Fig. 4), whereas thorn length and TL was weaker and was best described by a logarithmic relationship ($r^2 = 0.67$) (Fig. 5).

Overall, band resolution on vertebral sections was reasonable. However variable stain uptake presented problems on occasion, necessitating de-staining and re-staining. In addition, locating the birth-mark near the focus and crowding of bands towards the periphery was evident for a small number of samples of vertebral sections. Although band resolution was generally very good on caudal thorns from juveniles (<30 cm TL), thorn-wear near the tip in larger specimens (>45 cm TL) tended to occlude resolution in this region and in addition bands near the periphery were, on occasion, split or crowded.

Precision and bias

Mean CV estimates were similar for both readers, being consistently lower for thorn readings, at 9.07% and 9.73%, than vertebral readings at,

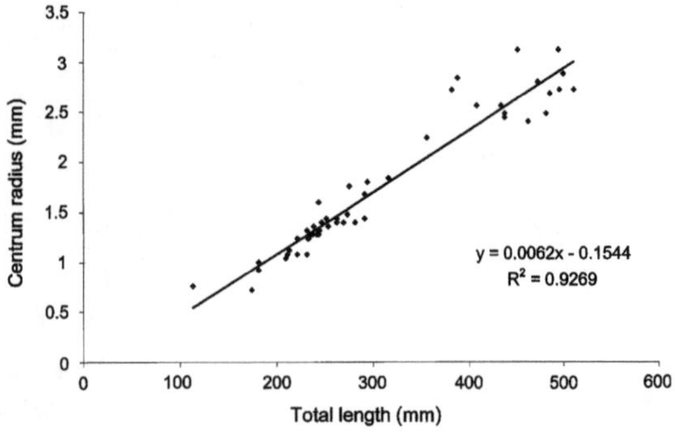

Fig. 4 The relationship of TL (mm) to centrum radius (mm) for combined sexes of thorny skate *A. radiata*. $n = 52$

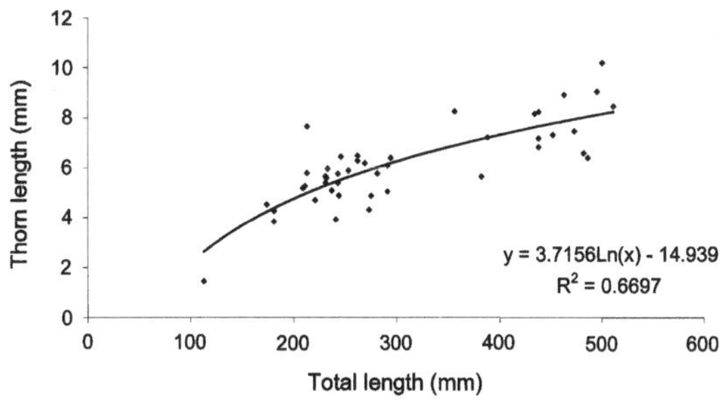

Fig. 5 The relationship of TL (mm) to thorn length (mm) for combined sexes of thorny skate *A. radiata*. $n = 52$

14.93% and 14.23%, for reader 1 and reader 2, respectively.

Age bias plots revealed no appreciable inter-structure bias (Figs. 6 and 7), apart from a higher average reading of 0.76 years for thorns compared to vertebrae for reader 1 between age classes 5–11 years (Fig. 6). Intra-reader bias for each structure was also minimal, with the largest bias evident for vertebral age readings, with reader 1 presenting 0.90 years higher ages than reader 2 between age classes 11–15 years (Fig. 7).

Discussion

Although relatively commonplace for commercial teleost species (Casselman 1983; Beamish and McFarlane 1987), inter-structure comparative age reading studies for elasmobranchs are rare (Gallagher and Nolan 1999; Francis and O'Maolagain 2005; Smith et al. 2005), mainly because few elasmobranchs, apart from skates, possess secondary calcified structures that can be used for age assessment purposes. Even when well established for certain teleosts, inter-structure age reading agreements across an entire size range for a given species are seldom achieved, especially for the larger size classes (Beamish and McFarlane 1987). Results from this study suggest that caudal thorns prove effective as an ageing alternative for thorny skates or serve as a useful form of age verification for vertebral age determination, given the similarity of age readings from both novel and conventional structures. Precision estimates for thorns were relatively reasonable for both readers, considering that over 90% of elasmobranchs studies returned CV estimates in excess of 10% (Campana 2001). CV estimates for vertebral age readings were higher for both readers, revealing less precision than thorns and were in line with precision estimates from other elasmobranch studies utilising vertebrae (Cailliet et al. 1990; Campana 2001; Francis et al. 2001). Consistent bias was not prevalent apart from slightly higher inter-reader vertebral age estimates for reader 1. This may in part be explained by the fact that unlike reader 2, reader 1 had previous experience at elasmobranch vertebral age reading, and generally more experienced age readers tend to resolve more bands in ageing structures (Officer et al. 1996). It is however difficult to ascertain which structure or indeed which reader presents the most accurate age estimates without knowing the true age of structures examined (Cailliet and Goldman 2004).

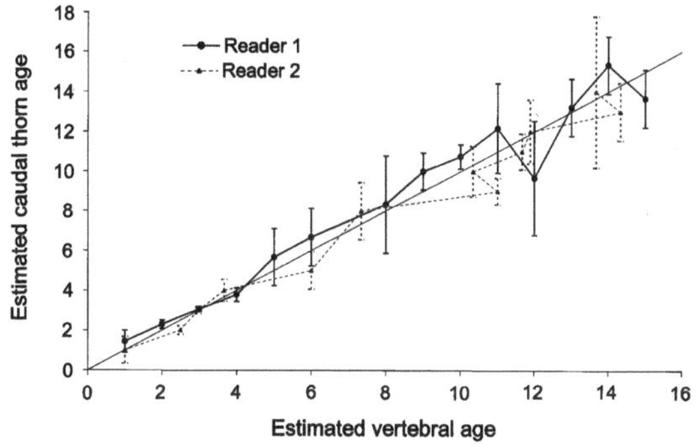

Fig. 6 Inter-structure age bias plot for the thorny skate *A. radiata*. Each error bar represents the 95% confidence interval for the mean caudal thorn age estimate, around that of a given vertebral age estimate by readers 1 and 2. The straight line represents the 1:1 equivalence line. $n = 52$

As with most elasmobranchs studied to date, the periodicity of band formation has not been comprehensively validated for the thorny skate (Sulikowski et al. 2005). However a comparison of tag-recaptures and vertebral age estimates in the North Sea (Walker 1999) and a study of marginal vertebral increments undertaken on samples collected in the Gulf of Maine (Sulikowski et al. 2005) support an annual band pair deposition cycle, as has been documented for other skate species (Holden and Vince 1973; Gallagher et al. 2005a). In addition, the oldest age

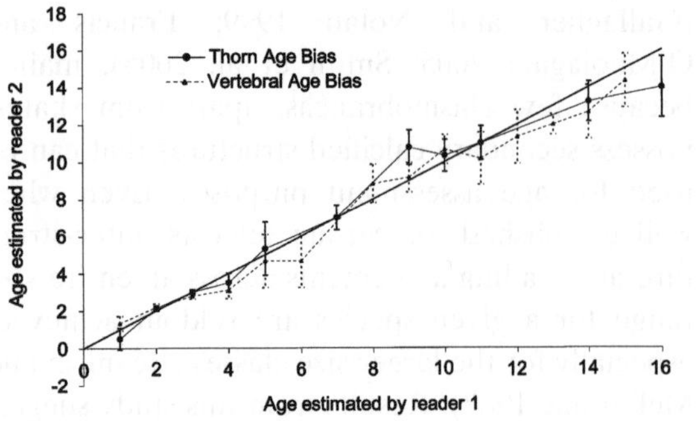

Fig. 7 Inter-reader caudal thorn and vertebral age bias plot for the thorny skate *A. radiata*. Each error bar represents the 95% confidence interval at the mean of reader 2's age estimate around that of a given age estimate by reader 1. The straight line represents the 1:1 equivalence line. $n = 52$

suggested by Sulikowski et al. (2005) for vertebral band counts of 16 years is comparable to 15 and 16 years for vertebral and caudal thorn band counts respectively in this study. Furthermore, Templeman's (1984) long-term tagging studies also revealed that the thorny skate has a longevity of at least 20 years.

Plotting of vertebral radius versus TL revealed a strong linear relationship ($r^2 = 0.93$), indicating that vertebral growth correlates well with somatic growth for the thorny skate. Gallagher et al. (2005b) demonstrated a strong logarithmic relationship between thorn height and TL for the white tipped skate *Bathyraja brachyurops*. However, thorn wear and proto-thorn damage was evident for several specimens in this study and precluded the effective use of this measurement. Thorn length versus TL was used instead, however the derived logarithmic relationship was relatively poor ($r^2 = 0.67$), largely due to the variability in intra specimen thorn size. It is probable that selection of a larger sample of specimens where thorn wear was not evident and measurement of thorn height would reveal allometric growth, as has previously been observed for caudal thorns (Gallagher et al. 2005b).

For elasmobranchs, there remains uncertainty about the causal factors of band formation processes and the link between environmental and physiological cues (Officer 1995; Cailliet and Goldman 2004). Natanson (1993) revealed that holding little skate, *R. erinacea*, under a constant temperature environment did not influence the rate of vertebral band formation, but did however influence band morphology. As initially postulated for dorsal fin spines (Holden and Meadows 1962), Gallagher et al. (2005b) suggested that surface defined ridges on caudal thorns are the external expressions of bases of successive cones that form with the cessation in somatic growth as a result of a marked annual cyclical reduction in seawater temperature. Although we collected samples within a narrow time period (October 2004), the majority of vertebral samples had either a fully formed opaque band or had initiated the formation of a translucent band, which concurs with the timing of vertebral band formation of other skate species in the north east Atlantic (Holden and Vince 1973; Gallagher et al. 2005a). In addition, for caudal thorns either a fully formed broad band or the initiation of a surface ridge was evident, which coincides with the initiation of the annual winter cycle of seawater temperature reduction (Rätz 1999), and also the movement of thorny skates into deeper colder water at this time of year in this area (>400 m) (Rätz 1999).

Not all skate species possess suitable caudal thorns for age assessment. The apparent absence of surface band patterns on caudal thorns from four temperate skate species from Irish coastal waters was attributed to a lack of a marked seasonal drop in seawater temperatures (Gallagher et al. 2005b). It has also been suggested that surface band patterns on caudal thorns were not evident on thorny skates collected off New England (Sulikowski personal communication). This was however a preliminary appraisal and it was acknowledged that further study was required to verify this (Sulikowski personal communication). It is clear that the cues for band formation warrant further investigation in this novel structure particularly if intra-specific regional differences exist. Of importance is the need for additional work on thorn growth processes through detailed analysis of the underlying band structures utilizing available technologies (e.g. histology, electron microscopy, laser ablation, electron microprobe, X-ray analysis) in conjunction with captive rearing trials.

The need for future research on thorn and band growth processes should not preclude their application as valid ageing tools for the thorny skate. Particularly since much remains unresolved regarding our understanding of vertebral growth band formation processes, the commonly used ageing tool for elasmobranchs (Cailliet and Goldman 2004). Caudal thorns offer a number of distinct advantages over conventional vertebral ageing including ease of removal and preparation for ageing (which results in minimal commercial loss). More importantly specimens do not need to be sacrificed, which is vital either in tag-recapture studies, where thorns can be removed at tagging and recapture for age validation, or where biological information is required for depleted stocks without further impacting on their status. This preliminary study has revealed that caudal thorns from thorny skates possess a surface band pattern that relates to vertebral band formation and thus can be used for age assessment purposes.

Acknowledgements We would like to express our thanks to the crew of the 'Walther Herwig III' for collection of samples and to Joerg Oehlenschlaeger and Hans Juergen-Knaack of the Federal Research Centre, Hamburg, Germany who arranged collection and delivery of samples. Gratitude is also extended to Peter Stafford and Alison Boyce of the Zoology Department, Trinity College Dublin, Ireland and Frances O'Dwyer of the Irish Sea Fisheries Board, Killybegs, Ireland who provided technical support. Thanks also to Coilín Minto from the Department of Biological Sciences, Dalhousie University, Halifax, Canada, for assistance in interpreting the data. We would like to thank the NOAA Fisheries Service—Southeast Fisheries Science Center for providing funding to present this paper at the 2005 Joint Meeting of Ichthyologists and Herpetologists, Tampa Bay, Florida, US.

References

Anderson ED (1990) Fishery models as applied to elasmobranch fisheries. In: Pratt HL, Gruber SH, Taniuchi T (eds) Elasmobranchs as living resources: advances in the biology, ecology, systematics and the status of the fishery. NOAA Technical Report NMFS 90, pp 473–484

Beamish RJ, McFarlane GA (1987) Current trends in age determination methodology. In: Summerfelt RC, Hall GE (eds) Age and growth in fish. Iowa State University Press, Ames, USA pp 15–42

Berestovskii EG (1994) Reproductive biology of skates of the family Rajidae in the seas of the far north. J Ichthyol 34:26–37

Bigelow HB, Schroeder WC (1954) Fishes of the western north Atlantic, part II. Sawfishes, guitarfishes, skates, rays, chimaeroids. Sears Foundation, Bingham Oceanographic Laboratory, New Haven, 588 pp

Cailliet GM (1990) Elasmobranch age determination and verification: an updated review. In: Pratt HL, Gruber SH, Taniuchi T (eds) Elasmobranchs as living resources: advances in the biology, ecology, systematics and the status of the fisheries. NOAA Technical Report NMFS 70, pp 157–165

Cailliet GM, Goldman KJ (2004) Age determination and validation in Chondrichthyan fishes. In: Carrier J, Musick JA, Heithaus MR (eds) Biology of sharks and their relatives. CRC Press LLC, Boca Raton, Florida pp 339–447

Cailliet GM, Yudin KG, Tanaka S, Taniuchi T (1990) Growth characteristics of two populations of *Mustelus manazo* from Japan based upon cross-readings of vertebral bands. In: Pratt HL, Gruber SH, Taniuchi T (eds) Elasmobranchs as living resources: advances in the biology, ecology, systematics, and the status of the fisheries. NOAA Technical Report NMFS 90, pp 167–176

Campana SE (2001) Accuracy, precision and quality control in age determination, including a review of the use and abuse of age validation methods. J Fish Biol 59:197–242

Campana SE, Annand MC, McMillan JI (1995) Graphical and statistical methods for determining the consistency of age determinations. Trans Am Fish Soc 124:131–138

Casselman JM (1983) Age and growth assessment of fish from their calcified tissue-techniques and tools. In: Prince ED, Pulos LM (eds) Proceedings of the international workshop on age determination of oceanic pelagic fishes: tunas, billfishes and sharks. NOAA Technical Report NMFS 8, pp 157–166

Chang WYB (1982) A statistical method for evaluating the reproducibility of age determination. Can J Fish Aquat Sci 39:1208–1210

Compagno LJV, Ebert DA, Smale MJ (1989) Guide to sharks and rays of southern Africa. New Holland Publishers, London, 158 pp

Del Río JL (2002) Some aspects of the thorny skate, *Amblyraja radiata*, reproductive biology in NAFO Division 3N. NAFO SCR Document 02/118, serial no. N4739, 14pp

Dolgov AV (2004) Feeding and food consumption by the Barents sea skates. J Northw Atl Fish Sci 35: http://www.journal.nafo.int/35/34–dolgov.html

Francis MP, O'Maolagain C, Stevens D (2001) Age, growth, and sexual maturity of two New Zealand endemic skates, *Dipturus nasutus* and *D. innominatus*. N Z J Mar Freshwater Res 35:831–842

Francis MP, O'Maolagain C (2005) Age and growth of the Antartic skate, *Amblyraja georgiana*, in the Ross Sea. CCAMLR Sci 12:183–194

Frisk MG, Miller TJ, Fogarty MJ (2001) Estimation and analysis of biological parameters in elasmobranch fishes: a comparative life history study. Can J Fish Aquat Sci 58:969–981

Gallagher MJ, Nolan CP (1999) A novel method for the estimation of age and growth in rajids using caudal thorns. Can J Fish Aquat Sci 56:1590–1599

Gallagher MJ (2000) The fisheries biology of commercial ray species from two geographically distinct regions. Ph.D. Thesis, Department of Zoology, University of Dublin, Trinity College, Dublin 2, Ireland

Gallagher MJ, Nolan CP, Jeal F (2005a) Age, growth and maturity of the commercial ray species from the Irish Sea. J Northw Atl Fish Sci 35:47–66

Gallagher MJ, Nolan CP, Jeal F (2005b) Structure and growth processes of caudal thorns. J Northw Atl Fish Sci 35:125–129

Henderson AC, Arkhipkin AI, Chtcherbich JN (2004) Distribution, growth and reproduction of the white-spotted skate *Bathyraja albomaculata* (Norman, 1937) around the Falkland Islands. J Northw Atl Fish Sci 35: http://www.journal.nafo.int/35/1-henderson.html

Holden MJ, Meadows PS (1962) The structure of the spine of the spur dogfish (*Squalus acanthias*) and its use for age determination. J Mar Biol Assoc U K 42:179–197

Holden MJ, Vince MR (1973) Age validation studies on the centra of *Raja clavata* using tetracycline. Journal du Conseil International pour l'Exploration de la Mer 35:13–17

Kulka DW, Miri CM (2003) The status of thorny skate (*Amblyraja radiata* Donovan, 1808) in NAFO Divisions 3L, 3N, 3O and subdivisions 3Ps. Canadian Science Advisory Secretariat Research Document—2003/031

Kulka DW, Mowbray FK (1999) An overview of the grand banks skate fishery. In: Shotton R (ed) Case studies of the management of elasmobranch fisheries. FAO Fisheries Technical Paper, Rome pp 47–73

Natanson LJ (1993) Effect of temperature on band deposition in little skate. Copeia 1:199–206

Officer RA (1995) Vertebral mineralisation patterns in gummy and school sharks and their utility for age determination. Ph.D Thesis, University of Melbourne, Victoria, Australia

Officer RA, Gason A, Walker TI, Clement JG (1996) Sources of variation in counts of growth increments in vertebrae from gummy shark, *Mustelus antarcticus*, and school shark, *Galeorhinus galeus*: implications for age determination. Can J Fish Aquat Sci 53:1765–1777

Rätz HJ (1992) Decrease in the fish biomass of West Greenland (Subdivisions 1B-1F), continued. NAFO SCR 92/40, serial no. N2088

Rätz HJ (1999) Structures and changes of the demersal fish assemblage off Greenland, 1982–96. NAFO Sci Coun Studies 32:1–15

Smith WD, Perez CR, Ebert DA (2005) Growth of the California skate, *Raja inornata*: assessment of multiple aging structures and somatic growth models. In: Abstract. American Society of Ichthyologists and Herpetologists/American Elasmobranch Society/Annual Meeting. Tampa, FL

Stehmann M, Burkel DL (1984) Rajidae. In: Whitehead PJP, Bauchot ML, Hureau JC, Nielson J, Tortonese E (eds) Fishes of the north-eastern Atlantic and the Mediterranean, vol 1. UNESCO, Paris, pp 163–196

Stehmann MFW, Merrett NR (2001) First records of advanced embryos and egg capsules of Bathyraja skates from the deep north-eastern Atlantic. J Fish Biol 59:338–349

Sulikowski JA, Kneebone J, Elzey S, Jurek J, Danley PD, Howell WH, Tsang PCW (2005) Age and growth estimates of the thorny skate (*Amblyraja radiata*) in the western Gulf of Maine. Fish Bull 103:161–168

Templeman W (1984) Migrations of thorny skate, *R. radiata*, tagged in the Newfoundland area. J Northw Atl Fish Sci 5:55–63

Templeman W (1987) Length–weight relationships, morphometric characteristics and thorniness of thorny skate (*Raja radiata*) from the northwest Atlantic. J Northw Atl Fish Sci 7:89–98

Vinther M (1989) Some notes on the biology of the starry ray *Raja radiata*, in the North Sea. Working Document for ICES study group on elasmobranch fisheries, April 1989, 20p

Walker PA, Hislop JRG (1998) Sensitive skates or resilient rays? Spatial and temporal shifts in ray species composition in the central and north-western North Sea between 1930 and the present day. ICES J Mar Sci 55:392–402

Walker PA (1999) Fleeting images dynamics of north sea ray populations. Ph.D. Thesis, University of Amsterdam

Environ Biol Fish (2006) 77:273–277
DOI 10.1007/s10641-006-9131-3

ORIGINAL PAPER

Terminology for the ageing of chondrichthyan fish using dorsal-fin spines

Maurice W. Clarke · Sarah B. Irvine

Received: 28 June 2006 / Accepted: 12 July 2006 / Published online: 17 August 2006
© Springer Science+Business Media B.V. 2006

Abstract Dorsal-fin spines are becoming increasingly popular as a suitable structure for ageing some chondrichthyan fishes. However, the terminology used to describe dorsal-fin structure is often inconsistent between studies, and is consequently unclear. Standardised terms and definitions of the dorsal-fin spine structure are proposed, with particular focus on those areas that are used for age and growth studies.

Keywords Squaliformes · Heterodontiformes · Chimaeriformes · Age · Growth

Introduction

Dorsal-fin spines are present in three extant orders of chondrichthyan fishes: Squaliformes (dogfishes), Heterodontiformes (horn sharks) and Chimaeriformes (chimaeras/ghost sharks). Spines are located at the dorsal midline over the vertebral column, and are not shed like teeth, denticles and stingray barbs. Chimaeras have a single spine anterior to the first dorsal fin that lies just beneath the skin and can be raised and lowered (Maisey 1979). Most dogfishes and horn sharks have two spines (one anterior to each dorsal fin) that are usually deeply embedded into the epaxial muscle tissue, and cannot be raised and lowered.

Spines of the spiny dogfish, *Squalus acanthias*, have been used to estimate age since the early 1900s (Kaganovskaia 1933), and the spines of other Chondrichthyes are receiving increasing attention as suitable structures for estimating age and growth (Clarke et al. 2002; Francis and ÓMaolagáin 2000; Irvine et al. 2006a). The structure of the dorsal-fin spine has been adequately detailed by Maisey (1979), Beamish and McFarlane (1985) and Tanaka (1990), although each study uses slightly different terms to define spine structure. Our objectives are to standardise the terms and definitions of dorsal-fin spines used in age and growth studies.

Spine structure

Standardised terms for the description of the external and internal spine structure are presented in Table 1. The spines of dogfishes and horn sharks are structurally similar, although most dogfish spines are generally curved posteriorly while horn sharks spines are almost vertical.

M. W. Clarke
Marine Institute, GTP, Parkmore, Galway, Ireland

S. B. Irvine (✉)
Department of Fisheries Western Australia,
Level 3 The Atrium, 168 St Georges Terrace, Perth,
WA 6001, Australia
e-mail: sbirvine@aapt.net.au

Table 1 Standardised definitions for dorsal-fin spine structure

Structure	Definition
External	
Anterior dentine portion	Thickened strip of dentine on anterior edge of external portion of some spines. Some deepwater dogfish species spine have enamel coating over portion. Sometimes referred to as the *anterior keel* on chimaera spines
Anterior–lateral faces	Forward faces of the spine
Enamel	Hard, sculptured tissue overlaying the stem
External spine point/base	Point where spine enters the body
Posterior face	Rearward face of spine, usually concave
Posterior–lateral dentine portions	Thickened, reduced rib of enamel present on posterior corners of the spines of some deepwater species
Mantle	Dentinous zone of the enamel cap
Spine base	Base of spine, point where stem ends within the body
Tip	Apex of spine
Stem	The main body of the spine often referred to as the *trunk*
Internal	
Cartilage rod	Rod of cartilage occupying the proximal part of lumen
Dentine layer	One of several layers of dentine that comprise the stem. Form separately, so age estimation must be based on only one layer
Discontinuity[a]	An irregular band between the inner and outer layers
Inner dentine layer	Inner concentric zone of dentine
Middle dentine layer[a]	Middle concentric zone of dentine
Outer dentine layer	Outer concentric zone of dentine
Lumen	Central cavity
Pulp cavity	Pulp filled part of the lumen, houses the nerves and blood vessels
Trunk primordium	Thin mineralised layer representing the beginning of dentinogenesis

[a] Authors differ on whether this is a part of the inner layer or a separate middle layer

The appearance of chimaera spines differs considerably (Freer and Griffiths 1993). The internal structure of dogfish and horn shark spines are similar, although there is disagreement as to the number of dentine layers in each spine. Several authors (Holden and Meadows 1962; Beamish and McFarlane 1985; McFarlane and Beamish 1987; Tanaka 1990; Irvine et al. 2006a) have reported three dentine layers. However, Guallart Furio (1998) and Clarke et al. (2002) reported only two layers (inner and outer), and defines the 'middle' layer reported in other studies as a discontinuity (Table 1). We do not seek to resolve this debate here; rather we hope that researchers of subsequent studies identify the inner dentine layer correctly by following the standardised terminology presented here to avoid further confusion.

Growth zones

Standardised terms for the growth zones of the spine and the related definitions for those regions used for ageing are presented in Table 2. This is based largely on the work of Wilson et al. (1983), and the definitions do not depart from that seminal glossary. There are three different areas that may exhibit growth bands; enamel, stem and internal dentine. Enamel (including anterior

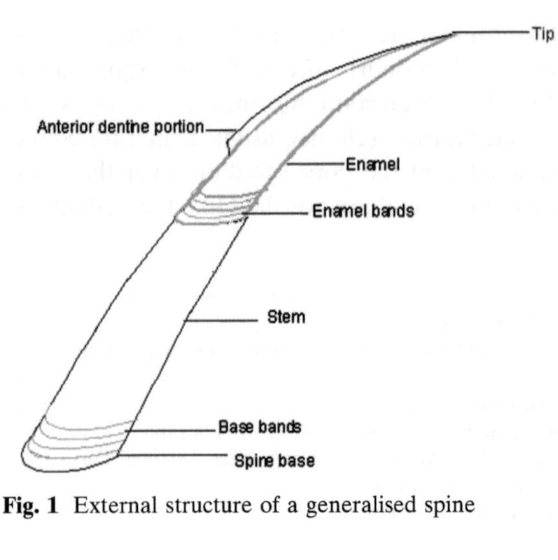

Fig. 1 External structure of a generalised spine

Table 2 Suggested definitions for age and growth studies using dorsal-fin spines

Ageing term	Definition
Annual mark	A feature that is formed annually. Often incorrectly termed *annulus*, which means a circular or ring-shaped structure
Band	A concentric zone or mark. One band generally refers to a dark and light ring/zone
Check	An abrupt discontinuity within or between structures
Cross-section	Transverse section sensu Wilson et al. (1983) to examine internal bands
Enamel band	Obvious band/ridge on the enamel. One band consists of either a ridge of enamel, or pigmented pair of dark and light rings
External band	A band that follows the shape of the spine based on the stem
Dorsal-fin spine	A spine of a squalid or heterodontid shark or a chimaera, being: • Open based • Unbranched distally • Unsegmented
Internal band	Growth band within the inner dentine layer. One band consists of dark and light ring that usually encircle the lumen or pulp cavity
Longitudinal section	Horizontal section sensu Wilson et al. (1983)
Marginal increment	The band/zone beyond the last identifiable band/zone at the margin of an ageing structure
Opaque	A zone inhibiting passage of light • Under transmitted light: appears dark • Under reflected light: appears bright
Primordium	A self-contained zone that represents the point of origin of growth
Ring	Light or dark concentric zone
Translucent	A zone allowing passage of light • Under transmitted light: appears bright • Under reflected light: appears dark
Validation	The process of showing that a method for age determination is accurate. Usually used for determining the annual formation of bands
Verification	The process of verifying the precision of band counts using multiple methods or ageing structures
No wear point	The most distal point of enamel/dentine on the anterior edge of a spine that has been abraded at the tip. Enamel bands past this point may be faint, difficult to read, or eroded away

dentine portion) bands (Fig. 1) are usually composed of a smoother area and ridge of enamel that is often pigmented (Kaganovskaia 1933; Clarke et al. 2002). Stem (external) bands generally follow the shape of the spine base (Irvine et al. 2006a; Irvine et al. 2006b). Bands in the

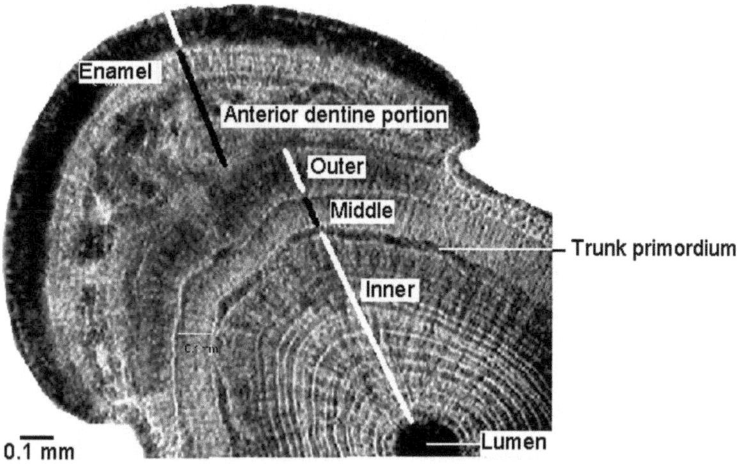

Fig. 2 Spine cross-section showing the internal structure [using the second dorsal spine of an Endeavour dogfish (*Centrophorus* cf. *uyato*) from southern Australia as an example]. Authors differ on whether the middle layer is a separate layer or part of the outer layer

Table 3 Definitions of morphometric measurements and the suggested acronyms

Measurement	Acronym	Definition
Total spine length[a]	TSL	Distance between tip to spine base
External spine length	ESL	Distance between tip and point of entry into body
External spine width	ESW	Width of spine at point of entry into the body
Wear point width	WPW	Width of spine at no wear point
Spine base width	SBW	Width of spine at spine base
Apex spine length	ASL	Distance between apex of the pulp cavity to spine base
Apex pulp cavity diameter	APC	Width of spine at the apex of the pulp cavity

[a] Unusable in worn spines

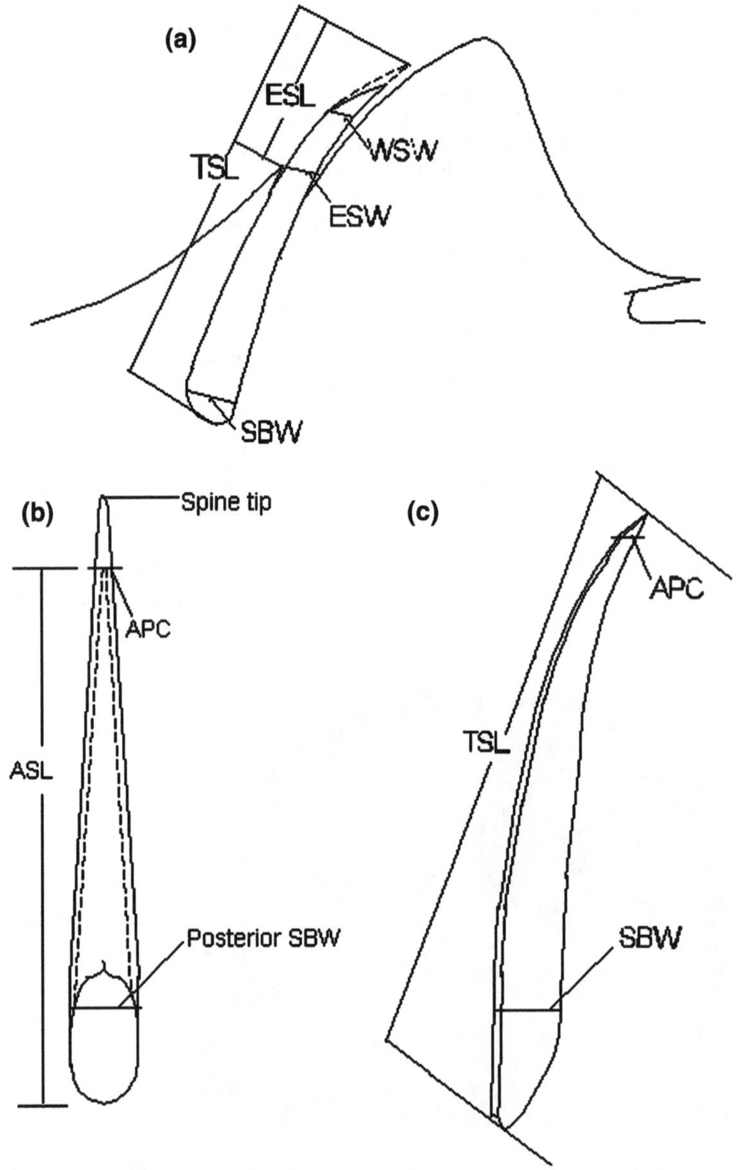

Fig. 3 Spine morphometric measurements. (**a**) Squalid or heterodontid spine (*lateral view*); (**b**) chimaera spine (*posterior view*); (**c**) chimaera spine (*lateral view*). Acronyms and definitions in Table 3

inner dentine layer (Fig. 2) are found in spine cross-sections, and one band comprises a translucent and opaque ring (Tanaka 1990; Freer and Griffiths 1993).

Morphometrics

Prior to the examination of growth bands, spine growth should be investigated by measuring spine morphometrics (Table 3; Fig. 3). The external spine length and external spine width must be measured prior to cleaning the spine. It is unknown if spines shrink when dried, therefore we suggest that all spines be measured whilst in the same condition (i.e. all fresh/frozen or all dried). The morphometric spine measurements are similar for dogfishes and horn sharks, although chimaera spines require different measurements (Ketchen 1975; Tanaka 1990; Francis and ÓMaolagáin 2000). Spine wear is an important issue when using enamel bands or trying to identify the optimum sectioning location when using internal dentine bands. We therefore suggest the method outlined in Ketchen (1975) be used for dogfish and horn shark spines, and the method of Sullivan (1977) for estimating the worn area in chimaera spines.

References

Beamish RJ, McFarlane GA (1985) Annulus development on the second dorsal spine of the spiny dogfish (*Squalus acanthias*) and its validity for age determination. Can J Fish Aquat Sci 42:1799–1805

Clarke MW, Connolly PL, Bracken JJ (2002) Age estimation of the exploited deepwater shark *Centrophorus squamosus* from the continental slopes of the Rockall Trough and Porcupine Bank. J Fish Biol 60:501–514

Francis MP, ÓMaolagáin C (2000) Age and growth of ghost sharks. Final research report for Ministry of Fisheries Research Project GSH1999/01 Objective 1, 29 pp

Freer DWL, Griffiths CL (1993) Estimation of age and growth in the St Joseph *Callorhinchus capensis* (Dumeril). S Afr J Mar Sci 13:75–81

Guallart Furio J (1998) Contribution to the knowledge of the biology and taxonomy of the deepsea shark, *Centrophorus granulosus* (Bloch & Schneider, 1801) (Elasmobranchii, Squalidae) in the Balearic Sea (western Mediterranean). Ph.D. thesis (in Spanish), Univertat de Valencia, Valencia, 290 pp

Holden MJ, Meadows PS (1962) The structure of the spine of the Spur Dogfish (*Squalus acanthias* L.) and its use for age determination. J Mar Biol Assoc UK 42:179–197

Irvine SB, Stevens JD, Laurenson LJB (2006a) Surface bands on deepwater squalid spines: an alternative method of ageing *Centroselachus crepidater*. Can J Fish Aquat Sci 63:617–627

Irvine SB, Stevens JD, Laurenson LJB (2006b) Comparing external and internal dorsal-spine bands to interpret the age and growth of the giant lantern shark, *Etmopterus baxteri* (Squaliformes: Etmopteridae) Environ Biol Fishes

Kaganovskaia SM (1933) A method for determining the age and composition of the catches of *Squalus acanthias*. Bulletin Far-East Br Acad Sci USSR 1(3):5–6 (translated from Russian by W.E. Ricker). Fisheries Research Board of Canada, Translation Series No. 281, vol. 1, pp. 5–6

Ketchen KS (1975) Age and growth of dogfish (*Squalus acanthias*) in British Columbia waters. J Fish Res Board Can 32:43–59

Maisey JG (1979) Finspine morphogenesis in squalid and heterodontid sharks. Zool J Linn Soc 66:161–183

McFarlane GA, Beamish RJ (1987) Validation of the dorsal spine method of age determination for spiny dogfish. In: Summerfelt RC, Hall GE (eds) The age and growth of fish. The Iowa State University Press, Ames, pp 287–300

Sullivan KJ (1977) Age and growth of the elephant fish *Callorhinchus milii* (Elasmobranchii: Callorhynchidae). NZ J Mar Freshw Res 11:745–753

Tanaka S (1990) The structure of the dorsal spine of the deep sea squaloid shark *Centrophorus acus* and its utility for age determination. Nippon Suisan Gakkai Shi (Bull Jpn Soc Sci Fish) 56:903–909

Wilson CA, Brothers EB, Casselman JM, Lavette-Smith C, Wild A (1983) Glossary. In: Prince ED, Pulos LM (eds) Proceedings of international workshop on age determination of oceanic pelagic fishes: tuna billfishes and sharks. US Department of Commerce, NOAA Technical Report No. 8, pp 207–208

Environ Biol Fish (2006) 77:279–292
DOI 10.1007/s10641-006-9129-x

ORIGINAL PAPER

Do differences in life history exist for blacktip sharks, *Carcharhinus limbatus*, from the United States South Atlantic Bight and Eastern Gulf of Mexico?

John K. Carlson · James R. Sulikowski · Ivy E. Baremore

Received: 12 June 2006 / Accepted: 3 July 2006 / Published online: 6 September 2006
© Springer Science+Business Media B.V. 2006

Synopsis We examined life history traits (e.g., mean length-at-age, growth rate, age-at-maturity) for blacktip sharks collected from two separate geographical areas (eastern Gulf of Mexico and South Atlantic Bight) to address the potential for separate stocks in southeastern US waters. Samples were obtained from fishery-dependent and independent sources. Growth and logistic models were fitted to observed length-at-age and reproductive data, respectively. von Bertalanffy growth parameters derived for blacktip shark from the Gulf of Mexico show that they attain a statistically smaller theoretical maximum length (L_∞ = 141.6 cm vs. L_∞ = 158.5 cm for female and L_∞ = 126.0 cm FL and L_∞ = 147.4 cm FL for male) and have a faster growth rate (k = 0.24 yr^{-1} vs. k = 0.16 yr^{-1} for female and k = 0.27 yr^{-1} vs. k = 0.21 yr^{-1} for male) than conspecifics in the South Atlantic Bight. Median length- and age-at-maturity were significantly different between sex and area. Length at which 50% of the population is mature was 117.3 cm FL for females and 103.4 cm FL for males in the Gulf of Mexico and 126.6 cm FL for females and 116.7 cm FL for males in the South Atlantic Bight. Median age-at-maturity was 5.7 yrs and 4.5 yrs for females and males in the Gulf of Mexico, respectively, while age-at-maturity was 6.7 yrs for females and 5.0 yrs for males for sharks from the South Atlantic Bight. Due to varying statistical results, temporal problems of sampling, and potential for gear bias, we could not definitively conclude that differences in life history characteristics exist.

J. K. Carlson (✉)
NOAA/National Marine Fisheries Service, Southeast Fisheries Science Center, 3500 Delwood Beach Road, Panama City, FL 32408, USA
e-mail: john.carlson@noaa.gov

J. R. Sulikowski
Florida Program for Shark Research, Florida Museum of Natural History, University of Florida, PO Box 117800, Gainesville, FL 32611, USA
e-mail: sulikowski@flmnh.ufl.edu

I. E. Baremore
Department of Fisheries and Aquatic Sciences, University of Florida, 7922 NW 71st St., Gainesville, FL 32653, USA
e-mail: ivalina@ufl.edu

Keywords Growth · Reproduction · Age · Stock

Introduction

Life history traits are the result of the strategy to which fish populations or stocks have evolved. Growth rates, age-at-maturity, and mortality reflect the underlying dynamics of the population. Estimates of these parameters are thought to be representative of individuals within a presumed stock and can be used to distinguish among separate stocks because these parameters are phenotypic expressions of the interaction of genotype and the environment (Begg 2005). Thus, differences in life history traits found between groups of individuals are assumed to be evidence that stocks of fish are geographically isolated and therefore are discrete stocks for management purposes (Ihssen et al. 1981).

Differences in life history between geographically separated stocks of elasmobranchs are becoming more widely documented. In waters off the United States (US), Carlson et al. (2003) noted a larger length-at-maturity for finetooth sharks, *Carcharhinus isodon*, from South Carolina than from the northeast Gulf of Mexico. Blacknose sharks, *Carcharhinus acronotus*, in the US south Atlantic Ocean have significantly lower growth rates (k) and reach maturity later than conspecifics in the Gulf of Mexico (Driggers et al. 2004). Neer and Thompson (2005) reported cownose rays, *Rhinoptera bonasus*, in the Gulf of Mexico have lower estimates of theoretical maximum length and growth rate than those from Chesapeake Bay, Virginia. However, these studies could not rule out that differences were artifacts of low sample sizes, methodology, or inter-annual comparisons.

Blacktip sharks, *Carcharhinus limbatus*, inhabit coastal waters off the United States from Massachusetts through Texas (Compagno 1984; Castro 1996). Blacktip sharks occur within two separate large marine ecosystems (Southeast US Continental Shelf and Gulf of Mexico, Musick et al. 2004), and conventional tagging evidence suggests little exchange between areas (Kohler et al. 1998; Carlson unpublished). However, blacktip sharks are managed as one stock under the current federal management plan (NMFS 2003). If sharks from separate geographic areas are assumed to share similar life history traits but actually differ, information used in the development of age-structured population models could result in errors in stock assessments and, possibly, overexploitation. To address the potential for separate stocks of blacktip sharks in US waters, we examined life history traits (e.g., mean length-at-age, growth rate, age-at-maturity) for sharks collected from two separate geographical areas, the eastern Gulf of Mexico and the South Atlantic Bight.

Materials and methods

Biological samples were collected during 1996–2002 through fishery-independent surveys (Hueter and Manire 1994; Grace and Henwood 1998; Carlson and Brusher 1999) and from fishery-dependent programs (Trent et al. 1997; Burgess and Morgan 2003). Additional reproductive data were also provided from a study by Castro (1996). Precaudal (PC), fork (FL), total (TL), and/or stretched total (STL) length (cm), sex, and maturity state were determined for each shark. When possible, weight was measured to the nearest kg (±0.1).

Age and growth

Vertebrae for age determination (3–6) were collected from either the column below the first dorsal fin or above the branchial chamber of each shark. Vertebral sections were placed on ice after collection and frozen upon return to the laboratory. Thawed vertebrae were cleaned of excess tissue and soaked in a 5% sodium hypochlorite solution for 5–30 min to remove remaining tissue. After cleaning, vertebrae were soaked in distilled water for 30 min and stored in 95% isopropanol. One vertebra was randomly selected, removed from alcohol, dried, fixed to a clear glass slide with resin, and sectioned using a Buehler 82 Isomet[1] low-speed saw.

To determine the most appropriate technique for enhancing visibility of growth bands, sagittal sections were cut from the vertebral centrum at 3 different thicknesses (±0.2 mm) and stained or left unstained (see review in Cailliet and Goldman 2004). Each section was mounted on a glass microscope slide with clear resin and examined using a dissecting microscope under transmitted light.

The annual periodicity of band pair formation was investigated using a marginal increment ratio analysis (MIR), which was calculated as the ratio between the final fully formed band pair and penultimate band (Cailliet and Goldman 2004). We observed that annuli in older adult specimens were compressed; therefore marginal increments

[1] Reference to trade names does not imply endorsement by NOAA Fisheries Service.

were calculated from randomly selected juvenile specimens below age 5 (e.g., Simpfendorfer 2000; Sulikowski et al. 2003, 2005). Measurements of the last complete band pair and the penultimate band from the centrum edge were taken using a compound microscope and optical micrometer. Mean marginal increment ratios were plotted by month of capture to identify trends in band formation, and a Kruskal–Wallis one-way analysis of variance on ranks was used to test for differences in marginal increment ratios by month.

In developing theoretical growth models, we assumed that (1) the birth mark is the band associated with a pronounced change in angle in the intermedialia and is formed on an arbitrary birth date of 1 May, (2) broad light bands are formed during summer, and (3) narrow dark bands are deposited in winter. Although vertebrae were taken from two different regions of the vertebral column (cervical and thoracic), an independent study found no significant difference in band counts between regions (B. Cullum, Florida Fish and Wildlife Commission, personal communication). Ages (yr) were calculated following the algorithm of Carlson et al. (1999): age = birth mark + number of winter marks –1.5. If only the birth mark was present, age was 0+ yrs. All age estimates from growth band counts were based on the hypothesis of annual growth band deposition (Branstetter 1987; Killam and Parsons 1989).

Vertebrae were read independently and randomly by two readers (JKC & IEB) without knowledge of location, sex, length or date of capture. Vertebral age estimates for which the readers disagreed were reviewed jointly by using a Meiji Techno R2 Dissecting Microscope equipped with a Hitachi KP-D50 Digital Camera and software.[1] Precision among age determinations was evaluated using percent agreement [(PA = Number agreed/Number read)*100] and percent agreement plus or minus 1 yr calculated for 10 cm FL (e.g., 76–85 cm) length intervals (Cailliet and Goldman 2004; Goldman 2004). Bowker's test of symmetry following Hoenig et al. (1995) was used to determine if differences between readers were systematic or due to random error. The Index of Average Percent Error (IAPE, Beamish and Fournier 1981) was calculated to compare the average deviation of readings from the means of all readings for each vertebral section:

$$\mathrm{IAPE} = \frac{1}{N}\sum_{j=1}^{N}\left[\frac{1}{R}\sum_{i=1}^{R}\frac{|x_{ij} - x_j|}{x_j}\right]$$

where N = number of sharks aged; R = number of readings; x_{ij} = ith age estimation of jth shark at ith reading, and x_j = mean age calculated for the jth shark.

Following Carlson and Baremore (2005), several growth models were fitted to the observed length-at-age data: the von Bertalanffy growth model (von Bertalanffy 1938; Beverton and Holt 1957; see also Cailliet et al. this issue), an alternate equation of the von Bertalanffy growth model with a length-at-birth intercept rather than the t_o parameter (Van Dykhuizen and Mollet 1992; Goosen and Smale 1997; Carlson et al. 2003), and a Gompertz growth model (Ricker 1975). All growth model parameters were estimated using Marquardt least-squares non-linear regression and SAS statistical software (PROC NONLIN, SAS Inst., Inc). Models were assessed based on a combination of examining residual mean square error (MSE), coefficient-of-determination (r^2), level of significance ($P < 0.05$), and standard residual analysis.

Several methods were employed to test for differences in age and growth within and between geographic areas. Following Kimura (1980), χ^2-tests of likelihood ratios were used to test for differences in combined parameters within the von Bertalanffy growth model. Comparisons of observed mean length-at-age between ages were made using a Welch modified two-sample t-test (Zar 1984). Growth rates were calculated from predicted fork lengths by the von Bertalanffy growth model. Growth intervals were represented by the time between band deposition (growth interval one is the time between band counts one and two). Sex-specific predicted growth rates (cm yr^{-1}) were compared between the areas.

Length-at-maturity estimation

Maturity was assessed following the guidelines of Castro (1996). To quantitatively assess length-at-maturity, the length at which 50% of the population is mature for male and female sharks was determined (i.e., median length-at-maturity). Data from the Gulf of Mexico and South Atlantic Bight were fitted separately to a logistic model:

$$Y = 1/(1 + e^{-(a+bX)})$$

where Y = the binomial maturity data (immature = 0, mature = 1) and X = length. Median length-at-maturity was expressed as $-a/b$. The model was fitted using maximum likelihood (PROC LOGISTIC, SAS Inst., Inc.) and the effects of area and sex were compared using χ^2-tests of likelihood ratios.

To assess age-at-maturity, lengths were back-transformed to age using each respective von Bertalanffy growth model. Similar to determination of length-at-maturity, ages were fitted to a logistic model using maximum likelihood and the effects of area and sex compared used χ^2-tests of likelihood ratios.

Results

A total of 628 biological samples were collected throughout the study (Fig. 1). Using data collected in this study and that from fishery-independent surveys (Carlson and Brusher 1999; Carlson 2003), several morphometric relationships to convert length measurements were developed. Linear regression formulae were determined as FL = 1.10(PC) + 0.29 (n = 1,096);

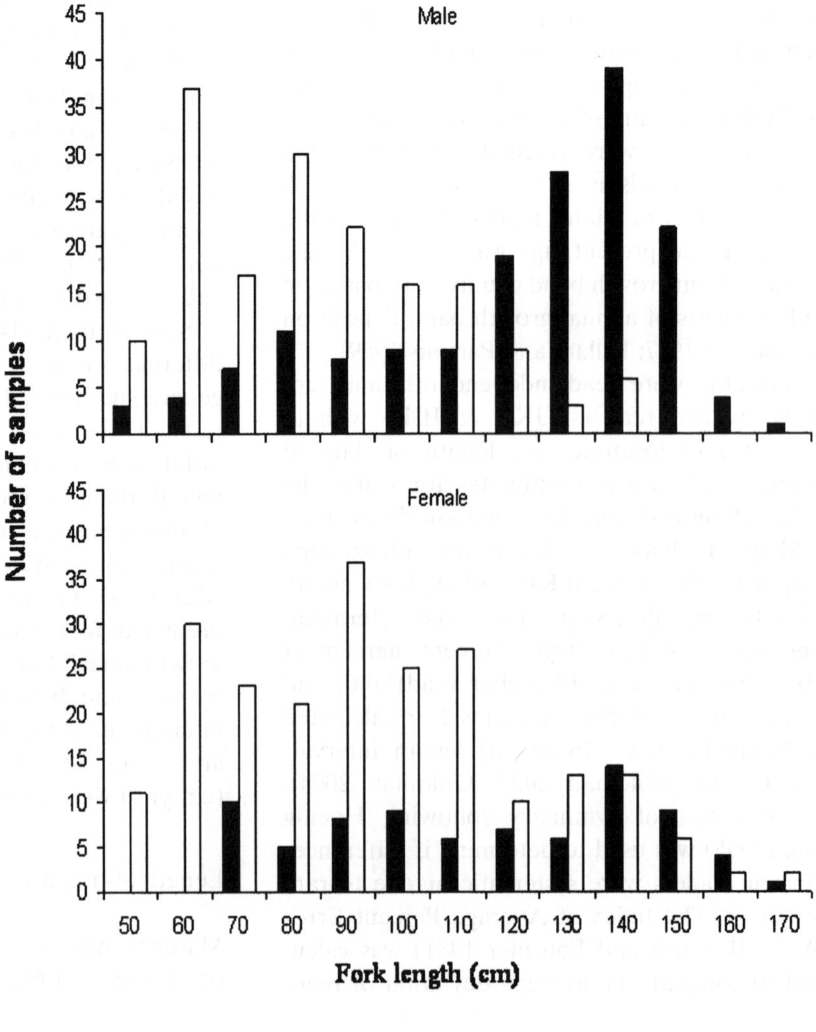

Fig. 1 Length frequency distributions for male and female blacktip sharks collected from the South Atlantic Bight (solid bar) and eastern Gulf of Mexico (open bar)

TL = 1.12(FL) + 1.12 (n = 1,248); and STL = 1.02(TL) + 0.99 (n = 926). All equations were highly significant ($P < 0.0001$) and had r^2 between 0.98 and 0.99.

Age and growth

Growth bands were found to be most apparent on unstained sagittal sections with a thickness of 0.3 mm (Fig. 2) and less apparent on 0.5 mm and 0.7 mm. The precision of band counts was high between readers and Bowker's test of symmetry (Hoenig et al. 1995) indicated that differences between readers were due to random error (χ^2 test, $P > 0.05$).

The first set of readings resulted in an index of average percent error of 3.9%. Initial percent agreement in all band counts between the readers was 76.5% within 1 band, 94.2% within 2 bands, 98.4% within 3 bands, 99.8% within 4 bands, 100.0% within 5 bands, and 99.8% within 6 bands. When grouped by 10 cm length intervals, agreement for combined sexes was reached for an average of 77.2% and 97.3% ±1 band for sharks less than 105 cm FL (Table 1). Above 105 cm FL, agreement was reached for 46.6% and 71.9% ±1 band of samples initially read. For those samples where band counts differed, consultation resulted in agreement for 608 out of 628 vertebrae. Samples with no resolution were discarded.

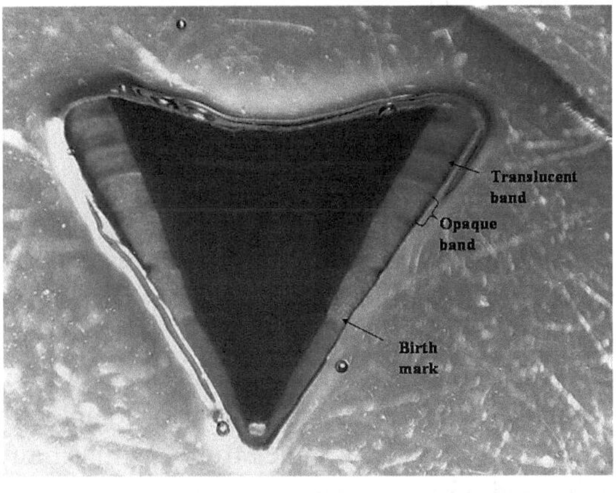

Fig. 2 Sagittal section from a blacktip shark vertebra used for age determination. Translucent bands (winter marks) correspond to thin areas under transmitted light, whereas opaque bands (summer marks) correspond to wide zones

Table 1 Percent agreement and percent agreement ±1 band from the initial set of readings for blacktip shark

FL interval	Total read	Percent agreement	Percent agreement ±1
Sexes combined			
36–45	3	66.7	100.0
46–55	61	100.0	100.0
56–65	62	96.8	100.0
66–75	63	81.0	96.8
76–85	76	68.4	100.0
86–95	58	65.5	93.1
96–105	68	61.8	91.2
106–115	48	52.1	85.4
116–125	39	56.4	89.7
126–135	66	60.6	90.9
136–145	62	54.8	82.3
146–155	18	55.6	83.3
156–165	4	0.0	0.0

A total of 79 vertebral samples were considered usable for marginal increment analyses. Six to 10 samples were available each month except for October and December, which were represented by a sample size of two and three, respectively, and April and November, when no specimens were collected. Marginal increment ratios were significantly different among those 8 months (Kruskal–Wallis, $P < 0.001$) with a distinct trend of increasing monthly increment growth that peaked in May, followed by a sharp decline to a numerical low in June (Fig. 3). Based on this information, the increment analyses support the likelihood that a single opaque band is formed annually on the vertebral centrum during the month of June.

All three growth models fit the data well (e.g., high r^2, low standard deviation of residuals). Because of the general similarity between the models and the ubiquitous use of von Bertalanffy model, we present and compare further age and growth results using only the von Bertalanffy equation. Growth parameters derived for blacktip sharks from the Gulf of Mexico indicated that they attain a smaller theoretical maximum length (L_∞) and that they reach L_∞ at a faster rate (k) than conspecifics in the South Atlantic Bight (Table 2, Fig. 4). von Bertalanffy growth parameters for sharks in the Gulf of Mexico were L_∞ = 141.6 cm FL, k = 0.24 yr^{-1}, t_o = –2.18 yr and L_∞ = 126 cm FL, k = 0.27 yr^{-1} t_o = –2.21 yr for females and males, respectively (Table 2). In the South

Fig. 3 Mean marginal increment ratio (MIR ± standard error) by month for combined sexes from sharks less than age 5. Numbers above each month represent sample size

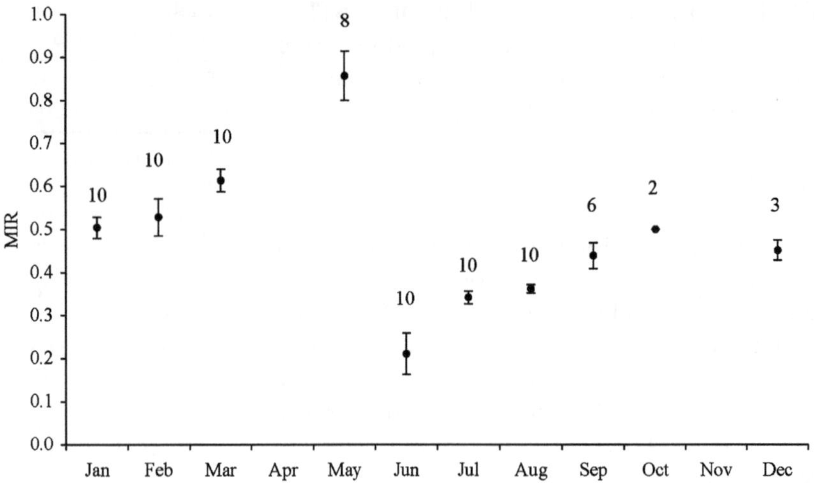

Table 2 Von Bertalanffy growth parameters for male, female, and sex combined blacktip sharks

	Male	SE	LCL	UCL	Female	SE	LCL	UCL	Combined	SE	LCL	UCL
Gulf of Mexico												
L_∞ (cm)	126.0	3.50	119.1	132.9	141.6	2.99	135.7	147.5	139.4	2.61	134.2	144.5
k (yr^{-1})	0.27	0.02	0.22	0.33	0.24	0.02	0.20	0.27	0.23	0.01	0.20	0.26
t_o (yr)	−2.21	0.18	−2.57	−1.84	−2.18	0.16	−2.49	−1.87	−2.33	0.13	−2.58	−2.07
N	161				207				368			
South Atlantic Bight												
L_∞ (cm)	147.4	2.60	142.2	152.5	158.5	5.71	147.1	169.8	150.9	2.51	145.9	155.8
k (yr^{-1})	0.21	0.02	0.17	0.24	0.16	0.02	0.11	0.21	0.19	0.01	0.16	0.22
t_o (yr)	−2.58	0.24	−3.06	−2.11	−3.43	0.50	−4.43	−2.43	−2.89	0.23	−3.34	−2.44
N	162				78				240			
Areas combined												
L_∞ (cm)	150.8	2.67	145.6	156.1	148.5	2.49	143.6	153.4	148.7	1.76	145.2	152.2
k (yr^{-1})	0.18	0.01	0.16	0.20	0.21	0.01	0.18	0.23	0.20	0.01	0.18	0.22
t_o (yr)	−2.76	0.15	−3.06	−2.46	−2.42	0.15	−2.74	−2.13	−2.57	0.11	−2.79	−2.36
N	323				285				608			

Estimates are provided for the Gulf of Mexico, South Atlantic Bight, and areas combined. Standard error = SE and 95% lower and upper confidence limits = LCL and UCL, respectively

Atlantic Bight, $L_\infty = 158.5$ cm FL, $k = 0.16$ yr^{-1}, $t_o = -3.43$ yr and $L_\infty = 147.4$ cm FL, $k = 0.21$ yr^{-1}, $t_o = -2.58$ yr for female and male blacktip sharks, respectively. Significant differences between von Bertalanffy growth curves of males and females were found within populations (Gulf log-likelihood ratio = 33.21, $P < 0.001$; Atlantic log-likelihood ratio = 9.32, $P < 0.05$) and between populations (females log-likelihood ratio = 18.65, $P < 0.001$; males log-likelihood ratio = 53.15, $P < 0.001$).

The maximum observed ages were 15.5+ yr (female) and 13.5+ yr (male) for sharks collected in the South Atlantic Bight and 12.5+ yr (female) and 11.5+ yr (male) in the Gulf of Mexico, based on vertebral band counts. Theoretical longevity estimates were 21.6 yr and 14.4 yr for females and 16.6 yr and 12.8 yr for males from South Atlantic Bight and Gulf of Mexico, respectively, using values obtained through von Bertalanffy growth models. The largest female aged in the Gulf of Mexico was 158 cm FL (11.5 yr) and 164 cm FL

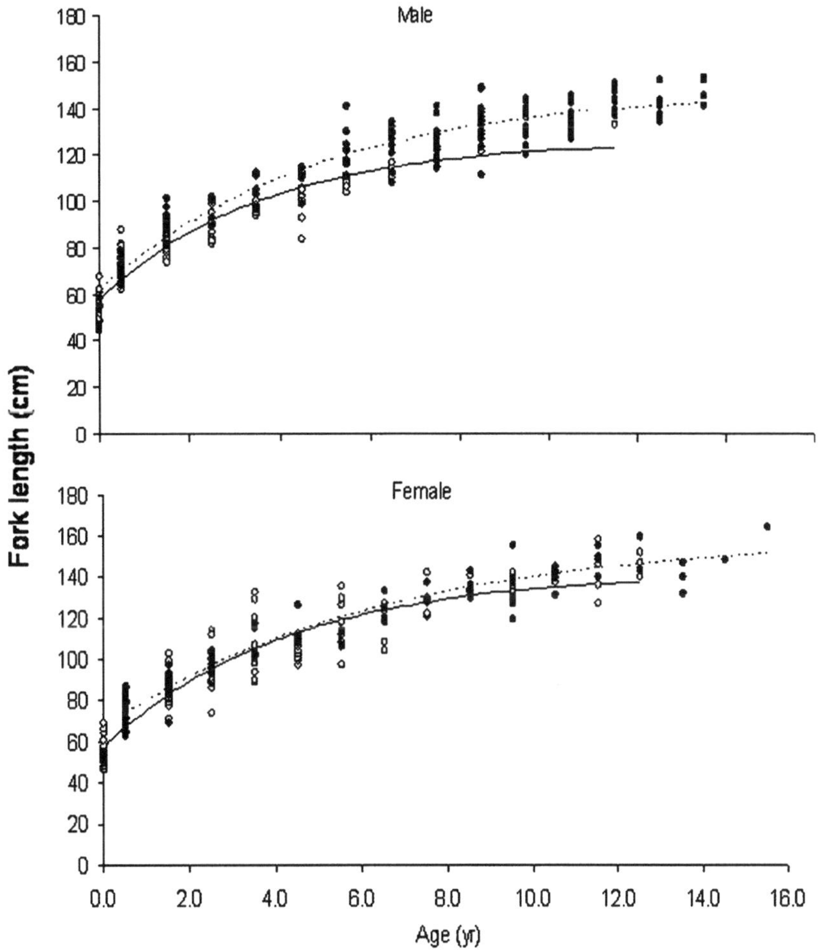

Fig. 4 Von Bertalannfy growth functions fitted to observed length-at-age data for male and female blacktip sharks. Solid circles for sharks and dashed lines = South Atlantic Bight while open circles and solid lines = Gulf of Mexico

(15.5 yr) from the South Atlantic Bight. For male blacktip sharks, the largest shark aged was 136 cm FL (9.5 yr) and 153 cm FL (age 13.5 yr) for the Gulf of Mexico and from the South Atlantic Bight, respectively.

Observed length-at-age and predicted growth rates were not significantly different ($P \geq 0.05$) between most ages for populations in the Gulf of Mexico and South Atlantic Bight (Table 3). Among females, mean observed length-at-age was only statistically different for one of out of 13 age classes (age 4.5). For male sharks, mean observed length-at-age was significantly differences in five of nine age classes (ages 1.5, 2.5, 3.5, 5.5, and 6.5). Predicted growth rates from age 0 yr to 12 yrs were similar among ages, averaging 6.6 cm yr^{-1} and 6.5 cm yr^{-1} for females and 5.5 cm yr^{-1} and 6.5 cm yr^{-1} for males from the Gulf of Mexico and South Atlantic Bight, respectively. Two-factor analysis of variance found no significant differences in growth rates (log transformed) between sexes ($F = 0.549$, df = 1, $P = 0.462$), area ($F = 1.060$, df = 1, $P = 0.308$) or their interaction ($F = 0.116$, df = 1, $P = 0.734$).

Length- and age-at-maturity

Median length- and age-at-maturity were different between sexes and areas. Length of the population at 50% maturity was 117.3 cm FL for females and 103.4 cm FL for males in the Gulf of Mexico (Fig. 5). The largest immature shark was 122 cm FL and 106 cm FL and the smallest mature shark was 109 cm FL and 102 cm FL for females and males, respectively. In the South

Table 3 Mean size-at-age for female and male blacktip sharks from the eastern Gulf of Mexico and South Atlantic Bight

Age	0.0	0.5	1.5	2.5	3.5	4.5	5.5	6.5	7.5	8.5	9.5	10.5	11.5	12.5	13.5	14.5	15.5
Female																	
Gulf	53.6	75.0	85.7	96.5	104.8	105.4	117.5	116.4	126.7	134.5	132.2	139.3	132.2	146.1	–	–	–
SE	0.8	1.1	1.2	1.5	2.6	1.5	3.8	4.4	4.0	3.2	2.5	1.4	5.2	3.4	–	–	–
LCL	52.0	72.7	83.2	93.4	99.4	101.9	108.8	104.1	115.5	120.9	126.4	133.2	126.9	131.6	–	–	–
UCL	55.2	77.4	88.1	99.5	110.3	108.8	126.2	128.7	137.8	148.1	138.0	145.4	155.8	160.6	–	–	–
N	46	27	34	28	19	11	10	5	5	3	8	3	5	3	–	–	–
$P > t$	–	0.17	0.22	0.55	0.28	0.03	0.48	0.24	0.36	0.87	0.47	0.93	0.30	0.29	–	–	–
Atlantic	–	72.4	88.8	98.4	111.9	114.9	110.5	123.9	126.8	134.4	134.6	136.6	144.6	146.1	144.3	148	164
SE	–	1.8	2.2	2.0	4.1	5.6	3.5	3.4	2.2	1.5	1.8	1.3	1.4	3.4	2.8	–	–
LCL		68.6	83.9	93.2	94.1	90.9	66.0	113.1	122.1	131.3	131.0	133.9	141.8	138.2	137.4	–	–
UCL		76.1	93.6	103.6	129.7	139.0	154.9	134.7	131.5	137.4	138.3	139.3	147.3	154.1	151.1	–	–
Male																	
Gulf	54.5	73.3	81.4	89.9	98.2	102.0	108.1	114.6	121.6	121.5	136	130.5	132.9	–	–	–	–
SE	0.6	0.9	0.9	1.9	1.1	3.1	1.2	2.5	7.4	–	–	–	–	–	–	–	–
LCL	53.2	71.3	79.3	85.5	95.7	94.9	105.3	108.7	27.5	–	–	–	–	–	–	–	–
UCL	55.8	75.2	83.4	94.4	100.7	109.1	110.9	120.5	215.6	–	–	–	–	–	–	–	–
N	51	32	26	10	10	9	9	8	2	1	1	1	1	–	–	–	–
$P > t$	0.17	0.43	<0.01	0.05	<0.01	0.17	<0.01	0.02	0.52	–	–	–	–	–	–	–	–
Atlantic	51.9	72.0	90.1	97.0	106.5	109.0	120.6	124.3	125.1	134.2	134.0	135.8	143.4	141.5	147.8	–	–
SE	2.2	1.0	1.8	2.5	2.4	2.7	2.3	2.7	1.9	1.8	1.9	1.4	1.4	3.1	2.9	–	–
LCL	45.5	69.9	86.0	89.2	100.3	101.6	115.5	118.3	120.9	130.4	130.0	132.8	140.4	132.9	5.7		
UCL	57.3	74.0	94.2	104.1	112.6	116.4	125.7	130.2	129.3	137.9	138.1	138.8	149.5	150.1	156.9		
N	7	18	12	5	6	5	12	10	14	21	15	16	12	5	4		

Standard error = SE, 95% lower and upper confidence limits = LCL and UCL, respectively and N = sample size. $P > t$ = the probability the t statistic is greater than the critical value of the t distribution

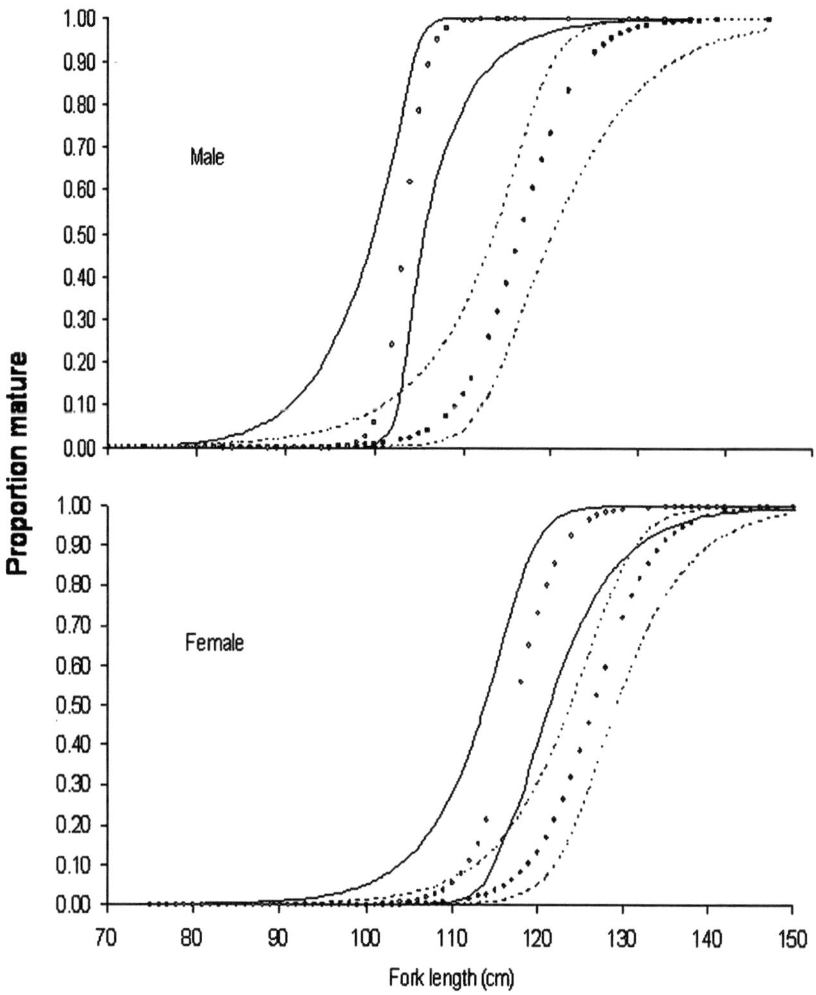

Fig. 5 Logistic models fitted to predicted length-at-maturity for male and female blacktip sharks. Lines = upper and lower 95% confidence intervals of the logistic curve. Solid circles and dashed lines = South Atlantic Bight while open circles and solid lines = Gulf of Mexico

Atlantic Bight, median length-at-maturity was 126.6 cm FL for females and 116.7 cm FL for males (Fig. 5). The largest immature shark was 134 cm FL and 119 cm FL and the smallest mature shark was 112 cm FL and 111 cm FL for females and males, respectively. Significant differences between logistic curves of males and females were found within populations (Gulf log-likelihood ratio = 310.19; $P < 0.0001$) (Atlantic log-likelihood ratio = 262.37; $P < 0.001$) and between populations (females log-likelihood ratio = 18.65; $P < 0.001$) (males log-likelihood ratio = 53.15; $P < 0.001$).

Converting length to age and fitting the logistic model resulted in an age-at-maturity of 5.7 yr and 4.5 yr for females and males in the Gulf of Mexico, respectively (Fig. 6). In the South Atlantic Bight, age-at-maturity was 6.7 yr for females and 5.0 yr for males.

Discussion

Life history parameters of many marine fish stocks have been shown to vary in response to environmental change and to the interaction between genotype and that particular environment (Begg 2005). Regional differences in annual sea surface temperatures are evident between the eastern Gulf of Mexico and the South Atlantic Bight (24.4°C vs. 22.5°C, respectively; NOAA/NOS/Center for Operational Oceanographic Products and Services, http://www.lternet.edu/technology/sensors/arrays.htm). Keeney et al.

Fig. 6 Logistic models fitted to predicted age-at-maturity for male and female blacktip sharks. Lines = upper and lower 95% confidence intervals of the logistic curve. Solid circles and dashed lines = South Atlantic Bight while open circles and solid lines = Gulf of Mexico

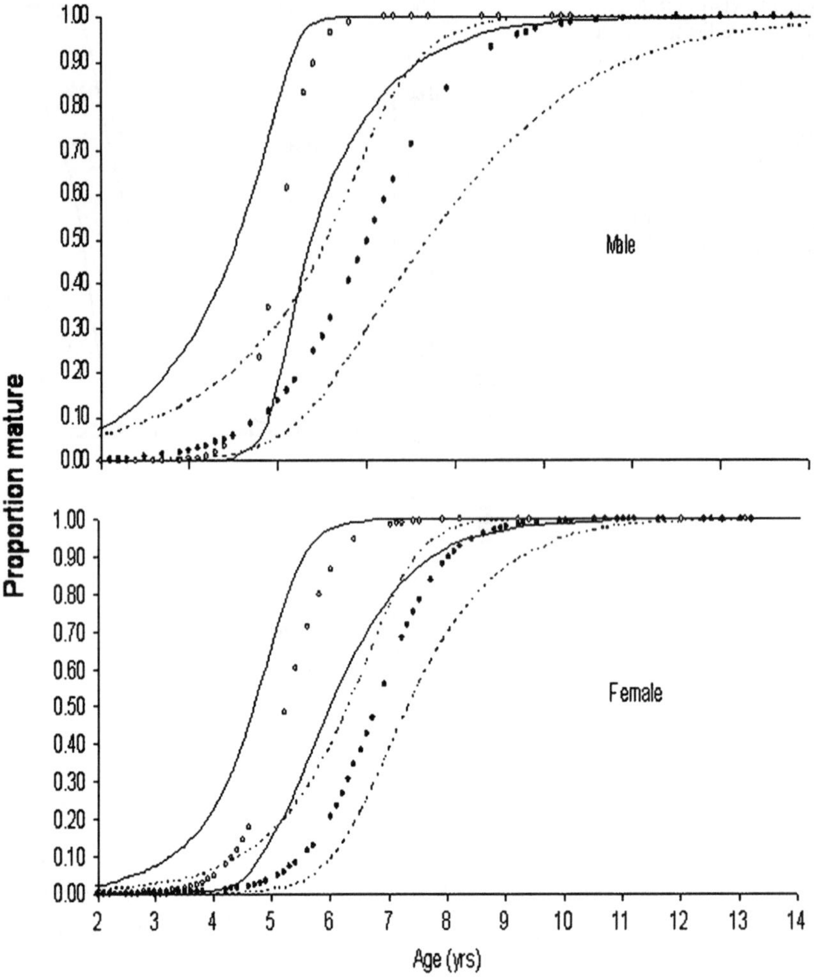

(2005) demonstrated genetic heterogeneity and female philopatry, which resulted in multiple genetic reproductive stocks among blacktip sharks in the Gulf of Mexico and South Atlantic Bight. Further, recoveries from conventional tagging of over 6,000 sharks since 1963 suggest very little mixing of sharks between these two areas (Kohler et al. 1998; Carlson unpublished data; J.P. Tyminski, Mote Marine Laboratory, personal communication). Despite these mechanisms that could potentially cause differences in life history traits in blacktip sharks between the South Atlantic Bight and eastern Gulf of Mexico, we could not definitively conclude that they exist. Although significant differences between sexes from each area were found in the overall von Bertalanffy growth models, mean length-at-age was not different for most ages and growth rates were similar. Length- and age-at-maturity differences could have been due to temporal disjunction, since most samples from the South Atlantic Bight (Castro 1996) were collected in 1981–1993 while sharks from the eastern Gulf of Mexico were captured during 1996–2002.

The temporal periodicity of growth zones should be evaluated to fully understand differences in shark stocks. In the current study, minimal marginal increment ratios occurred in sharks captured in June and maximal ratios occurred in sharks captured in May. These results support the hypothesis of that one band pair form annually in the vertebral centra of this species. These results compare favorably to cycles in marginal increments (Cailliet and Goldman 2004) and to annual vertebral band patterns in other shark species examined in both the Gulf of Mexico (Carlson et al. 1999, 2003) and the Atlantic Ocean (Natanson et al. 1995; Conrath et al. 2002). However, Killam and Parsons (1989) suggest ring deposition occurs in January for blacktip sharks captured in Tampa Bay, Florida. The reason for this difference in marginal increment formation between the studies is unknown but may be due to differing techniques. For example, Killam and Parsons (1989) used vernier calipers on whole vertebrae whereas the present study utilized an ocular micrometer on sectioned vertebrae.

Since band counts of the largest and oldest animals in the present study were compressed (too small to discern marginal increments from their widths), marginal increment analysis was only conclusive for juvenile animals (sharks <5 yrs in age). Thus, the annular nature of growth bands was verified for only those age groups. Nevertheless, we assumed that as sharks grew larger and older, the annual nature of growth ring deposition continued throughout their lifetimes (Cailliet and Goldman 2004).

Geographic variation in growth for at least two stocks of starspotted dogfish, *Mustelus manazo*, was proposed by Taniuchi et al. (1983), but a re-examination of their data using cross-exchange and comparative reader analysis found no significant differences (Cailliet et al. 1990). Similarly, Tanaka et al. (1990) reported that differences among band readers and methodologies produce variations in growth for blue shark, *Prionace glauca*, that were artifacts. We attempted to control these factors and feel they had little effect on our results. We suspect the most significant factor in our study that could affect our results was inherent bias associated with gear selectivity. An ideal study would ensure all samples would be collected using a similar fishery-independent gear. Growth models fitted to fishery-dependent data can have biases due to length-selective fishing where more fast-growing sharks, fewer large young sharks, and small old sharks are differentially removed from the stock (Kimura 1980; Walker et al. 1998). In our study, blacktip sharks were collected with a variety of sampling gears from scientific gillnets to commercially fished longlines. This is a common problem when attempting to model growth on sharks with wide ranges in lengths. Blacktip sharks range in length from 35 cm to 170 cm. Gear utilized for catching smaller sharks would not be appropriate for larger sharks. For example, longlines designed to capture smaller sharks (i.e., small hooks, low gangion leader strength) would not be appropriate for sampling larger sharks due to differential catchability as a result of bite-offs from low leader strength (Beerkircher et al. 2003). Conversely, gillnets of larger mesh lengths designed for adult sharks would not be suitable for juvenile sharks as smaller sharks would simply pass through the net. Carlson and Cortés (2003) documented gillnet selectivity for small coastal sharks in US waters, but few controlled experimental studies are available on how gear variation affects catches of sharks or resulting growth models (Walker et al. 1998).

Despite these fundamental biases associated with determining variation in life history traits, at least one study has illustrated that differences in life history traits do occur in shark populations. Lombardi-Carlson et al. (2003) demonstrated increasing length-, age-at-maturity, and mass of near-term embryos in bonnetheads, *Sphyrna tiburo*, with increasing latitude in the eastern Gulf of Mexico. Further, these authors documented an increase in shark growth rate with an increase in latitude (termed countergradient variation, Conover and Present 1990). Along with controlling for sample preparation and reading, Lombardi-Carlson et al. (2003) collected sharks simultaneously in all areas with similar sampling gear to minimize sampling bias.

Estimates of age, growth, and length- and age-at-maturity for male and female blacktip sharks in the Gulf of Mexico were different than those reported by Killam and Parsons (1989) for sharks collected off Tampa Bay, FL. von Bertalanffy growth parameters were $L_\infty = 160$ cm FL, $k = 0.19$, and $L_\infty = 137$ cm FL, $k = 0.28$ during 1985–1987 for females and males, respectively. These indicate an increase in growth and a decrease in theoretical maximum length for sharks collected for our study 11–14 yrs later. Observed maximum age also increased from 10 yrs and 9 yrs in 1985–1987 to 12.5 yrs and 11.5 yrs in the present study. Length- and age-at-maturity decreased from about 110 (age = 4–5 yrs) and 132 cm FL (age = 6–7 yrs) in 1985–1987 to 103 (age = 4.5 yrs) and 117 cm FL (age = 5.7 yrs) in 1996–2001 for males and females, respectively. Blacktip sharks have been heavily harvested in the Gulf of Mexico since the 1980's (NMFS 2003), thus the observed decrease in length- and age-at-maturity and increased growth rate lends support to the potential for a density-dependent compensatory response. Compensatory growth and reproductive responses have been documented in a few species of sharks (Sminkey and Musick 1995; Carlson and Baremore 2003). For reasons previously outlined, it could not be determined if these temporal changes in age and growth were due to differences in methodology, anthropogenic influences, or natural causes.

Given the caveats observed in this study and the current data, we could not definitively conclude that differences in life history characteristics exist between blacktip sharks in the eastern Gulf of Mexico and South Atlantic Bight. A synoptic study sampling the entire geographic range of blacktip sharks (i.e., entire Gulf of Mexico and northwest Atlantic Ocean) would be required to fully resolve the question of separate stocks. The application of archival satellite tags could also help to define spatial distributions and long-term movement patterns, information that can assist in stock discrimination (Punt 2001).

Acknowledgements Ken Goldman, Pete Sheridan and Linda Lombardi provided helpful comments on an earlier version of this manuscript. Lori Hale (NOAA Fisheries Service-Panama City Laboratory) assisted with the processing and reading of samples and Linda Lombardi developed the SAS code for the maturity estimates. George Burgess and Matt Callahan (University of Florida) obtained samples from the directed shark longline fishery. John Tyminski (Mote Marine Laboratory) provided samples off Tampa Bay and Charlotte Harbor, FL. Armando de ron Santiago, C.J. Greene, Matt Rayl, Bill Habich, Mike Farni, Jacques Hill, and Jeff Pulver collected samples from the directed shark gillnet fishery. Mark Grace and Lisa Jones (NOAA Fisheries-Mississippi Laboratory) helped to provide samples during R/V Oregon II longline surveys. Thanks also go to Jose Castro (NOAA Fisheries-Miami Laboratory) who provided data from his blacktip reproduction study.

References

Beamish RJ, Fournier DA (1981) A method for comparing the precision of a set of age determinations. Can J Fish Aquat Sci 38:982–983

Beerkircher L, Shivji M, Cortés E (2003) A Monte Carlo demographic analysis of silky shark (*Carcharhinus falciformis*): implications of gear selectivity. Fish Bull 101:168–174

Begg GA (2005) Life history parameters. In: Cadrin SX, Friedland KD, Waldman JR (eds) Stock identification methods. Elsevier Academic Press, New York, pp 119–150

Beverton RJH, Holt SJ (1957) On the dynamics of exploited fish populations. Ministry of Agriculture. Fisheries and Food Fishery Investigation Series II XIX, 533 pp

Branstetter S (1987) Age and growth estimates for the blacktip, *Carcharhinus limbatus*, and spinner, *C. brevipinna*, sharks from the northwestern Gulf of Mexico. Copeia 1987:964–974

Burgess GH, Morgan A (2003) Commercial shark fishery observer program. Renewal of an observer program to monitor the directed commercial shark fishery in the Gulf of Mexico and south Atlantic: 2002(2) and 2003(1) fishing seasons. Final Report, National Marine Fisheries Service, Highly Migratory Species Management Division Award NA16FM1598, 15 pp

Cailliet GM, Goldman KJ (2004) Age determination and validation in chondrichthyan fishes. In: Carrier JC, Musick JA, Heithaus M (eds) The biology of sharks and their relatives, CRC Press, Boca Raton, FL, pp 399–447

Cailliet GM, Yudin KG, Tanaka S, Taniuchi T (1990) Growth characteristics of two populations of *Mustelus manazo* from Japan based upon cross reading of vertebral bands. In: Pratt HL Jr, Gruber SH, Taniuchi T (eds) Elasmobranchs as living resources: advances in the biology, ecology, systematics, and the status of the fisheries. NOAA Technical Report, NMFS 90. Silver Spring, MD, pp 167–176

Carlson JK (2003) Shark nurseries in the northeastern Gulf of Mexico. In: McCandless CT, Pratt HL Jr (eds) Gulf of Mexico and Atlantic States shark nursery overview. U.S. Department of Commerce, National Oceanic and Atmospheric Administration. Silver Spring, MD, pp 165–182

Carlson JK, Baremore IE (2003) Changes in biological parameters of Atlantic sharpnose shark *Rhizoprionodon terraenovae* in the Gulf of Mexico: evidence for density-dependent growth and maturity? Mar Freshwater Res 54: 227–234

Carlson JK, Baremore IE (2005) Growth dynamics of the spinner shark, *Carcharhinus brevipinna*, off the United States Southeast and Gulf of Mexico coasts: a comparison of methods. Fish Bull 103:280–291

Carlson JK, Brusher JH (1999) An index of abundance for coastal species of juvenile sharks from the northeast Gulf of Mexico. Mar Fish Rev 61(3):37–45

Carlson JK, Cortés E (2003) Gillnet selectivity of small coastal sharks off the southeastern United States. Fish Res 60:405–414

Carlson JK, Cortés E, Bethea D (2003) Life history and population dynamics of the finetooth shark, *Carcharhinus isodon*, in the northeast Gulf of Mexico. Fish Bull 101:281–292

Carlson JK, Cortés E, Johnson AJ (1999) Age and growth of the blacknose shark, *Carcharhinus acronotus*, from the eastern Gulf of Mexico. Copeia 1999:683–690

Castro JI (1996) Biology of the blacktip shark, *Carcharhinus limbatus*, off the southeastern United States. Bull Mar Sci 59:508–522

Compagno LJV (1984) FAO species catalogue: sharks of the world. An annotated and illustrated catalogue of shark species known to date. FAO Fisheries Synopsis (125), Vol. 4, part 1: Hexanchiformes to Lamniformes. Food and Agriculture Organization of the United Nations, Rome, 249 pp

Conover DO, Present TMC (1990) Countergradient variation in growth rate: compensation for length of the growing season among Atlantic silversides from different latitudes. Oecologia 83:316–324

Conrath CL, Gelsleichter J, Musick JA (2002) Age and growth of the smooth dogfish (*Mustelus canis*) in the northwest Atlantic Ocean. Fish Bull 100:674–682

Driggers WB, Carlson JK, Oakley D, Ulrich G, Cullum B, Dean JM (2004) Age and growth of the blacknose shark, *Carcharhinus acronotus*, in the western North Atlantic Ocean with comments on regional variation in growth rates. Environ Biol Fish 71:171–178

Goldman KJ (2004) Age and growth of elasmobranch fishes. In: Musick JA, Bonfil R (eds) Elasmobranch fisheries management techniques. Asia Pacific Economic Cooperation, Singapore, pp 97–132

Goosen AJJ, Smale MJ (1997) A preliminary study of the age and growth of the smoothhound shark *Mustelus mustelus* (Triakidae). S Afr J Mar Sci 18:85–91

Grace M, Henwood T (1998) Assessment of the distribution and abundance of coastal sharks in the U.S. Gulf of Mexico and eastern seaboard, 1995 and 1996. Mar Fish Rev 59(4):23–32

Hoenig JM, Morgan MJ, Brown CA (1995) Analysing differences between two age determination methods by tests of symmetry. Can J Fish Aquat Sci 52:364–368

Hueter RE, Manire CA (1994) Bycatch and catch-release mortality of small sharks in the Gulf coast nursery grounds of Tampa Bay and Charlotte Harbor. Mote Marine Laboratory Technical Report 368. Sarasota, FL, pp 1–183

Ihssen PE, Booke HE, Casselman JM, McGlade JM, Payne NR, Utter FM (1981) Stock identification: materials and methods. Can J Fish Aquat Sci 38:1838–1855

Keeney DB, Heupel M, Hueter RE, Heist EJ (2005) Microsatellite and mitochondrial DNA analyses of the genetic structure of blacktip shark (*Carcharhinus limbatus*) nurseries in the northwestern Atlantic, Gulf of Mexico, and Caribbean Sea. Mol Ecol 14:1911–1923

Killam KA, Parsons GR (1989) Age and growth of the blacktip shark, *Carcharhinus limbatus* near Tampa Bay, Florida. Fish Bull 87:845–857

Kimura DK (1980) Likelihood methods for the von Bertalanffy growth curve. Fish Bull 77:765–776

Kohler NE, Casey JG, Turner PA (1998) NMFS cooperative shark tagging program 1962–93: An atlas of shark tag and recapture data. Mar Fish Rev 60(2):1–87

Lombardi-Carlson LA, Cortés E, Parsons GR, Manire CA (2003) Latitudinal variation in life-history traits of bonnethead shark, *Sphyrna tiburo*, (Carcharhiniformes:Sphyrnidae) from the eastern Gulf of Mexico. Mar Freshwater Res 54:875–883

Musick JA, Harbin MM, Compagno LJV (2004) Historical zoogeography of the selachii. In: Carrier JC, Musick JA, Heithaus M (eds) The biology of sharks and their relatives. CRC Press, Boca Raton, FL, pp 33–78

Natanson L, Casey JG, Kohler NE (1995) Age and growth estimates for the dusky shark, *Carcharhinus obscurus*, in the western Atlantic Ocean. Fish Bull 93:116–126

Neer JA, Thompson BA (2005) Life history of the cownose ray, *Rhinoptera bonasus*, in the northern Gulf of Mexico, with comments on geographic variability in life history traits. Environ Biol Fish 73:321–331

NMFS (National Marine Fisheries Service) (2003) Final amendment 1 to the fishery management plan of the Atlantic tunas, swordfish and sharks. U.S. Department of Commerce, National Oceanic and Atmospheric Administration. Silver Spring, MD, 412 pp

Punt AE (2001) Review of the assessments of and management advice for Atlantic large coastal sharks. An independent review document for the National Marine Fisheries Service. U.S. Department of Commerce, National Oceanic and Atmospheric Administration. Silver Spring, MD, 28 pp

Ricker WE (1975) Computation and interpretation of biological statistics of fish populations. Bull Fish Res Board Can 191, 382 pp

Simpfendorfer CA (2000) Age and growth of the whiskery shark, *Furgaleus macki*, from south-western Australia. Environ Biol Fish 58:335–343

Sminkey TR, Musick JA (1995) Age and growth of the sandbar shark, *Carcharhinus plumbeus*, before and after population depletion. Copeia 1995:871–883

Sulikowski JA, Kneebone J, Elzey S, Jurek J, Danley P, Howell WH, Tsang PWC (2005) Age and growth estimates of the thorny skate (*Amblyraja radiata*) in the western Gulf of Maine. Fish Bull 103:536–543

Sulikowski JA, Morin MD, Suk SH, Howell WH (2003) Age and growth of the winter skate, *Leucoraja ocellata*, in the Gulf of Maine. Fish Bull 101:405–413

Tanaka S, Cailliet GM, Yudin KG (1990) Differences in growth of the blue shark, *Prionace glauca*: technique or population? In: Pratt HL, Gruber SH, Taniuchi T (eds) Elasmobranchs as living resources: advances in the biology, ecology, systematics, and the status of the fisheries. NOAA Technical 395 Report. NMFS 90. Silver Spring, MD, pp 177–187

Taniuchi T, Kuroda N, Shimizu M, Nose Y (1983) Age, growth, reproduction, and food habits of the star-spotted dogfish *Mustelus manazo* collected from Chosi. Bull Jpn Soc Sci Fish 49:1325–1334

Trent L, Parshley DE, Carlson JK (1997) Catch and bycatch in the shark drift gillnet fishery off Georgia and Florida. Mar Fish Rev 59(1):19–28

Van Dykhuizen G, Mollet HF (1992) Growth, age estimation, and feeding of captive sevengill sharks, *Notorynchus cepedianus*, at the Monterey Bay Aquarium. Austr J Mar Freshwater Res 43:297–318

Von Bertalanffy L (1938) A quantitative theory of organic growth (inquiries on growth laws. II). Hum Biol 10:181–213

Walker TI, Taylor BL, Hudson RJ, Cottier JP (1998) The phenomenon of apparent change of growth rate in gummy shark (*Mustelus antarcticus*) harvested off southern. Austr Fish Res 39:139–163

Zar JH (1984) Biostatistical analysis, 2nd edn. Prentice-Hall, New Jersey, 243 pp

Environ Biol Fish (2006) 77:293–300
DOI 10.1007/s10641-006-9110-8

ELASMOBRANCHS

Evidence of two-phase growth in elasmobranchs

Miguel Araya · Luis A. Cubillos

Received: 5 June 2006 / Accepted: 23 June 2006 / Published online: 8 August 2006
© Springer Science+Business Media B.V. 2006

Abstract It is often assumed that the von Bertalanffy growth model (VBGM) is appropriate to describe growth in length-at-age of elasmobranchs. However, a review of the literature suggests that a two-phase growth model could better describe growth in elasmobranchs. We compare the two-phase growth model (TPGM) with the VBGM for 18 data sets of elasmobranch species, by fitting the models to 36 age-length-at-age data pairs available. The Akaike Information Criteria (AIC) and the difference in AIC between both models revealed that in 23 cases the probability that the TPGM was true ≥50%. The VBGM tends to estimate larger L_∞ values than the two-phase growth model, while the k parameter tends to be underestimated. The growth rate in length-at-age appears tends to decrease near the age at first maturity in several species of elasmobranch. The importance of the TPGM lies in that it may better describe this aspect of the life history of many elasmobranchs. In this context, we conclude that the TPGM should be used along with other growth models in order to precisely estimate elasmobranch life history parameters.

Keywords Elasmobranchs · von Bertalanffy · Two-phase growth · Sharks · Skates

M. Araya (✉)
Departamento de Ciencias del Mar, Universidad Arturo Prat, Avda. Arturo Prat 2120, 111-0939 Iquique, Chile
e-mail: miguel.araya@unap.cl

L. A. Cubillos
Departamento de Oceanografía, Facultad de Ciencias Naturales y Oceanográficas, Universidad de Concepción, Casilla 160–C, Concepción, Chile

Introduction

Knowledge of the age structure of a population and the k parameter of the von Bertalanffy growth model (VBGM, von Bertalanffy 1938) are central to understanding the responses of an exploited elasmobranches population. In fact, k has been used as an index of the vulnerability of a stock subject to excessive mortality and is useful for comparing life history strategies and limitations among species (Pratt and Casey 1990; Musick 1999). Those groups having a k coefficient value of <0.1 seem to be particularly vulnerable. Most elasmobranchs fall under that category (Cailliet and Goldman 2004).

The VBGM is one of the most used models to describe growth in elasmobranchs. It is based on the premise that an organism is analogous to a chemical reaction that obeys the mass action law and is described by the familiar equation:

$$L_t = L_\infty\left(1 - e^{-k(t-t_0)}\right)$$

where L_∞ is the asymptotic length-at-age, which represents the average length-at-age of individuals in a stock would attain if they grew indefinitely, k is a curvature parameter determining the rate at which the fish reach the asymptotic length-at-age, and t_0 is a position parameter defining the initial condition on the time axis when mean fish length-at-age is zero.

Caillet et al. (1992) proposed a variant of the VBGM replacing t_0 by L_0 because length-at-age at-birth is generally more easily obtainable in elasmobranchs. This model is expressed as:

$$L_t = L_\infty - (L_\infty - L_0)e^{kt}$$

where L_0 is the length-at-age-at-birth.

Other models used to describe growth in elasmobranchs include the generalized VBGM (Pauly 1981) and Schnute's model (Schnute 1981) used by Goosen and Smale (1997) for *Mustelus mustelus* and a seasonal growth model for *M. lenticulatus* (Francis and Francis 1992). A modification of the VBGM known as the two-phase growth model (TPGM, Soriano et al. 1992) has recently been used to describe the growth of *Isurus oxyrinchus*, *Prionace glauca*, and *Lamna nasus*[1]. The importance of this TPGM resides in that the model predicts the decrease in growth rate observed for different species between the ages of four and seven. Natanson et al. (2002) also observed a decrease in growth rate when male and female *Lamna nasus* from the Northeast Atlantic Ocean first reached sexual maturity, but they fitted the VBGM to the data when a TPGM might have been more appropriate. Similarly, Natanson and Caillet (1990; see their Fig. 4) identified a change in growth between 800 and 1000 mm TL (approximately 28 bands) in *Squatina californica*, which was coincident with the length-at-first maturity of this species, but they did not fit any model to describe the growth. Skomal and Natanson (2003) described a two-phase growth for *Prionace glauca* by using a non-parametric fit (LOESS), whose change in phase was coincident with the age-at-maturity of females, but suggested that it could have been a problem due to low sample size for females.

A review of published papers on age and growth of elasmobranchs, reveals that in many of the figures in which length-at-age is plotted against age, a decrease in growth rate in length-at-age is observed, coinciding with the attainment of the age at first maturity. This pattern in growth is evident in several elasmobranch species, but it is usually ignored because the VBGM is adopted when the TPGM might be more appropriate, for example; *Carcharhinus obscurus* (Natanson and Kohler 1996), *Dasyatis chrysonota chrysonota* (Cowley 1997), *Squalus acanthias* (Avsar 2001), and *Prionace glauca* (Skomal and Natanson 2003). The problem is compounded in some studies that have fitted growth curves to back-calculated data, which tends to obscure the decrease in growth rate when first reaching maturity. Other studies do not present the distribution of observed length-at-age against age at all (Natanson et al. 1995; Sminkey and Musick 1995).

It seems warranted to propose that elasmobranchs may follow a pattern of growth in length-at-age different from that predicted by the VBGM. The most appropriate growth model for elasmobranchs (or groups of elasmobranchs) has not been clearly established, and no studies have yet attempted to explain the decrease in growth rate near the age at first maturity within a growth model. Walker et al. (1998) and Walker (1998) provide evidence of length-at-age selective fishing mortality as a cause for distorting von Bertalanffy growth curves. These authors also stress the potential effects of length-at-age selective sampling bias and overestimation of age for distorting growth curves. Is two phase growth a common characteristic in the growth of elasmobranchs? We addressed this question by comparing the fit of the TPGM and the VBGM to 36 length-at-age data sets available for 18 populations of elasmobranchs.

[1] Acuña, E., L. Cid, E. Pérez, I. Kong, M. Araya, J. Lamilla & J. Peñailillo. 2001. Estudio biológico de tiburones (marrajo dentudo, azulejo y tiburón sardinero) en la zona norte y central de chile. Informe FIP N° 2000-23. Subsecretaría de Pesca. 128 pp. Available from the Internet URL http://www.fip.cl/pdf/informes/inffinal%202000-23.pdf

Materials and methods

Data and growth models

Length-at-age data were requested and made available by several authors that previously had published their results in peer-reviewed journals. In a few cases, data was extracted directly from published figures. In other cases, we used unpublished data[1]. Data based in the analysis belonged to 16 species and 4 orders of elasmobranches (Table 1). Depending on the available information, the analyses were carried out for male, female, or combined sexes.

Growth parameters of both the VBGM and the two-phase growth model were estimated by fitting the model to the observed data through non-lin ear regression by considering an additive error structure. In the case of the VBGM, this is expressed as:

$$L_t = L_\infty(1 - e^{-k(t-t_0)}) + \varepsilon_t$$

where ε_t is the random error, which is assumed normally distributed.

The TPGM of Soriano et al. (1992), which consists of a modified version of the VBGM, is expressed as:

$$L_t = L_\infty(1 - e^{-kA_t(t-t_0)}) + \varepsilon_t$$

This equation is the second variant in Soriano et al. (1992), where A_t is a factor that modifies k when the age is increased, and can be defined by:

$$A_t = 1 - \frac{h}{(t - t_h)^2 + 1}$$

where t_h is the age at which the transition between the two phases occur, and h determines the magnitude of the maximum difference in length-at-age between the VBGM and the TPGM in the point t_h. The loss-function in both models used was least-squares.

Model selection

A maximum likelihood method was used to select the model that best fitted elasmobranch growth (Burnham and Anderson 2002; Motulsky and Christopoulos 2003). The objective of the method is to evaluate the relative power of the evidence supporting a given model. Relative support was evaluated by considering likelihood theory combined with Akaike's Information Criterion (AIC) (Burnham and Anderson 2002). A single AIC

Table 1 Elasmobranchs species considered in the present study

Species	Order	F	M	Both	Source
Squalus acanthias	Squaliformes	X	X	X	Avsar (2001)
Deania calceus	Squaliformes	X	X		Clarke et al. (2002)
Raja clavata	Rajiformes	X	X		Holden (1972)
Raja batis	Rajiformes			X	Du Buit (1976)
Raja naevus	Rajiformes			X	Du Buit (1976)
Rhinobatos productus	Rajiformes		X	X	Timmons and Bray (1997)
Dasyatis chrysonota chrysonota	Rajiformes	X	X		Cowley (1997)
Leucoraja ocellata	Rajiformes	X	X	X	Sulikowski et al. (2003)
Dipturus chilensis	Rajiformes	X	X	X	Araya et al.[1]
Alopias vulpinus	Lamniformes			X	Cailliet et al. (1983)
Lamna nasus	Lamniformes	X	X	X	Natanson et al. (2002)
Lamna nasus	Lamniformes			X	Acuña et al. (2001)
Isurus oxyrinchus	Lamniformes	X	X	X	Acuña et al. (2001)
Negaprion brevirostris	Carcharhiniformes	X	X	X	Brown and Gruber (1988)
Triakis semisfaciata	Carcharhiniformes			X	Kusher et al. (1992)
Carcharhinus leucas	Carcharhiniformes			X	Cruz-Martínez et al. (2002)
Prionace glauca	Carcharhiniformes	X			Skomal and Natanson (2003)
Prionace glauca	Carcharhiniformes	X	X	X	Acuña et al. (2001)

F: females; M: males

[1] M. Araya, H. Arancibia & P. Ortiz (Unpublished data)

value does not have an interpretation by itself, but comparisons between different values of AIC allow evaluation of the relative support of the data for two or more models. The AIC penalizes the complexity of the model, given by the number of parameters, by attaining an optimum between parsimony and accuracy. The expression used is:

$$\mathrm{AIC} = n\,\ln(\hat{\sigma}^2) + 2p$$

where p is the number of estimated parameters, n is the number of observations, and σ^2 is given by:

$$\hat{\sigma}^2 = \frac{\sum \hat{\varepsilon}_i^2}{n}$$

where $\hat{\varepsilon}_i$ are the residuals for a given model. The model with the lowest AIC value is selected as the most probable and true model for the data. How much more probable? According to Burnham and Anderson (2002) and Motulsky and Christopoulos (2003) the difference between the AIC with the higher number of parameters and the model with fewer parameters, ΔAIC, allows computation of the AIC weight (w), which corresponds to the probability of choosing the true model, i.e., in this case the TPGM:

$$w = \frac{e^{-0.5\,\Delta AIC}}{1 + e^{-0.5\,\Delta AIC}}$$

Results

The values of AIC's, ΔAIC and w, were obtained for the different species (Table 2). There were significant differences between models in the L_∞ parameter estimates as suggested by the slope of the regression (=0.866), which was significantly different from unity ($p < 0.0001$; Fig. 1a). The VBGM tends to estimate larger values of L_∞ than the TPGM. In 15 cases, the difference in L_∞ values between the two models was between 10 and 141 cm, in 19 cases the difference was between 9.4 and 9.7 cm, and in only two cases the difference was less than –26 cm (Fig. 1b).

The opposite occurred in the case of the k parameter, i.e., the VBGM tends to estimate a lower k compared with the TPGM. The slope of the regression (=0.866) was not significantly different from unity ($p = 0.318$), but the intercept (=0.036) was significantly different from zero ($p < 0.01$). The majority of k values estimated by the TPGM was above the based or line of 45° (Fig. 2a). In 13 cases, the difference in k values was between –0.026 and –0.165 year^{-1}, while in 23 cases the difference ranged between –0.017 and 0.018 year^{-1} (Fig. 2b).

Length-at-age L_{th} of different species at age t_{h} when the difference between the models is maximum was always higher for the VBGM than for the TPGM (Fig. 3). This means that there is a time in the life history of elasmobranchs when the rate of change in growth in length-at-age tends to decrease. In fact, in 23 of 36 cases, the probability that the TPGM is true was ≥50% (Fig. 4), representing 16 species (9 cases were females, 7 males, and 7 sexes combined).

Discussion

The AIC is a good criterion for selecting the most parsimonious model for explaining the observed variation in the data while using fewer parameters (Burnham and Anderson 2002). However, which model is better or worse will always depend on the context. Wang et al. (1995) pointed out that choosing a growth curve is often subjective and recommend that in some cases a pragmatic decision based on previous studies and experience rather than goodness-of-fit of the data should be used. In contrast, Haddon (2001) stated that the model best explaining the growth process should be used, but unfortunately it is not so simple to decide what the "better" description of the process is since statistical results and biological interpretation can sometimes conflict (see Cailliet et al. this issue). Generally, it is not possible to maximize all of the attributes of a particular model simultaneously, because often parsimony, precision, accuracy, and biological realism are not independent attributes.

Ricker (1979) considered that growth may be divided into a series of stages in the life history of a fish, and the changes between stages are characterized by some crisis or discontinuity in development, such as maturity, changes in behavior, or changes in habitat. This is in agreement

Table 2 Values of AIC for the two models (AIC_{VB}: von Bertalanffy model growth; AIC_{TP}: two-phase model); ΔAIC: difference between the AIC with the higher number of parameters and the model with fewer parameters; w_{TPGM}: AIC weight corresponds to the probability of the TPGM being correct

Species	Female				Male				Both			
	AIC_{VB}	AIC_{TP}	ΔAIC	w_{TPGM}	AIC_{VB}	AIC_{TP}	ΔAIC	w_{TPGM}	AIC_{VB}	AIC_{TP}	ΔAIC	w_{TPGM}
S. acanthias	42.35	43.04	0.68	41.5%	27.11	30.51	3.39	15.5%	38.52	38.89	0.38	45.3%
D. calceus	337.21	341.08	3.87	12.6%	235.46	237.56	2.10	25.9%				
R. clavata	105.45	101.85	–3.60	85.8%	48.34	46.36	–1.97	72.9%				
R. batis									275.85	269.31	–6.54	96.3%
R. naevus									96.00	95.96	–0.04	50.5%
R. productus					103.77	100.16	–3.61	85.9%	179.76	181.49	1.73	29.6%
D. ch. chrysonota	74.82	73.67	–1.15	64.0%	47.69	37.39	–10.30	99.4%				
L. ocellata	892.36	892.30	–0.07	50.8%	682.64	681.45	–1.19	64.5%	1576.19	1578.76	2.57	21.6%
D. chilensis	228.18	222.47	–5.71	94.6%	136.70	140.41	3.71	13.6%	368.30	371.18	2.88	19.1%
A. vulpinus									949.06	952.85	3.79	13.1%
L. nasus	1485.81	1487.28	1.47	32.4%	1412.45	1408.85	–3.59	85.8%	2981.25	2977.23	–4.02	88.2%
L. nasus									313.66	312.86	–0.80	59.8%
I. oxyrinchus	1373.13	1345.83	–27.30	100.0%	1537.23	1534.70	–2.52	77.9%	3110.73	3133.11	22.38	0.0%
N. brevirostris	183.44	178.66	–4.78	91.6%	187.70	186.45	–1.25	65.2%	360.25	370.79	10.54	0.5%
T. semifasciata									1519.49	1516.22	-3.27	83.7%
C. leucas									124.31	121.84	-2.47	77.5%
P. glauca	603.23	592.51	–10.72	99.5%								
P. glauca	993.23	992.88	–0.35	54.4%	553.97	560.07	6.08	4.6%	1584.36	1593.63	9.27	1.0%

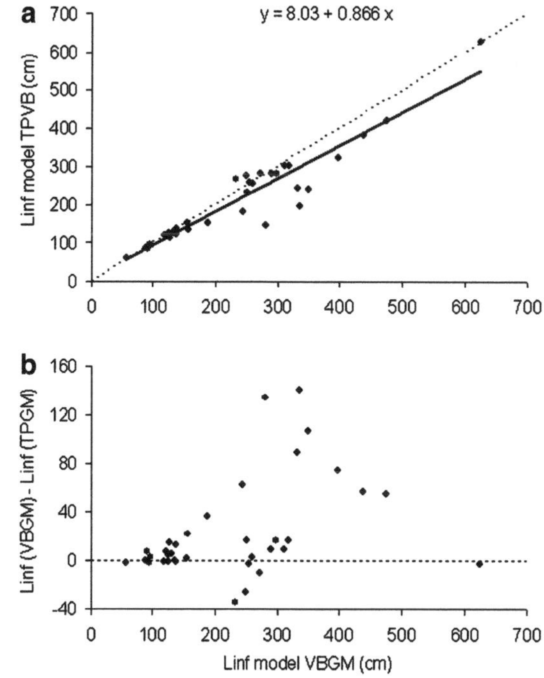

Fig. 1 (a) Relationship between parameter L_∞ (Linf) estimated by the models. The dotted line corresponds to 1:1. (b) Relationship between Linf estimated by VBGM and the difference between both models. VBGM: von Bertalanffy growth model; TPGM: two-phase growth model

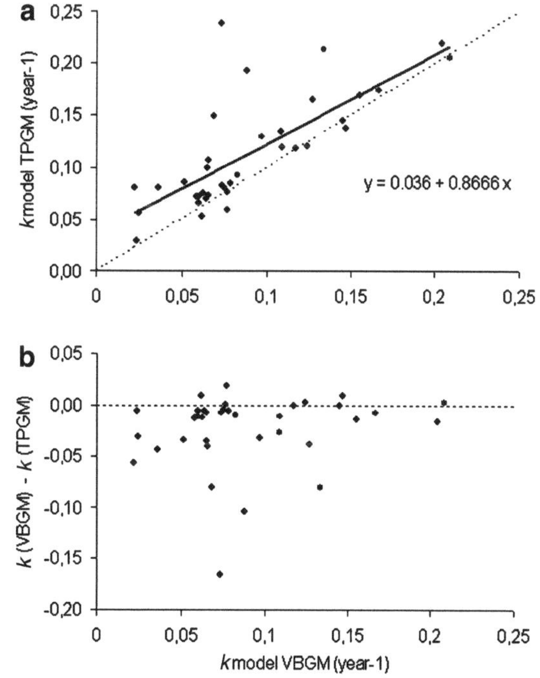

Fig. 2 (a) Relationship between parameter k estimated by both models. The dotted line corresponds to 1:1. (b) Relationship between k estimated by VBGM and the difference between both models. VBGM: von Bertalanffy growth model; TPGM: two-phase growth model

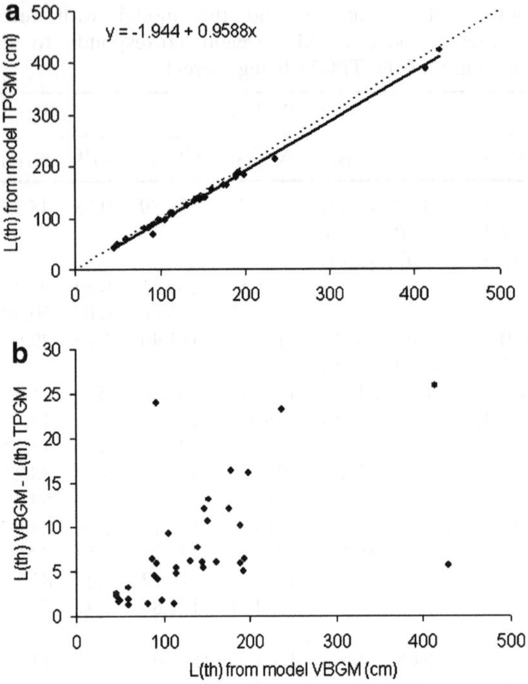

Fig. 3 (a) Relationship between the length-at-age at age t_h, $L(t_h)$ as estimated by both models. The dotted line corresponds to 1:1. (b) Relationship between $L(t_h)$ estimated with VBGM and the difference between both models

with the suggestion of Day and Taylor (1997), who stated that the growth trajectory should be specified by two separate equations: one describing pre-maturity, in which essentially no surplus energy is destined to reproduction and post-maturity equation, in which all (determinate growth) or some (indeterminate growth) surplus energy is used for reproduction. These authors in fact suggested that there are not reasons for the use of the VBGM when studying the relationship between growth and maturity because growth in pre- and post-maturity seems to be different. The maturity principle involves a reduction in the energy destined to growth because the energy is allocated to reproduction (Jensen 1985). Consequently, the growth trajectory should exhibit a fundamental change at the onset of maturity, and growth models should be able to distinguish between the process of growth before and after maturity (Day and Taylor 1997).

Attaining maturity would likely be recognized as a new growth phase if individual growth data were available, but growth in wild populations is inferred by using averages of individuals from different year classes (Ricker 1979). As such, changes in growth rate at the onset of maturity are obscured by data from fish that come from different year classes.

According to Soriano et al. (1992), the two-phase growth would be apparent when individuals are sampled only in a single episode and the average length-at-age is computed from various annual classes. If one-year class has had a lower growth rate, then a decline will appear in the growth curve. However, most elasmobranch growth studies surveyed here had samples collected during one or more years, so lower growth rates at intermediate ages should not be related to the sampling procedure. Conversely, when the two-phase models fitted to length-at-age data pairs of combined sexes is considered, e.g., *Leucoraja ocellata, Negaprion brevirostris, Isurus oxyrinchus, Prionace glauca*, the probability (w) in choosing the TPGM is very low. However, when we take into account male and female individually the conclusion is completely different. There is a higher probability favoring the TPGM. This situation could be because usually female and male attain their first maturity at different ages, and therefore the reduction in the length-at-age growth rate predicted by the two-phase model for both sexes could be masked, affecting the fit of the model when both sexes are combined. Of course, this explanation is right only if the reduction in the length-at-age growth rate is related to the age at first maturity.

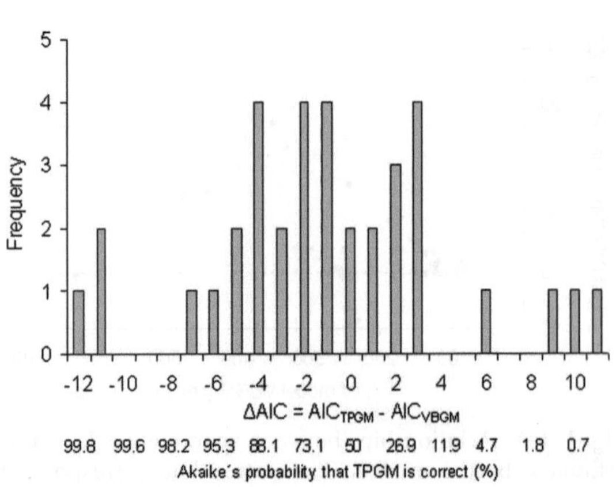

Fig. 4 Frequency distribution of the differences in the AIC between models VBGM and TPGM (ΔAIC) and corresponding Akaike's probability that the TPGM is true

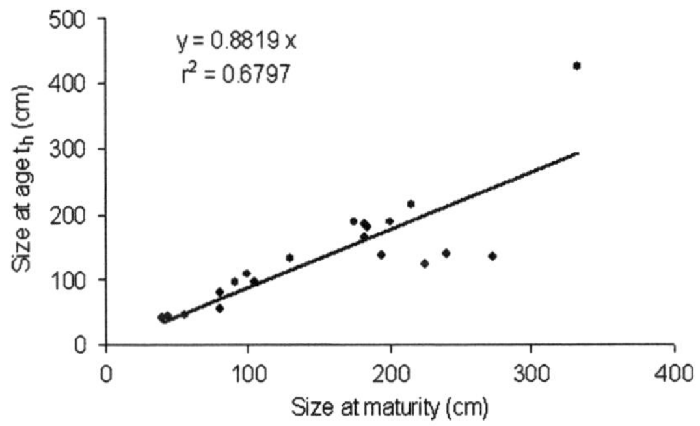

Fig. 5 Relationship between the size at maturity and size at age t_h. The slope is not significantly different from one. The size at maturity of Lamniformes and Carcharhiniformes was obtained from Cortés (2000) and size of Squaliformes and Rajiformes from Froese and Pauly (2004)

The TPGM could be related to the age at first maturity in elasmobranchs. Indeed, length at first maturity of most elasmobranchs coincides with the average length-at-age (i.e. t_h in Fig. 5), where the slope was not significantly different from the unity. In 7 of 10 species, t_h of females was higher than that of males, since females reach maturity at a larger size than males (Cortés 2000).

According to our results, growth in elasmobranchs should be analyzed carefully because the VBGM does not always apply (Cailliet et al. 2006). Many of the elasmobranchs considered in this study seem to follow a different growth pattern, characterized by a decrease in the growth rate in length-at-age at a time that apparently coincides with the onset of maturity. This could result in errors in stock assessments that make use of growth parameters derived from the VBGM, as well as in age-structured models that take into account the average weight and length-at-age.

Acknowledgements The authors would like to thank Enric Cortés and Terry Walker for commentaries and revision of the manuscript. In addition, we would like to thank the authors who facilitated their information to us to conduct the analyses and to two anonymous referees for reviewing the manuscript. We would particularly like to thank the California Sea Grant for providing funding to travel to the meeting and present the results of this study.

References

Avsar D (2001) Age, growth, reproduction and feeding of the spurdog (*Squalus acanthias* Linnaeus, 1758) in the South-eastern Black sea. Estuar Coast Shelf Sc 52:269–278

Brown CA, Gruber SH (1988) Age assessment of the lemon shark, *Negaprion brevirostris*, using tetracycline validated vertebral central. Copeia 1988:747–753

Burnham KP, Anderson DR (2002) Model selection and multimodel inference: a practical information-theoretic approach 2 ed Springer-Verlag, New York, NY

Cailliet GM, Martin LK, Harvey JK, Kusher D, Welden BA (1983) Preliminary studies on the age and growth of blue, *Prionace glauca*, common thesher, *Alopias vulpinus*, and shortfin mako, *Isurus oxyrinchus*, sharks from California waters. In: Proceedings of the international workshop on age determination of oceanic pelagic fishes: tunas, billfishes, and sharks. pp 179–188. In: Prince ED, Pulos LM (eds) NOAA Tech. Rep. NMFS 8, U.S. Dept. Comm., Washington, DC

Cailliet GM, Mollet HF, Pittenger GG, Bedford D, Natanson LJ (1992) Growth and demography of the pacific angel shark (*Squatina californica*), based upon tag returns off California. Aust J Mar Freshw Res 43:1313–1330

Cailliet GM, Goldman KJ (2004) Age Determination and Validation in Chondrichthyan Fishes. In: Carrier JC, Musick JA, Heithaus MR (eds), Biology of sharks and their relatives. CRC Press, Boca Raton, FL, pp 399–446

Cailliet GM, Smith WD, Mollet HF, Goldman KJ (2006). Chondrichthyan growth studies: an updated review, stressing terminology, sample size sufficiency, validation, and curve fitting. pp. 000–000. In: Carlson JK, Goldman KJ (eds), Special volume from symposium of the American Elasmobranch Society.

Clarke MW, Connolly PL, Bracken JJ (2002) Catch, discarding, age estimation, growth and maturity of the squalid shark *Deania calceus* west and north of Ireland. Fish Res 56:139–153

Cortés E (2000) Life history patterns and correlations in sharks. Rev Fish Sci 8:299–344

Cowley PD (1997) Age and growth of the blue stingray *Dasyatis chrysonota chrysonota* from the south-east ern cape coast of South Africa. S Afr J Mar Sci 18:31–38

Cruz-Martínez A, Chiappa-Carrara X, Arenas-Fuentes V (2002) Age and growth of the bull shark *Carcharhinus leucas* from the southern Gulf of Mexico. NAFO SCR Doc. 02/88, 1–11

Day T, Taylor PD (1997) von Bertalanffýs growth equation should not be used to model age and size at maturity. Am Nat 149:381–393

Du Buit MH (1976) Age et croissance de *Raja batis* et de *Raja naevus* en Mer Celtique. J Cons Per Int Expl Mer 37:261–265

Francis MP, Francis RICC (1992) Growth rate estimates for New Zealand rig (*Mustelus lenticulatus*). Aust J Mar Freshw Res 43: 1157–1176

Froese R, Pauly D (eds) (2004) FishBase-world wide web electronic publication. www.fishbase.org, version (03/2004)

Goosen AJJ, Smale MJ (1997) A preliminary study of age and growth of the smooth-hound shark *Mustelus mustelus* (Triakidae). S Afr J Mar Sci 18:85–91

Haddon M (2001) Modelling and Quantitative Methods in Fisheries. Chapman & Hall/CRC, Boca Raton, FL. p 406

Holden MJ (1972) The growth rates of *Raja brachyura*, *R. clavata* and *R. montagui* as determined from tagging data. J Cons Per Int Expl Mer 34:161–168

Jensen AL (1985) Relations among net reproductive rate and life history parameters for lake Whitefish (*Coregonus clupeaformis*). Can J Fish Aquat Sci 42:164–168

Kusher DI, Smith SE, Cailliet GM (1992) Validated age and growth of the leopard shark, *Triakis semifasciata*, with comments on reproduction. Environ Biol Fish 35:187–203

Motulsky H, Christopoulos A (2003) Fitting models to biological data using linear and nonlinear regression. A practical guide to curve fitting. GraphPad Sofware Inc., San Diego CA, www.graphpad.com

Musick JA (1999) Life in the slow lane: ecology and conservation of long-lived marine animals. American fisheries society symposium 23, Bethesda, Maryland

Natanson LJ, Cailliet GM (1990) Vertebral growth zone deposition in Pacific Angel sharks. Copeia 1990:1133–1145

Natanson LJ, Casey JG, Kohler NE (1995) Age and growth estimates for the dusky shark, *Carcharhinus obscurus*, in the western North Atlantic Ocean. Fish Bull 93:116–126

Natanson LJ, Kohler NE (1996) A preliminary estimate of age and growth of the dusky shark *Carcharhinus obscurus* from the South-West Indian Ocean, with comparisons to the Western North Atlantic population. S Afr J Mar Sci 17:217–224

Natanson LJ, Mello JJ, Campana SE (2002) Validated age and growth of the porbeagle shark (*Lamna nasus*) in the western North Atlantic Ocean. Fish Bull 100:266–278

Pauly D (1981) The relationships between gill surface area and growth performance in fish: a generalization of von Bertalanffýs theory of growth. Meeresforschung 28:251–282

Pratt HL, Casey JG (1990) Shark reproductive strategies as a limiting factor in directed fisheries, with a review of Holden's method of estimating growth parameters. NOAA Tech. Rep. NMFS 90: 97–109

Ricker WR (1979) Growth rates and models. In: Hoar WS, Randall DJ, Brett JR (eds) Fish physiology. Acad. Press, New York, NY, pp 677–743

Schnute J (1981) A versatile growth model with statistically stable parameters. Can J Fish Aquat Sci 38:1128–1140

Skomal GB, Natanson LJ (2003) Age and growth of the blue shark (*Prionace glauca)* in the North Atlantic Ocean. Fish Bull 101:627–639

Sminkey TS, Musick JA (1995) Age and growth of the sandbar shark, *Carcharhinus plumbeus*, before and after population depletion. Copeia 1995:871–883

Soriano M, Moreau J, Hoenig JM, Pauly D (1992) New functions for the analysis of two-phase growth of juvenile and adult fishes, with application to nile perch. Trans Am Fish Soc 121:486–493

Sulikowski JA, Morin MD, Suk SH, Howell WH (2003) Age and growth estimates of the winter skate (*Leucoraja ocellata*) in the western Gulf of Maine. Fish Bull 101:405–413

Timmons M, Bray RN (1997) Age, growth, and sexual maturity of shovelnose guitarfish, *Rhinobatos productus* (Ayres). Fish Bull 95: 349–359

von Bertalanffy L (1938) A quantitative theory of organic growth. Hum Biol 10:181–213

Walker TI (1998) Can shark resources be harvested sustainably? A question revisited with a review of shark fisheries. Mar Freshw Res 49:553–572

Walker TI, Taylor BL, Hudson RJ, Cottier JP (1998) The phenomenon of apparent change of growth rate in gummy shark (*Mustelus antarcticus*) harvested off southern Australia. Fish Res 39:139–163

Wang Y-G, Thomas MR, Somers IF (1995) A maximum likelihood approach for estimating growth from tag-recapture data. Can J Fish Aquat Sci 52:252–259

Environ Biol Fish (2006) 77:301–308
DOI 10.1007/s10641-006-9112-6

SI - ELASMOBRANCHS

Two Bayesian methods for estimating parameters of the von Bertalanffy growth equation

Kate I. Siegfried · Bruno Sansó

Received: 21 June 2006 / Accepted: 3 July 2006 / Published online: 19 August 2006
© Springer Science+Business Media B.V. 2006

Abstract The von Bertalanffy growth equation (VBGE) is commonly used in ecology and fisheries management to model individual growth of an organism. Generally, a nonlinear regression is used with length-at-age data to recover key life history parameters: L_∞ (asymptotic size), k (the growth coefficient), and t_0 (a time used to calculate size at age 0). However, age data are often unavailable for many species of interest, which makes the regression impossible. To confront this problem, we have developed a Bayesian model to find L_∞ using only length data. We use length-at-age data for female blue shark, *Prionace glauca*, to test our hypothesis. Preliminary comparisons of the model output and the results of a nonlinear regression using the VBGE show similar estimates of L_∞. We also developed a full Bayesian model that fits the VBGE to the same data used in the classical regression and the length-based Bayesian model. Classical regression methods are highly sensitive to missing data points, and our analysis shows that fitting the VBGE in a Bayesian framework is more robust. We investigate the assumptions made with the traditional curve fitting methods, and argue that either the full Bayesian or the length-based Bayesian models are preferable to classical nonlinear regressions. These methods clarify and address assumptions made in classical regressions using von Bertalanffy growth and facilitate more detailed stock assessments of species for which data are sparse.

Keywords Markov Chain Monte Carlo · Elasmobranch · Stock assessment · Asymptotic size

K. I. Siegfried (✉)
NOAA-Fisheries, Panama City Laboratory, 3500 Delwood Beach Road, Panama City, FL 32408, USA
Tel.: +1-850-234-6541; Ext: 215
e-mail: kate.siegfried@noaa.gov

B. Sansó · K. I. Siegfried
The Center for Stock Assessment Research (CSTAR), Santa Cruz, USA

B. Sansó
Department of Applied Mathematics and Statistics, University of California, Santa Cruz, USA

Introduction

The von Bertalanffy growth equation (VBGE) can be used to describe the way individual organisms grow (von Bertalanffy 1938, 1957). The form used by Beverton and Holt (1959) is the following

$$L(a) = L_\infty\left(1 - e^{-k(a-t_0)}\right) \quad (1)$$

where a is age, k is the growth coefficient, t_0 is a value used to calculate size when age is zero, and L_∞ is asymptotic size. Generally, to obtain estimates of the parameters, the VBGE is fit to length-at-age data using classical nonlinear regression techniques (Grafen and Hails 2002). This curve-fitting technique somewhat changes the assumptions of the VBGE; using a least squares estimate assumes an average maximum size for the population rather than a truly asymptotic size—often termed L_{max} in the literature rather than L_∞. However, there is a physiological maximum size a fish can attain. Bayesian methods presented in this paper allow for the use of the VBGE while maintaining its biological assumptions.

Individual growth is the basis of many stock assessment models (Jennings et al. 2001). However, length-at-age data—required for parameter estimation—are not available or abundant for many populations, particularly for long-lived species. Length data are easier to gather, and more available for fished species. We developed a Bayesian model to estimate the maximum size using only length data.

We also present a full Bayesian model designed to fit the VBGE to the length-at-age data. We include literature-derived priors in our analysis and compare our results from both Bayesian models to the results from a classical nonlinear regression.

Methods

We have length-at-age data with $a = 0,\ldots,i$ indexing age and r indexing the number of length data per age. The number of data varies per age, and the following is a visualization of our data:

$$\begin{matrix} a=1 & L_{1,1} & \ldots & L_{1,r} \\ \vdots & \vdots & & \vdots \\ a=i & L_{i,1} & \ldots & L_{i,r} \end{matrix} \quad (2)$$

There are likely two sources of error in our data: process error and observation error (Hilborn and Mangel 1997). We expect the process error to be the variability of vital rates such as growth and mortality within each age class. Observation error will manifest in the way each population is sampled; we expect to see variation in the selectivity of the gear (i.e. the distribution of sizes caught by the gear type). However, for this paper, we include one multiplicative error term to represent all error.

For each of the following models, we used the same data set from the literature to test the ideas (Acuña et al. 2001).[1] The length-at-age data are for female blue shark, *Prionace glauca*, from the Eastern Tropical Pacific, representing ages 0–14. We used only female shark data, since many shark species have different growth trajectories for each sex (Cortés 2000). Although we have length-at-age data, we only use the age data for the least squares regression and the full Bayesian model.

Least squares regression

Least squares regression is a common technique used to fit curves to data (Grafen and Hails 2002). We used the VBGE (Eq. 1) to calculate predicted length at age and then compared those values with the observed length data. The best fit parameters minimize the sum of the squared differences between predicted and observed lengths. For this analysis, we used the solver in Microsoft Excel to find the best combination of parameters minimize the sum of the squared errors.

Full Bayesian model

Our full Bayesian model fits the VBGE to the length-at-age data using Markov Chain Monte Carlo (MCMC) methods (Hastings 1970; Gelman et al. 2004). These methods require the following four steps:

(1) Find the likelihood of the data,
(2) Establish priors for all parameters,

[1] See the reference for sampling and aging methods.

(3) Find the full conditional probabilities for parameters, when possible, and
(4) Decide whether to use the Metropolis Hastings or the Gibbs sampling algorithm to sample the posterior distribution for each parameter (Gelman et al. 2004).

We start with the VBGE, including multiplicative error

$$L(a) = L_\infty\left(1 - e^{-k(a-t_0)}\right)e^{\varepsilon_a} \tag{3}$$

and then find the log transform

$$l_a = l_\infty + \log\left(1 - e^{-k(a-t_0)}\right) + \varepsilon_a \tag{4}$$

$$\varepsilon_a \sim N(0, \sigma^2) \tag{5}$$

where the lower case l indicates the log of the value. Thus, our likelihood is the following:

$$L(l_{ar}|l_\infty, k, t_0, \sigma^2) \propto \prod_{a=0}^{14}\prod_{r=1}^{R_a}\frac{1}{\sqrt{2\pi\sigma^2}}\exp\left\{-\frac{1}{2\sigma^2}(l_{ar} - (l_\infty + \log(1 - \exp\{-k(a-t_0)\})))^2\right\} \tag{6}$$

where R_a is the number of length data for each age.

We constructed informative priors for k and t_0 based on published estimates of the same parameters in the literature for blue shark (Holden 1973; Cailliet et al. 1983; Hoenig and Gruber 1990; Cortés 2000):

$$p(k) = \text{Gamma}(15, 100)^2 \tag{7}$$

$$p(-t_0) = \text{Gamma}(14, 4) \tag{8}$$

We show the probability density functions (pdfs) of the priors for k and t_0 in Figs. 1 and 2 respectively. We used diffuse priors for l_∞ and σ^2, giving the full power of estimation to the data:

$$p(l_\infty) \propto 1 \tag{9}$$

$$p(\sigma^2) \propto \frac{1}{\sigma^2} \tag{10}$$

Our prior for σ^2 is a Jeffreys prior (a form of non-informative prior, see Gelman et al. 2004). Marginalizing the joint posterior—the likelihood multiplied by the four priors—for each parameter gave two full conditionals:[3]

$$L_\infty \sim \log\text{normal}\left(\bar{l} - \frac{\sum_{a=0}^{14}\sum_{r=1}^{R_a}\log(1-\exp\{-k(a-t_0)\})}{n}, \frac{\sigma^2}{n}\right) \tag{11}$$

$$\sigma^2 \sim \text{Inverse-gamma}\left(\frac{n}{2}, 2\sum_{a=0}^{14}\sum_{r=1}^{R_a}\left(\left(l_{ar} - \left(l_\infty + \log\left(1-\exp\left\{-k\left(a-t_0\right)\right\}\right)\right)\right)^2\right)\right) \tag{12}$$

where $\bar{l}$ is the average logged lengths and R_a is the number of lengths for each age (Appendix).

Since the conditional probabilities for l_∞ and σ^2 are known pdfs (Eqs. 9 and 10), we used a Gibbs sampler for these parameters in the MCMC. However, we had to use a Metropolis step for k and t_0 because we could not find their full conditional distributions. We used a random walk jumping distribution[4] for both Metropolis steps.

Length-based Bayesian model

The full Bayesian model is not only computationally intensive but also requires length-at-age data. For cases where only length data are available—which is frequently the case in developing fisheries—we developed a method to estimate the maximum size of the fish in the stock using the

[2] We used the following form of the gamma distribution: $\text{Gamma}(\alpha, \beta) \sim \frac{\beta^\alpha}{\Gamma(\alpha)}x^{(\alpha-1)}e^{-x\beta}$ for $x>0$

[3] For the functional forms of the lognormal or inverse-gamma distributions, see Gelman et al. (2004)

[4] A random walk jumping distribution samples the probability space around the initial point following a normal distribution, with the initial point as the mean of the normal. The variance is subjectively changed at the start of each MCMC run to achieve a desirable acceptance rate, between 0.3 and 0.5. See Hastings (1970)

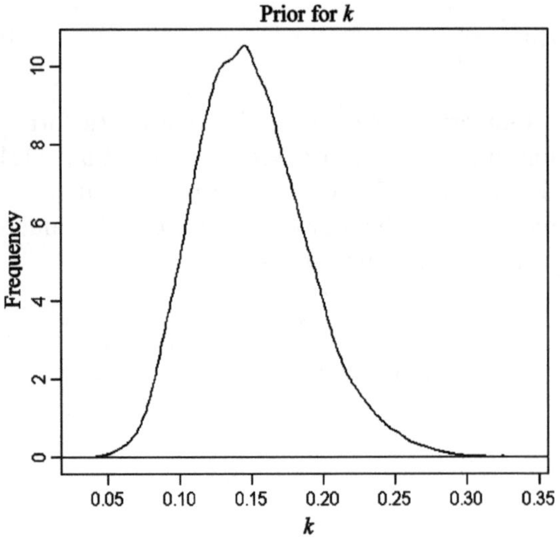

Fig. 1 The frequency of values for k, the Brody growth coefficient, given the prior probability density. Most of the density lies between 0.09 and 0.19

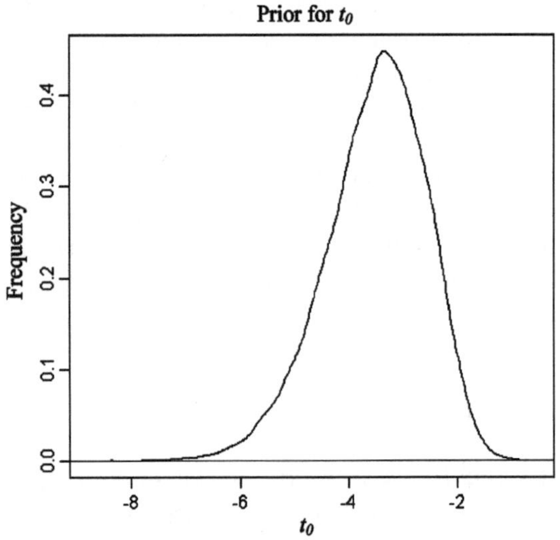

Fig. 2 The frequency of values for t_0 given the prior probabilities. Most of the density lies between –2 and –5

available data. The new method employs the following Bayesian model. Let $\{x_1,..., x_n\}$ be individual lengths and

$$Y_i = \log\left(\frac{\frac{x_i}{\theta}}{1-\frac{x_i}{\theta}}\right) \quad \text{for } x_i < \theta \tag{13}$$

where θ is our proxy for $L_{\max}$ and we assume $Y_i \sim N(\mu, \sigma^2)$. A plot of the transformed data was used to determine if the Y_is were distributed

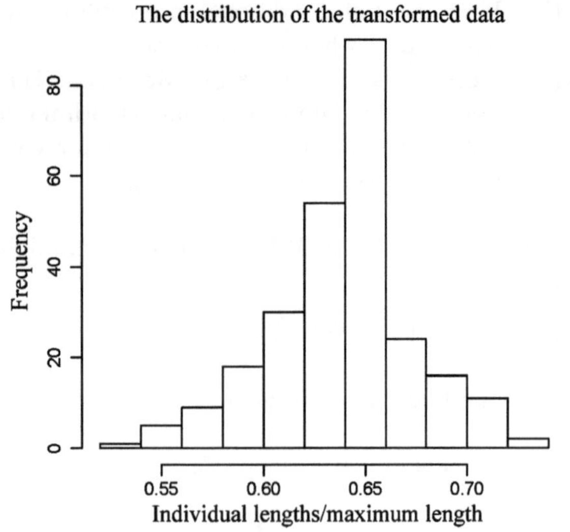

Fig. 3 A histogram of the transformed data shows an approximate normal distribution

normally (Fig. 3). We used the logit transformation of the individual lengths divided by the maximum datum. The parameter θ is our proxy to the maximum attainable size a fish in the population can achieve. Our key condition is that all data be less than θ.

Our likelihood is the following, based on the normal distribution of the Y_is:

$$L(\mu, \sigma^2, \theta) = \exp\left\{\frac{1}{2\sigma^2}\sum_{i=1}^{n}\left(\log\left(\frac{\frac{x_i}{\theta}}{1-\frac{x_i}{\theta}}\right)-\mu\right)^2\right\} \left(\frac{1}{\sigma^2}\right)^{\frac{n}{2}}\left(\prod_{i=1}^{n}\frac{1}{x_i\left(1-\frac{x_i}{\theta}\right)}\right) \tag{14}$$

We include no information from the literature for the mean and variance parameters of the normal distribution in Eq. 14 by assuming Jeffreys priors for both parameters

$$p(\mu, \sigma^2) \propto \frac{1}{\sigma^2} \tag{15}$$

However, we used information from the literature for our prior for $L_{\max}$ (θ)

$$p(\theta) \sim \text{Normal}(310, 80) \tag{16}$$

where 310 cm TL is the mean and 80 is the variance of the normal prior. From the full joint posterior, we found two full conditionals

$$p(\mu|\sigma^2, \theta, \text{data}) \propto \text{Normal}\left(\frac{1}{n}\sum_i \left(\frac{\frac{x_i}{\theta}}{1-\frac{x_i}{\theta}}\right), \frac{\sigma^2}{n}\right) \quad (17)$$

$$p(\sigma^2|\mu, \theta, \text{data}) \propto \text{Inverse} - \text{gamma} \left(\frac{n}{2}, \frac{2}{\sum_{i=1}^{n}\left(\log\left(\frac{\frac{x_i}{\theta}}{1-\frac{x_i}{\theta}}\right) - \mu\right)^2}\right) \quad (18)$$

Again, since we have recognizable pdfs for μ and σ^2, we used a Gibbs sampler for these parameters. We used the Metropolis Hastings algorithm with a random walk jumping distribution[3] to find θ (Hastings 1970).

Results

Least squares method

The combination of parameters that minimized the sum of the squared error was 252 cm, 0.13, and –3.09 for L_∞, k, and t_0 respectively (Table 1).

Full Bayesian model

We used a burn-in period of 1,000 samples and generated posteriors for the three parameters of the VBGE with the remaining samples. Our mean results are as follows: 419 cm TL, 0.06, and –3.25 for L_∞, k, and t_0 respectively. We include 95% probability intervals for each estimate in Table 1.

Length-based Bayesian model

We initialized the model with a number slightly larger than the maximum data point. The chain converged quickly to an estimate of 308 cm (Table 1). Using our estimate for L_∞ in the least squares regression, we generated a value for k and t_0 to use for comparison (0.08 and –4.13 respectively).

Convergence

We tested the convergence of our Bayesian models using both the Z-scores of the Geweke test, and by starting the runs at various initial values to see if the sample chains converge on the same number. Both tests were successful, showing convergence in the models.[5]

Discussion

Comparing estimates

First, we compare the least squares regression and the full Bayesian model (Figs. 4, 5 & Table 1). The estimates of t_0 are very similar; however, the difference in assumptions is very obvious when comparing the estimates of L_∞ and k. The full Bayesian model estimates a much larger asymptote. However, since L_∞ and k are inversely proportional, we expect, and find, a much smaller k estimate.[6] To ensure the difference between the model estimates was not only due to that correlation, we checked the correlation between parameters and the influence of the priors on the estimates (Fig. 6). The scatter plots have no structure in any of the other parameters, therefore we ruled out other correlations. We then examined the sensitivity of the full Bayesian model to the informative prior for k. Recall the mean of the gamma prior for k was 0.15. Our full Bayesian model estimate is 0.06, which means that there was enough information in the data to overwhelm any influence of the prior on the estimate (Fig. 7). In fact, when we used a non-informative prior for k, the posterior estimate was not significantly different. We also removed the ten largest data points and ran the full Bayesian model and classical regression again. The Bayesian model output did not change significantly, but the regression results changed drastically. Therefore, we are confident with our model estimates for the full Bayesian model.

Next, we compare the least squares regression with the length-based Bayesian model. The esti-

[5] See Gelman et al. (2004). For information on convergence diagnostics.

[6] The VBGE comes from the following differential equation: $\frac{dL}{d_t} = k(L_\infty - L)$. When most of the data are small lengths, it becomes $\frac{dL}{d_t} \approx kL_\infty$ if $L_\infty \gg L$, exacerbating the correlation problem.

Table 1 A summary of the results from the three different models

Model	Estimate of L_∞	Estimate of k	Estimate of t_0
Least squares estimation (VBGE)	252 cm [192, 312]	0.13	–3.09
Full Bayesian model (VBGE)	419 cm [378, 457]	0.06 [0.05, 0.07]	–3.25 [–3.23, –3.29]
Length-based Bayesian model	308 cm [304, 312]	0.08	–4.13

All three parameters are estimated using the two models that use age data. Although our length-based model only estimates L_∞, it is possible to draw a von Bertalanffy growth curve using the inverse relationship between k and L_∞

mates of these two models overlap significantly (Table 1). In fact, the mean of the length-based model is well within the confidence interval of the regression.

Finally, we compare the two Bayesian models (Table 1). The length-based model is essentially a compromise between the physiological basis of the VBGE and the availability of length-at-age data. It performs similarly to the least squares regression while maintaining the maximum size assumption (recall that the logit treats all data as a proportion relative to the maximum datum). However, we argue that the full Bayesian model is the most desirable model, if the data are available. The maximum reported size of blue shark is 396 cm TL (Smith 2006) and our full Bayesian estimate is closer than the length-based model estimate. However, when length-at-age data are unavailable, this new method provides a sound estimate of asymptotic size. As data availability and sampling coverage improve, we expect the length-based model estimate to converge on the full Bayesian estimate.

Conclusion

Our full and length-based Bayesian models are designed to solve two problems: estimate asymptotic size in the absence of length-at-age data, and maintain the biological assumptions of the VBGE when fitting the model to data.

The length-based method uses the distribution of the data to determine what the maximum size is for individuals in the population. In contrast, the least squares regression calculates an average

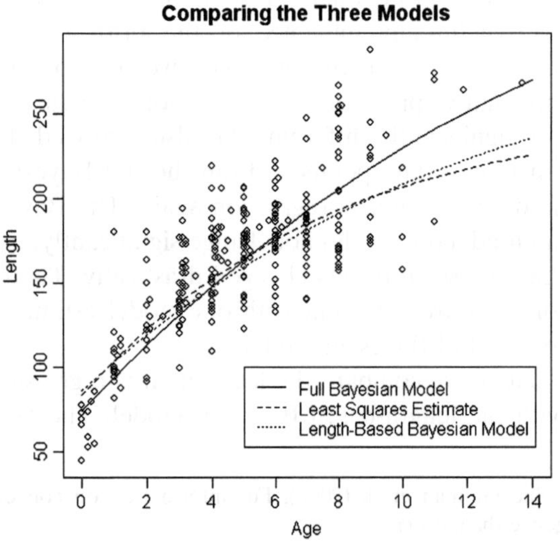

Fig. 4 The full Bayesian model and the least squares regression are fit to the length-at-age data for the female blue shark. The Bayesian model assumes L_∞ is a true asymptote

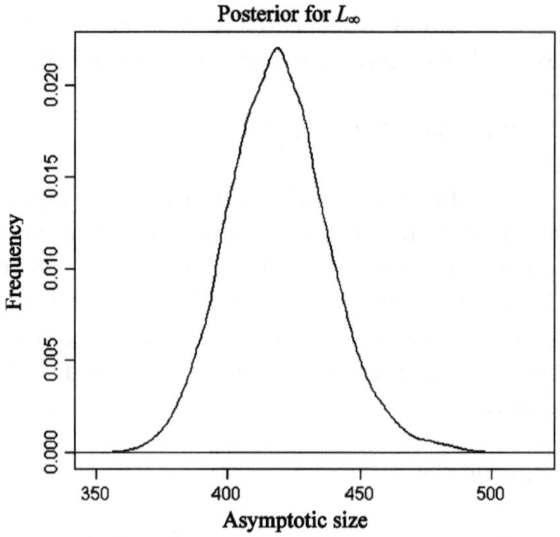

Fig. 5 The posterior of L_∞ for the full Bayesian model

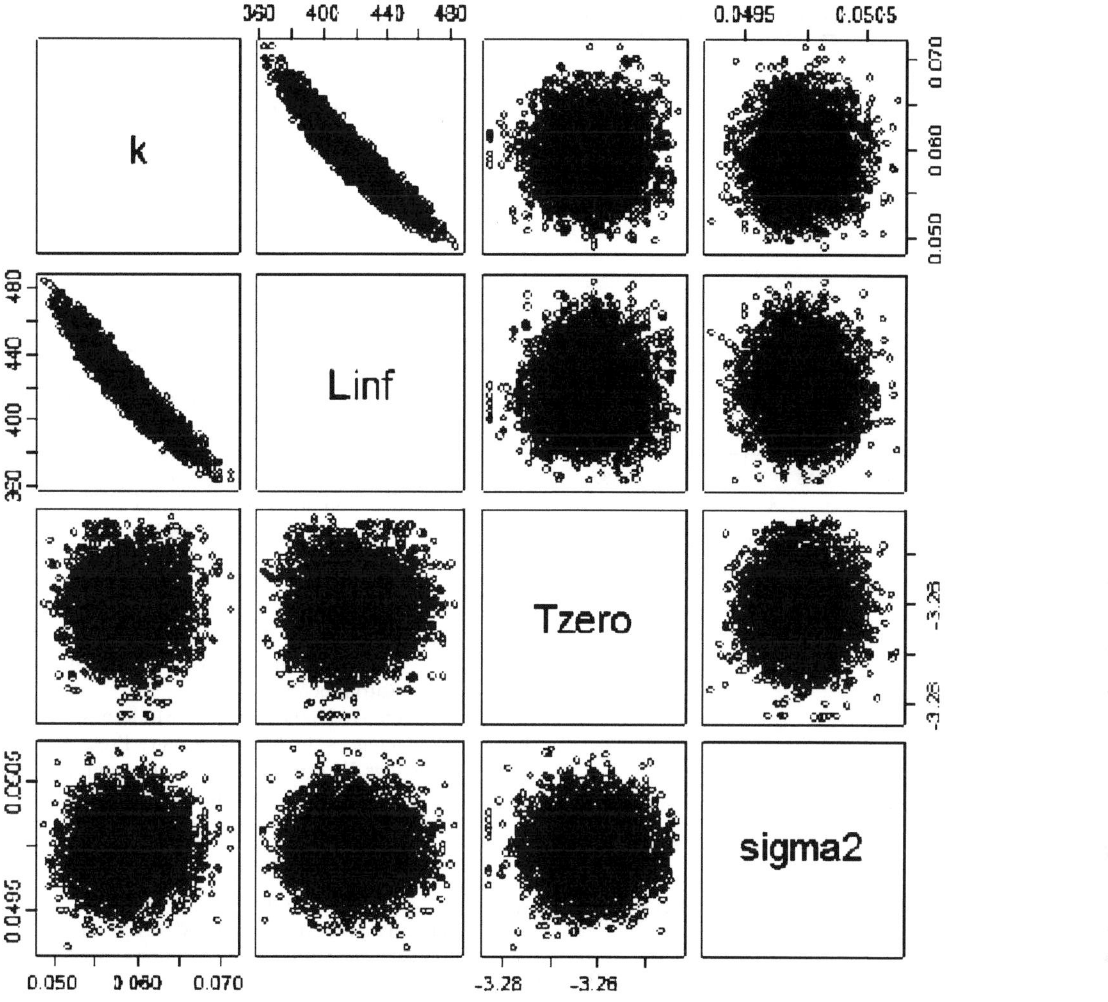

Fig. 6 This plot illustrates the correlation between each of the estimated parameters. Each row and column is one parameter; in order from top to bottom we plotted k, L_∞, t_0, and σ^2. For little or no correlation, we expect a structureless scatterplot. There is an apparent inverse correlation between k and L_∞

maximum size across individuals in the population. The distinction is a subtle, but important one. That is, how should one think about asymptotic size? Is it an individual or population-level parameter? Data are gathered to assess population dynamics, and we offer our length-based and full Bayesian models to assess parameters on the population scale. Our methods are generally applicable to species for which we expect asymptotic growth, and they may be particularly useful for assessments of developing fisheries or of long-lived fishes.

Acknowledgements This work was partially supported by The Center for Stock Assessment Research (CSTAR), a partnership between UCSC and NOAA Fisheries Southwest Fishery Science Center, Santa Cruz Laboratory. Financial support for travel to this symposium was also provided by California-Sea Grant and NOAA Fisheries Service-Southeast Fisheries Science Center. We thank Marc Mangel, Enric Cortés, Miguel Araya, Raquel Prado, Grant Thiltgen, and Greg Cailliet for helpful discussions, data, and support. We appreciate the comments of the anonymous reviewers, and we are especially thankful to John Carlson and Ken Goldman for organizing the symposium and editing this special issue.

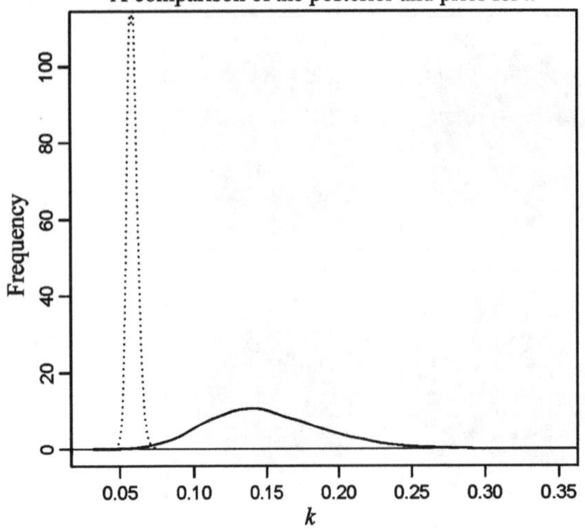

Fig. 7 The posterior of k for the full Bayesian model (dashed line) plotted against its prior (solid line)

Appendix

The full Bayesian model

In our computations, we replaced t_0 with q because it is easier to work with a strictly positive distribution. At the end of the model runs, we take the negative results for q and set them equal to t_0.

Our joint posterior is the combination of the four priors—for k, q, L_∞, and σ^2—and the likelihood:

$$p(l_\infty, k, q, \sigma^2 | l_{ar}) \propto \frac{1}{\sigma^2} q^{13} \exp\{-4q\} k^{-85}$$
$$\exp\{-k(100)\} \times \prod_{a=0}^{15} \prod_{r=1}^{R} \frac{1}{\sqrt{2\pi\sigma^2}}$$
$$\exp\left\{-\frac{1}{2\sigma^2}(l_{ar} - (l_\infty + \log(1 - \exp\{-k(a+q)\})))\right\}$$

Marginalizing, or multiplying out each of the parameters one at a time, give us either a formula to use in a Metropolis Hastings algorithm or a full conditional to use with a Gibbs sampler as detailed in the main text.

References

Acuña, E, Cid L, Pérez E, Kong I, Araya M, Lamilla J, Peñailillo J (2001) Estudio biológico de tiburones (marrajo dentudo, azulejo y tiburón sardinero) en la zona norte y central de chile. 128 pp, Subsecretaría de Pesca

Beverton RJH, Holt SJ (1959) A review of the lifespans and mortality rates of fish in nature and the relation to growth and other physiological characteristics. Ciba Foundation colloquia in ageing; the lifespan of animals, Churchill, London, pp 147–177

Cailliet, GM, Martin LK, Harvey JT, Kusher DI, Welden BA (1983) Preliminary studies on the age and growth of blue, *Prionace glauca*, common thresher, *Alopias vulpinus*, and shortfin mako, *Isurus oxyrinchus*, sharks from California waters. In: Prince ED, Pulos LM (eds) Proceedings of the international workshop on age determination of oceanic pelagic fishes: Tunas, billfishes, and sharks. NOAA Tech. Rept. NMFS 8, pp 179–188

Cortés E (2000) Life history patterns and correlations in sharks. Reviews in Fisheries Science 8:299–344

Gelman A, Carlin JB, Stern HS, Rubin DB (2004) Bayesian data analysis. Chapman and Hall, Boca Raton, 668 pp

Grafen A, Hails R (2002) Modern statistics for the life sciences. Oxford University Press, New York, 351 pp

Hastings W (1970) Monte Carlo sampling methods using Markov chains and their applications. Biometrika 57:97–109

Hilborn R, Mangel M (1997) The ecological detective: confronting models with data. Princeton University Press, Princeton, 315 pp

Hoenig, JM, Gruber SH (1990) Life history patterns in the elasmobranchs: implications for fisheries management. In elasmobranchs as living resources: advances in the biology ecology, and systematics, and the status of the fisheries. National Oceanic and Atmospheric Administration, Washington, DC, pp 1–16

Holden MJ (1973) Are long-term sustainable fisheries for elasmobranchs possible? Rapports et Proces-Verbaux des Reunions du Conseil International pour l'Exploration de la Mer 164:360–367

Jennings S, Kaiser MJ, Reynolds JD (2001) Marine fisheries ecology. Blackwell Science Ltd., London, 417 pp

Smith S, Holts D, Ramon D, Rassmussen R, Show C (2006) Shark Research-Blue shark (*Prionace glauca*), Southwest Fisheries Science Center, La Jolla, California, http://swfsc.nmfs.noaa.gov/frd/HMS/Large%20Pelagics/Sharks/species/blue.htm

von Bertalanffy L (1938) A quantitative theory of the organic growth (inquires on growth laws. II). Human Biology 10:181–213

von Bertalanffy L (1957) Quantitative laws in metabolism and growth. The Quarterly Review of Biology 32:217–231

Environ Biol Fish (2006) 77:309–315
DOI 10.1007/s10641-006-9111-7

ELASMOBRANCHS

A critical appraisal of marginal increment analysis for assessing temporal periodicity in band formation among tropical sharks

Rosângela Lessa · Francisco Marcante Santana · Paulo Duarte-Neto

Received: 6 June 2006 / Accepted: 27 June 2006 / Published online: 28 July 2006
© Springer Science+Business Media B.V. 2006

Abstract Marginal increment ratio (MIR) analyses were conducted as part of age and growth studies on three coastal/semi-oceanic species, the smalltail shark, *Carcharhinus porosus*, daggernose shark, *Isogomphodon oxyrhynchus* and the night shark *C. signatus*, and two ubiquitous oceanic species, blue shark, *Prionace glauca*, and whitetip shark, *C. longimanus*, collected in equatorial areas off Brazil with the aim of establishing the interspecific temporal nature of vertebral band formation. Monthly variations in marginal bands were analyzed using mean MIR on the entire sample as the standard method for all species. Reasons for the inconclusive results regarding these species are critically appraised with respect to three main sources of bias that are associated with marginal increment analysis (MIA). Bias due to insufficient sample sizes may have hampered the analysis for *I. oxyrhynchus* and *C. longimanus* due to movements from shallow waters to seamounts for the former species and to extensive migrations for the latter. Bias due to data collection over too long a period is thought to have influenced monthly mean MIR for *C. porosus* and *P. glauca*. For the latter, individuals from different age groups lay down rings at different times, making band deposition inconsistent between individuals. Finally, bias due to births occurring over too long a period was the prevalent cause for confounding MIR values among *I. oxyrhynchus* and *C. signatus* species, whose birth period lasts several months and leads to different ages within the same cohort. Other approaches used for MIA in *C. signatus* and *P. glauca* led to distinct times of band formation by age-groups when compared to MIR applied on the entire sample. For the daggernose shark, delays in events related to the reproductive cycle from one year to the next were also found to confound MIR. Requirements for the use of MIR implying a band width that displays a sinusoidal cycle when temporally plotted (month or season) were not fulfilled for any of these species. The method has been of little utility for detecting the periodicity of band deposition among sharks from the tropics. This emphasizes the need for supplying information on the temporal periodicity of pair deposition based on other methods.

R. Lessa (✉) · F. M. Santana · P. Duarte-Neto
Laboratório de Dinâmica de Populações Marinhas—DIMAR, Departamento de Pesca, Universidade Federal Rural de Pernambuco—UFRPE, Av. D. Manoel de Medeiros, s/n, Dois Irmãos, PE CEP 52171-900, Brazil
e-mail: rlessa@ufrpe.br

P. Duarte-Neto
Unidade Acadêmica de Garanhuns—UAG, Universidade Federal Rural de Pernambuco—UFRPE, Rua Ernesto Dourado, 82, Heliópolis, Garanhuns, PE CEP 55296-190, Brazil

Keywords Age verification · Age and growth · Southwestern Equatorial Atlantic

Introduction

Information on elasmobranch growth has been derived from counts of opaque and translucent bands in the vertebrae (Cailliet and Goldman 2004). Marginal increment analysis (MIA) involves the characterization or measurement of band on the margin of growth structures and is one of the most commonly applied age verification methods used for elasmobranchs to establish the periodicity of band formation (Cailliet 1990). Such an evaluation is essential for the understanding the growth process and is critical when using hard parts for age estimates (Wilson et al. 1983; Cailliet et al. 1986). Although several verification methods are available, MIA has become the most popular due to its moderate sampling requests and low costs. This technique, however, has proven problematic, particularly due to technical difficulties related to resolving the margins of growth bands (Campana 2001).

Several authors have postulated various reasons for MIA failure to indicate the timing of annulus formation (Brothers 1983; Cailliet et al. 1986; Campana 2001), particularly in migratory oceanic sharks. Brothers (1983) and Cailliet (1990) identified three main sources of bias generally associated with, insufficient sample sizes, excessively extended sample periods and a lack of a defined reproductive period. In an attempt to improve the resolution of MIA, specimens have been separated by age class, and relative (MIR) rather than the absolute size of the marginal band has been used in order to reduce bias (Brothers 1983; Campana 2001).

Strict requirements for the use of MIA methods imply that periodicity verification of growth increments carried out on younger (and faster growing) individuals should not be extended to older age groups, as the number of bands and time of deposition may vary with age (Beamish and McFarlane 1983; Brothers 1983; Cailliet et al. 1986; Campana 2001).

When band pairs are formed on an annual basis (the general pattern for elasmobranchs), band width should display a sinusoidal cycle when temporally plotted (month or season) (Carlson et al. 1999), reaching values close to one when the band is completed, and values close to zero when a new band starts to form. Age verification through MIA requires knowledge on life cycles and an understanding of species distribution, for a correlation is necessary between the time of band formation and both environmental and life history events (Brothers 1983). However, verification through MIA often proves inconclusive even with the associated distribution and life cycle information.

As part of age and growth studies MIA, using mean MIR, were conducted on three coastal/semi-oceanic species: smalltail shark, *Carcharhinus porosus*, daggernose shark, *Isogomphodon oxyrhynchus* and night shark, *C. signatus* and two ubiquitous oceanic species: blue shark, *Prionace glauca* and whitetip shark, *C. longimanus*, to establish the interspecific temporal nature of vertebral band formation. The findings of these analyses are presented in this paper and discussed using results from published works.

Methods

Carcharhinus porosus was sampled from 1984 to 1987 and *I. oxyrhynchus* from 1984 to 1987 and in 1998 in the western coast of the state of Maranhão in northern Brazil. The third species is the semi-oceanic *C. signatus*, caught from 1995 to 1997 on seamounts off northeastern Brazil. Samples of the migratory oceanic *P. glauca* and the oceanic *C. longimanus* were caught by commercial vessels operating in the southern Atlantic from 1992 to 2000 and by research vessels from 1992 to 1997 (Lessa and Santana 1998; Lessa et al. 1999a, 2000, 2004; Santana and Lessa 2004). For all the species mentioned above the gestation period lasts 1 year and the protracted birth season takes several months. Life history data are presented in Table 1.

Previous analyses carried out on the study species have used different approaches for marginal increment (MI) assessments. Currently, monthly variations in marginal bands were analyzed using mean MIR on the entire sample as the standard method for all species (Natanson et al. 1995, see also Cailliet et al., this issue).

$$\mathrm{MIR} = (\mathrm{VR} - R_n)/(R_n - R_{n-1}),$$

where VR = the vertebral radius; R_n = the last complete band; and R_{n-1} = the next to last (penultimate) complete band.

Mean MIR, with confidence interval, and standard deviation (s.d.) was plotted monthly in order to locate periodic trends in annulus formation (Fig. 1). Data were tested for normality and variance analyses (ANOVA) were performed to detect significant differences throughout the year. All statistical inferences were made at a significance level of 0.05.

Results

Bias due to small sample

'The small sample size in any particular month or for any age group weakens MIA' (Brothers 1983; Campana 2001).

For *I. oxyrhynchus*, low sample sizes may have been caused by movements from shallow waters to seamounts outside the fishing area. Thus, highest MIR value obtained in October (n = 1) hampered MIA interpretation. Monthly mean MIR values were around 0.5 and did not result in the expected sinusoidal shape (Fig. 1).

Also, for *C. longimanus* significant differences found in MIR during the July peak corresponded to a small number of individuals (n = 3). Although differences were significant (P < 0.0062), the sinusoidal shape was not exhibited. For both species MIR could not be performed by age groups due to the limited size of samples (Fig. 1).

Bias due to data collection over too long a period

'Data collection over too long a period causes variability on account of annual marks that are

Table 1 Birth time, size at birth, size and age at maturity and area of collection for five shark species caught in equatorial areas off northern and northeastern Brazil

Species	Birth time/size (cm TL)	Size (cm TL)/age-at-maturity (years)	Latitude	Author
Carcharhinus porosus	Sep–Nov/31.1	71 M; 70 F/6	1°05′–2°45′ S/43°30′–45°30′ W	Lessa et al. (1999b)
Isogomphodon oxyrhynchus	Oct–Feb 1990/42.3; Feb–Apr 1991	103 M/5–6; 115 F/6–7	1°05′–2°45′ S/43°30′–45°30′ W	Stride et al. (1992), Lessa et al. (1999c), Lessa et al. (2000)
Carcharhinus signatus	Undefined/67, At least Jan–June	200–205 F/10; 185–190 M/8	02°16′–04°05′ S/33°43′–37°30′ W	Hazin et al. (2000), Santana and Lessa (2004)
Prionace glauca	Sep–Jan/56.4, Dec–Jan, Oct–Jan	225 M; 228 F/5	Southern Hemisphere	Amorim (1992), Lessa et al. (2004), Vooren and Montealegre (2004)
Carcharhinus longimanus	Undefined/71.1, At least July–Nov	180–190/6–7	South eastern Africa, Southern Brazil, 1°–9° S/30°–40° W	Bass et al. (1973), Amorim et al. (1998), Lessa et al. (1999d)

M, Males; F, females

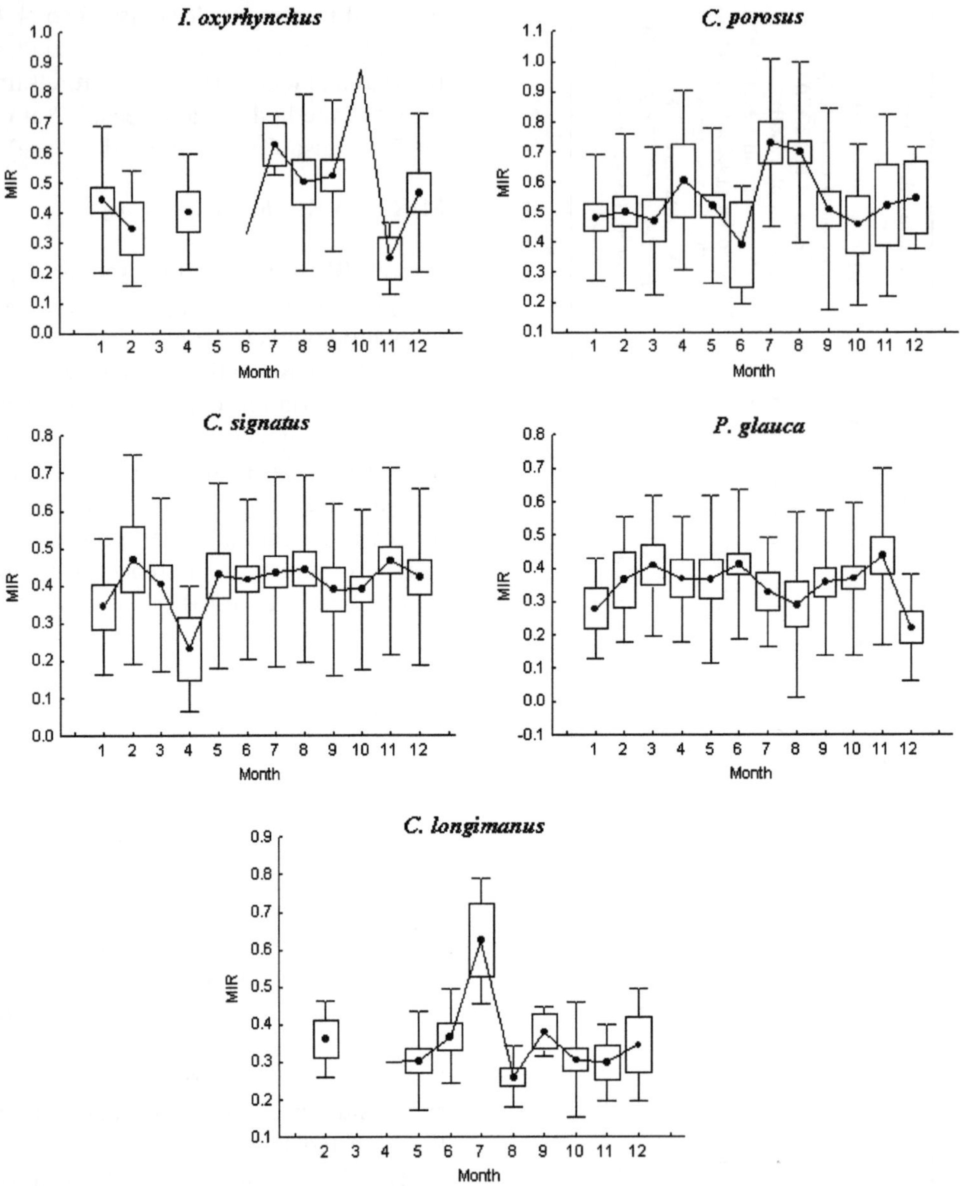

Fig. 1 Mean vertebral marginal increment ratios (MIR) ± s.d. by month for *I. oxyrhynchus*, *C. porosus*, *C. signatus*, *P. glauca* and *C. longimanus* caught in northern and northeastern Brazil. Boxes are the confidence interval for the means

not formed at the same time in consequence of life-cycle events (reproduction, migration, mating, etc...) that do not occur at the same time every year. Thus, when data are collected over a long span of time, considerable variability is included into the system' (Brothers 1983).

Monthly mean MIR for *C. porosus* showed maximal values in July–August and lowest values in October, oscillating from 0.4 to 0.7. Throughout most of the evaluated period, low and high mean MIR occurred, as suggested by the confidence intervals (Fig. 1). Lower MIR values were displayed for *C. signatus* from February to April. However, a small sample size was obtained in April. Throughout the year MIR was around 0.4–0.5, but standard deviations of means indicated that high and low values were present year round (Fig. 1).

For *P. glauca*, monthly mean MIR ranged from 0.3 to 0.5 throughout the year, with high standard deviations of means indicating that narrow and large edges were present during the entire period (Fig. 1). The absence of clear periods of low and high increments throughout the year yielded no temporal MIA trends.

Bias due to births occurring over a too long period

'Protracted parturition period may lead to variation in MI of the same cohort as individuals are born in different birth dates varying by several months' (Santana and Lessa 2004).

For *I. oxyrhynchus*, birth was estimated to take place from the halfway point of the dry season/beginning of the raining season (October) to the end of rainy season, over a period of about 6 months (Lessa et al. 1999c). This context leads to different ages within the same cohort, with differences spanning 6 months. Similarly, *C. signatus* embryos measuring 10–40 cm TL were collected in February, whereas 31.8–37.2 cm TL embryos were found in June, indicating that the birth period lasts several months (Fig. 1).

Discussion

A parturition period spreading over a long season represents an adaptation to the exploitation of food by juveniles over a long growing period; it maximizes species survival and tends to be common in fish from low latitudes (Nikolsky 1969). This adaptation is true for the equatorial species analyzed here, which appear to have long birth periods. Such a context is considered to have confounded MIA by generating biases due to fish from the same cohort being born at different times. Regarding coastal/semi-oceanic species, although *I. oxyrhynchus* partially shares the environment with *C. porosus*, the former seems to present a more complex behavior, dwelling in turbulent banks during the rainy season (Lessa et al. 1999b). Delays in events related to the reproductive cycle from 1 year to the next have been recorded for the daggernose shark by Stride et al. (1992), who observed that the movement to bays, which took place from October to February in 1991, began 4 months later in 1992. This implies a postponing of mating and gestation periods perhaps due to changes in environmental conditions. Thus, the species may be capable of compensating for interannual climate variability that delays annual cycle events. This kind of adaptation has never been clearly described for tropical sharks and merits further investigation. Furthermore, the small size of the vertebral sample impeded the annual analyses both for the entire sample and by age groups.

Overall, sampling time ranged between 4 and 5 years for all analyzed species. The one exception was *C. signatus* for which vertebral collection took place from 1995 to 1997 when vertebrae were more evenly sampled. In a previous study, MIA for the night shark was carried out using different approaches (Santana and Lessa 2004). It was expressed: (1) as the percentage marginal increment—PMI (Crabtree and Bullock 1998), and (2) as characterizing the vertebral edge (Carlson et al. 1999). Results of these analyses coincided and significant differences were found for the former ($P < 0.0463$), indicating that bands are completed from June to October when the highest frequency of broad-light edges (MI 0.6–1) occurred. A third analysis using absolute MI by quarter for juveniles (4 and 5 years) and adults (6 to >8 years) led to significant differences for juveniles with highest values in the third and fourth quarters ($P < 0.05$). Contrarily, no significant difference was detected for adults as a result of the smaller sample size. Currently, no significant differences were detected using mean MIR, which highlights the need for different approaches, as contradictory results were obtained for PMI and MIR. Moreover, the use of absolute MI allowed a better understanding of the differences in band deposition by age groups. These differences may have hindered the detection of the time of ring deposition using MIR.

Variability in embryo size suggests a protracted birth season for the night shark, with births recorded over a period of at least 6 months (Hazin et al. 2000). Furthermore, individuals measuring ~40 cm TL in February are born long before individuals measuring 37.2 cm TL in June. This leads to differences in MI within the same cohort. Thus, after a 12-month gestation period, individuals are born with birth dates varying by several months. Such a pattern was assumed to have led to inconclusive MIA (Santana and Lessa 2004). Nikolsky (1969) related the adaptation discussed here to unstable conditions, which seems to be the case for *C. signatus*, as the species inhabits seamounts where the physical regime is most variable due to the proximity of the Northern-Brazilian Current, causing turbulence to the environment.

Concerning more highly mobile species that undergo extensive migrations over tropical and temperate regions of the southern hemisphere, the traditional MIA approach is complicated, as individuals of different age groups may lay down rings at different times, making band deposition inconsistent between individuals. Moreover, referring to large pelagics the analysis of data collected over a large time span (as is the case here, 4–5 years) is obscured, as fish do not spawn at the same time every year (Brothers 1983). This introduces high variability when data are analyzed on an annual basis. Results obtained in the current study are in accordance with criticisms by the above-quoted author indicating reasons why MIR for these species failed in the study area.

For *P. glauca*, absolute MI was used in a previous study showing a majority of individuals with recently formed bands from November onward and significant differences occurring in mean MI throughout the year for the 3–5 age group and for the overall sample. However, MI analysis did not conclusively supported the annual periodicity of bands, as it does not depict a clear trend of low and high periods with low standard deviations of means during the months when the new zone is forming. In the current study, an identical pattern was depicted using mean MIR, the values of which ranged between 0.2 and 0.4 throughout the year, with no significant difference found.

A review of papers addressing age validation in recent decades makes it clear that MIA methods, which were regarded as powerful tools in 1983 (Brothers 1983; Casselman 1983), are currently deemed problematic as an analysis method for certain species (Campana 2001). The most definitive validation method for elasmobranchs is tagging and release with chemical markers (mostly OTC) used on specimens of known-age (Cailliet 1990), which should ideally be carried out for all age classes, as the periodicity of band formation may change with age (Cailliet and Goldman 2004) as shown above. However, results obtained thus far make it clear that the lack of oxytetracycline-injected recaptured fish over the entire size range has not allowed for full age and growth validation, such as for *Lamna nasus*, validated in juveniles up to an age of 11 years (Natanson et al. 2002); *P. glauca*, validated up to 4 years (n = 2, Skomal and Natanson 2003); and *Carcharodon carcharias* (n = 1, Wintner and Cliff 1999).

In conclusion, MIR method for the study species has been of little utility in validating the periodicity of band deposition among tropical sharks from the southern Atlantic Ocean. This emphasizes the need for supplying information based on tagging and chemical marking (OTC). Such information is required for management and is particularly crucial for the coastal species that were recently listed as endangered (MMA 2004).

Acknowledgements The "Elasmobranch Project" was funded by the Secretaria da Comissão Interministerial para os Recursos do Mar-SECIRM/UFMA from 1984 to 1989. Data were collected from 1992 to 1995 through the "ECOTUNA Project" IBAMA/UFRPE. The Ministerio do Meio Ambiente-MMA/SECIRM in the scope of the "Programa Nacional de Avaliação do Potencial Sustentável dos Recursos Vivos—REVIZEE" funded research from 1997 to 2002. We are grateful to the Conselho Nacional de Desenvolvimento Científico e Tecnológico—CNPq for a Research Grant provided to the senior author (Procs: 301048/83-OC) and scholarships for students. We are indebted to Norte Pesca S/A and to the fishermen involved. Remarks by anonymous reviewers led to substantial improvements to the manuscript; we greatly appreciate their careful and constructive comments.

References

Amorim AF (1992) Estudo da pesca e reprodução do cação-azul *Prionace glauca* L. 1758, capturado no Sudeste e Sul do Brasil. D.Sc. Thesis, Universidade Estadual Paulista, Rio Claro, 205 pp

Amorim AF, Arfelli CA, Fagundes L (1998) Pelagic elasmobranchs caught by longliners off Southern Brazil during 1974–97: an overview. Mar Freshwater Res 49:621–632

Bass AJ, D'Aubrey JD, Kitnasamy N (1973) Sharks of the east Coast of southern Africa. I. Genus *Carcharhinus* (Carcharhinidae). South African Association of Marine Biology Research Investment, Report 38, 100 pp

Beamish RJ, McFarlane GA (1983) Validation of age determination estimates: the forgotten requirement.

In: Prince ED, Pulos LM (eds) Proceedings of the international workshop on age determination of oceanic pelagic fishes: tunas, billfishes, and sharks. NOAA Technical Report NMFS 8, pp 29–33

Brothers EB (1983) Summary of round table discussions on age validation. In: Prince ED, Pulos LM (eds) Proceedings of the international workshop on age determination of oceanic pelagic fishes: tunas, billfishes, and sharks. NOAA Technical Report NMFS 8, pp 35–44

Cailliet G (1990) Elasmobranch age determination and verification: an updated review. In: Pratt HL Jr, Gruber SH, Taniuchi T (eds) Elasmobranchs as living resources: advances in biology, ecology, systematics and the status of the fisheries. NOAA Technical Report NMFS 8, pp 157–165

Cailliet GM, Goldman KJ (2004) Age determination and validation in chondrichthyan fishes. In: Carrier J, Musick JA, Heithaus MR (eds) Biology of sharks and their relatives, chapter 14. CRC Press LLC, Boca Raton, Florida pp 339–447

Cailliet GM, Radke RL, Welden BA (1986) Elasmobranch age determination and verification: a review. In: Uyeno T, Arai R, Taniuchi T, Matsuura K (eds) Indo-Pacific fish biology: proceedings of the second international conference on Indo-Pacific fishes. Ichthyology Society of Japan, Tokyo, pp 345–360

Campana SE (2001) Accuracy, precision and quality control in age determination, including a review of the use and abuse of age validation methods. J Fish Biol 59:197–242

Carlson JK, Cortés E, Johnson AG (1999) Age and growth of the blacknose shark *Carcharhinus acronotus* in the eastern Gulf of Mexico. Copeia 3:684–691

Casselman JM (1983) Age and growth assessment of fish form their calcified structures techniques and tools. National Oceanic and Atmospheric Administration Technical Report NMFS 8, pp 1–17

Crabtree RE, Bullock LH (1998) Agem growth and reproduction of the black grouper *Mycteroperca bonaci*, in Florida waters. Fish Bull 96:735–753

Hazin FH, Lucena F, Souza TSL, Boeckman C, Broadhurst M, Menni R (2000) Maturation of the night shark, *Carcharhinus signatus*, in the south-western equatorial Atlantic Ocean. Bull Mar Sci 66:173–185

Lessa R, Santana FM (1998) Age determination and growth of the smalltail shark *Carcharhinus porosus* from northern Brazil. Mar Freshwater Res 49:705–711

Lessa R, Santana FM, Paglerani R (1999a) Age, growth and stock structure of the oceanic whitetip shark, *Carcharhinus longimanus*, from the southwestern Equatorial Atlantic. Fish Res 42:21–30

Lessa R, Santana FM, Menni R, Almeida Z (1999b). Population structure and reprodutive biology of the smalltail shark (*Carcharhinus porosus*) off Maranhão. Mar Freshwater Res 50:383–388

Lessa R, Batista V, Almeida Z (1999c) Occurrence and biology of the daggernose shark *Isogomphodon oxyrhynchus* (Chondrichthyes: Carcharhinidae) off the Maranhão Coast (Brazil). Bull Mar Sci 64:115–128

Lessa RP, Paglerani R, Santana F (1999d) Morphometry and reproductive biology of *Carcharhinus longimanus* from southwestern equatorial Atlantic. Cybium (Paris, França) 23(4):353–368

Lessa R, Santana FM, Batista V, Almeida Z (2000) Age and growth of the daggernose shark, *Isogomphodon oxyrhynchus*, from northern Brazil. Mar Freshwater Res 51:339–347

Lessa R, Santana FM, Hazin FH (2004) Age and growth of the blue shark *Prionace glauca* (Linnaeus, 1758) off northeastern Brazil. Fish Res 66:19–30

Natanson LJ, Casey JG, Kohler NE (1995) Age and growth estimates for the dusky shark, *Carcharhinus obscurus*, in the western North Atlantic Ocean. Fish Bull 93:116–126

Natanson LJ, Mello JJ, Campana SE (2002) Validated age and growth of the porbeagle shark (*Lamna nasus*) in the western North Atlantic Ocean. Fish Bull 100:266–278

Nikolsky GV (1969) Theory of fish population dynamics. Oliver & Boyd, Edinburgh, 323 pp

Santana FM, Lessa R (2004) Age determination and growth of the night shark (*Carcharhinus signatus*) off the northeastern Brazilian coast. Fish Bull 102:156–167

Skomal GB, Natanson LJ (2003) Age and growth of the blue shark (*Prionace glauca*) in the North Atlantic Ocean. Fish Bull 101:627–639

Stride RK, Batista VS, Raposo LAB (1992) Pesca experimental de tubarão com redes de emalhar no litoral maranhense. São Luis, ODA/FINEP/UFMA, vol III, 160 pp

Vooren CM, Montealegre SQ (2004) Biologia, ecologia, pesca e identificação de estoque do tubarão azul (*Prionace glauca*) no Atlântico Sul. Workshop Report. IV Reunião da SBEEL, Recife, Brazil, 16 pp

Wilson CA, Brothers ED, Casselman JM, Smith CL, Wild A (1983) Glossary. In: Prince ED, Pulos LM (eds) Proceedings of the international workshop on age determination of oceanic pelagic fishes: tunas, billfishes, and sharks. NOAA Technical Report NMFS 8, pp 207–208

Wintner S, Cliff G (1999) Age and growth determination of the white shark. *Carcharodon carcharias*, from the east coast of South Africa. Fish Bull 97:153–159

Environ Biol Fish (2006) 77:317–325
DOI 10.1007/s10641-006-9124-2

SI-ELASMOBRANCHS

Elemental signatures in the vertebral cartilage of the round stingray, *Urobatis halleri*, from Seal Beach, California

Loraine F. Hale · John V. Dudgeon · Andrew Z. Mason · Christopher G. Lowe

Received: 6 June 2006 / Accepted: 14 July 2006 / Published online: 16 August 2006
© Springer Science+Business Media B.V. 2006

Abstract Although numerous studies have utilized elemental analysis techniques for age determination in bony fishes, little work has been conducted utilizing t'hese procedures to verify age assessments or temporal periodicity of growth band formation in elasmobranchs. The goal of this study was to determine the potential of laser ablation inductively coupled plasma-mass spectrometry (LA-ICP-MS) to provide information on the seasonal deposition of elements in the vertebrae of the round stingray collected from Seal Beach, California. Spatially resolved time scans for elements across the round stingray vertebrae showed peaks in calcium intensity that aligned with and corresponded to the number of seasonal growth bands identified using standard light microscopy. Higher signals of calcium were associated with the wide opaque bands while lower signals of calcium corresponded to the narrow translucent bands. While a close alignment between the numbers of calcium peaks and annual growth bands was observed in round stingray samples aged 5 years or younger, this relationship was less well defined in vertebral samples from round stingrays over 11 years old. To the best of our knowledge, this is the first study of its kind to utilize ICP-MS to verify age assessments and seasonal band formation in an elasmobranch. The results from this preliminary study indicate that LA-ICP-MS elemental analysis of the vertebral cartilage of the round stingray may have potential to independently verify optically derived age assessments and seasonal banding patterns in elasmobranch vertebrae.

Keywords Round stingray · Elemental composition · TOF-ICP-MS · Laser ablation

L. F. Hale · J. V. Dudgeon · A. Z. Mason · C. G. Lowe
Department of Biological Sciences, California State University Long Beach, 1250 Bellflower Blvd., Long Beach, CA 90840, USA

L. F. Hale (✉)
NOAA Fisheries Service, Southeast Fisheries Science Center, 3500 Delwood Beach Rd., Panama City, FL 32408, USA
e-mail: Loraine.Hale@noaa.gov

J. V. Dudgeon · A. Z. Mason
Institute for Integrated Research in Materials, Environments and Society, California State University Long Beach, 1250 Bellflower Blvd., Long Beach, CA 90840, USA

Introduction

Age and growth analyses are the first step in examining fundamental biological processes of a population of fish species (Casselman 1983). Estimations of the age structure of a population and growth rates of a species are used to determine longevity, mortality, productivity, yield, and population dynamics which in turn are essential for management of populations (Casselman 1983). Ages are typically assessed by enumerating the growth bands in calcified structures such as the otoliths and vertebrae using light microscopy, which correlate closely with the age of the fish (Casselman 1983). While this approach is applicable for most bony fishes, it cannot always be utilized for some species such as elasmobranchs which do not have comparable calcified structures (Cailliet et al. 1986).

A number of alternative approaches have been advocated for determining age in cartilaginous fishes. The use of disequilibrium analyses between radioisotopes through bomb radiocarbon dating has provided a method for age validation for many hard-to-age fish such as rockfish (Kastelle et al. 2000; Cailliet et al. 2001; Andrews et al. 2005) and has been applied to elasmobranchs (Campana et al. 2002). Welden et al. (1987) used radiometric dating to determine ages of the leopard shark, *Triakis semifasciata*, the Pacific angel shark, *Squatina californica*, the common thresher shark, *Alopias vulpinas*, and the white shark, *Carcharodon carcharias*, but found that the age estimates based on radiometric dating were highly variable and in some cases predicted negative ages. However, Campana et al. (2002) subsequently utilized radiometric dating techniques to successfully age the porbeagle, *Lamna nasus*, and the shortfin mako shark, *Isurus oxyrinchus*.

Another more generally applicable approach that has been used to age elasmobranchs has been to utilize spatially resolved elemental analyses to determine the periodicity of annual growth bands in cartilaginous structures such as vertebrae. Energy dispersive X-ray microanalysis was used by Jones and Geen (1977) to determine the age of the spiny dogfish, *Squalus acanthias.* They found the peaks of calcium and phosphorus in the dorsal fin spine corresponded with age estimations from length–frequency analysis and suggested that elemental analysis of the cartilage of elasmobranchs could be useful in ageing elasmobranchs when no alternate ageing structure can be found (Jones and Geen 1977). Cailliet and Radtke (1987) used electron microprobe analysis to spatially analyze the temporal peaks in calcium and phosphorus in the vertebrae of the gray reef shark, *Carcharhinus amblyrhynchos*, and the common thresher shark, *Alopias vulpinus*, and found a strong correlation between the periodicity of deposition of these elements and optically determined annual growth band counts. They suggested the use of elemental analysis of elasmobranch vertebral cartilage as a way to verify the temporal periodicity of growth bands; however, no further studies have been published using this procedure for age verification in this group of fishes.

The current study aims to investigate the utility of laser ablation inductively coupled plasma mass spectroscopy (LA-ICP-MS) as a new methodology for studying the spatial distribution of elements in elasmobranch cartilage. The major advantage of LA-ICP-MS for studying the spatial distribution of elements in a solid sample over other established techniques such as energy dispersive (EDX) and wavelength dispersive X-ray (WDX) spectroscopy is its sensitivity. Thus, the relative elemental detection limits for interference free isotopes by LA-ICP-MS are generally in the sub-μg g^{-1} range while those for EDX and WDX are typically in the 0.01–1% range (Campana et al. 1997; Durrant and Ward 2005).

The utility of this technique has been tested by identifying the seasonal and annual peaks in calcium and phosphorus in the vertebrae of the round stingray, *Urobatis halleri*, from Seal Beach, California and determining if these elemental peaks correspond with the annual periodicity of optically determined growth band pairs in the vertebrae. Recently, Hale (2005) utilized standard light microscopy to assess the calcified concentric growth bands in round stingray vertebral cartilage and determined that these fish live to be at least 14 years old and becomes sexually mature at 4 years of age. To the best of our knowledge, this is the first study of its kind to use elemental analysis by LA-ICP-MS for age assessment in an elasmobranch (Campana 2005).

Methods

Vertebrae preparation and ageing

Vertebrae were dissected from 12 specimens of round stingrays collected in 2003 and 2004

(Table 1). The vertebrae were mounted in Tap Plastics resin and a 0.5 mm wide bow-tie section was made using a Buehler isomet saw. Vertebral sections were viewed and growth bands were counted using a stereo-microscope with transmitted light field at 4× magnification (Olympus™ model SZX-12). Two readers independently assessed the age of each round stingray by counting the bands in individual vertebral centra. Annual periodicity of growth band formation was validated by maintaining 10 oxytetracycline-injected round stingrays in captivity for 2 years (Hale 2005).

Elemental composition

The round stingray vertebral sections were analyzed for elements using a GBC Optimass orthogonal time of flight inductively coupled plasma mass spectrometer (oTOF-ICP-MS) with an attached New Wave Research (Fremont, CA) LUV-213 high performance Nd:YAG UV laser operated at 213 nm (Reish and Mason 2003). Sectioned vertebrae samples were mounted on plexiglass slides and inserted into the ablation chamber of the laser system. Prior to analysis, the sections were pre-ablated with the laser to remove external contamination, after which the specimen chamber and transmission line was purged with high-purity argon gas to remove residual analytes in the sample introduction system. For data acquisition, a laser spot size of 30 μm with a constant energy density of 8.95 J $(cm^2)^{-1}$ and a repetition rate of 20 Hz was used to ablate the sample. The laser was scanned at a rate of 10 μm s^{-1} and elemental data was collected along a linear transect running through the focal region and from edge to edge of the whole vertebral corpus calcareum. The ICP-MS was operated in the time resolved mode. The instrumental conditions of operation were as follows: nebulizer flow 1.1 l min^{-1}; plasma flow 10 l min^{-1}; auxiliary flow 0.5 l min^{-1}; RF 1,600 W; skimmer –1,300 V; extraction lens –1,250 V; pushout plate 590 V; pushout grid –400 V; blanker 200 V; and reflectron 650 V. Calcium (^{44}Ca), phosphorus (^{31}P), magnesium (^{24}Mg) and strontium (^{88}Sr) were the main elements screened in the vertebral sections and the isotopic data were acquired at an acquisition rate of 1 point s^{-1}, representing a total ion pushout of 14,705.

Table 1 Round stingray samples used in ICP-MS elemental analysis. It is unknown when UH-094 was caught, so it was excluded from any seasonal analyses

Sample number	Sex	Disc width (mm)	Age (yr)	Month caught (mm/dd/yr)
19	M	250	11	05/16/03
54	M	82	0	10/24/03
69	F	151	5	08/15/03
74	F	253	12	09/19/03
86	F	104	1	01/30/04
94	F	95	1	–
100	F	143	5	03/19/04
105	M	84	0	03/19/04
114	M	150	3	05/19/04
116	F	155	4	06/09/04
149	F	154	4	08/27/04
159	M	254	12	08/27/04

Analytical precision was determined under these operating conditions for a suite of elements including sodium (Na), magnesium (Mg), aluminum (Al), silicon (Si), calcium (Ca), scandium (Sc), titanium (Ti), vanadium (V), chromium (Cr), manganese (Mn), iron (Fe), nickel (Ni), cobalt (Co), copper (Cu), zinc (Zn), arsenic (As), rubidium (Rb), strontium (Sr), yttrium (Y), zirconium (Zr), niobium (Nb), tin (Sn), antimony (Sb), cesium (Cs), barium (Ba), lanthanum (La), cerium (Ce), praseodymium (Pr), neodymium (Nd), samarium (Sm), europium (Eu), gadolinium (Gd), terbium (Tb), dysprosium (Dy), holmium (Ho), erbium (Er), thulium (Tm), ytterbium (Yb), lutetium (Lu), hafnium (Hf), tantalum (Ta), lead (Pb), thorium (Th), and uranium (U) using a NIST 612 glass standard reference material both before and after sample analyses (40 μg g^{-1}). The recorded %RSD values were typically below 6% over five analysis replicates with the exception of As and Zr which were 8.2 and 7.2%, respectively. Air analyses were used for blanks. Theoretical detection limits, based upon the 3σ value of the blank signal and the responses from the NIST 612, were typically <1 μg g^{-1} as reported by others (Sanborn and Telmer 2003; Trejos et al. 2002).

The presence of growth bands was determined from the peaks in the time resolved ion profiles and the spatial distribution of elemental signatures was compared to the annual and seasonal periodicity of growth bands used for age assessment. Statistical tests to identify peaks from background noise were conducted using a statistical integration program (Origin™, Microcal Software Inc. 1997). An 11-point running average was applied to smooth the data, and peaks were identified using prescribed parameters within the software that included a minimum height of 3% of the total amplitude of the data in the range and a minimum width of 3% of the total number of points in the data range. For the elements of interest these values were approximate to the recommended 3σ of the smoothed blank signal (Sinclair et al. 1998; Sanborn and Telmer 2003).

Additionally, the spatial patterns in the ion intensity ratios of calcium to strontium, phosphorus and magnesium across the vertebrae were compared to identify changes in the relative distribution of these elements. Expression of the ion intensities in the form of ratios normalizes the ion intensities of the analytes to compensate for variations in laser ablation efficiency and topographical effects to provide a better index of the enrichment of these elements relative to ^{44}Ca along the transect (Reish and Mason 2003). Average Sr:Ca ratios were calculated by averaging the intensity ratio at each time point.

Results

A representative photomicrograph of a sectioned vertebrae across a transect, together with the corresponding smoothed spatial ion signal for ^{44}Ca, is shown in Fig. 1. Oscillations in the relative signal intensity of ^{44}Ca were observed along the transect that closely correlated with the identified growth bands with increased signal intensities occurring at the mid point of each yearly band and peaks in ^{44}Ca that were statistically identified above the background noise corresponded with growth bands (Fig. 1). Elevated signals were obtained from the edges of the specimen while a marked depression was noted in the ^{44}Ca signal at the foci of each vertebrae (Fig. 1). Analyses of other elements showed strong discernible signals for ^{88}Sr and ^{31}P that closely tracked the signal for ^{44}Ca (Fig. 2). Other

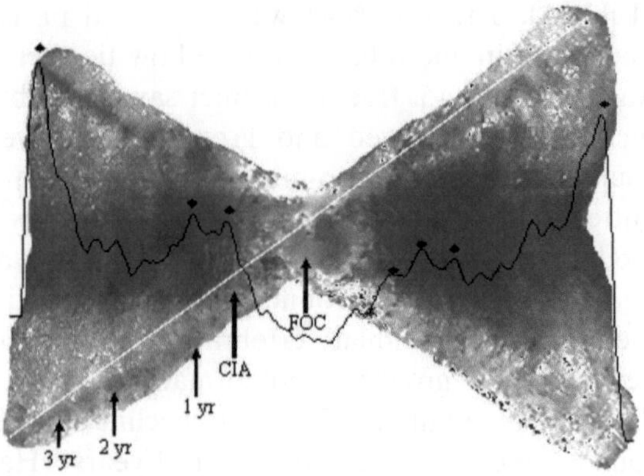

Fig. 1 Laser scan across a sectioned vertebrae of round stingray sample UH-114 showing the relative distribution of ^{44}Ca. The focus (FOC), change in angle (CIA), and year marks are indicated with arrows. The statistically determined peaks are indicated with a diamond (♦). The position of the laser transect is indicated with a diagonal line across the whole left corpus calcareum

elements producing pronounced signals correlating with ^{44}Ca included ^{24}Mg and ^{23}Na (not shown). Barium (^{137}Ba) was found primarily within the foci region while lead (^{208}Pb) was located at the edges of the vertebrae (not shown). No significant signals were obtained for other elements including uranium (^{238}U), niobium (^{93}Nb), and cerium (^{140}Ce) which have been recorded at relatively high concentrations within the biomineralized shells of various marine polychaetes (Reish and Mason 2003). Plots of the ion intensity ratios of ^{31}P and ^{24}Mg to ^{44}Ca showed that the relative abundances were conservative along the transect while those for ^{88}Sr showed

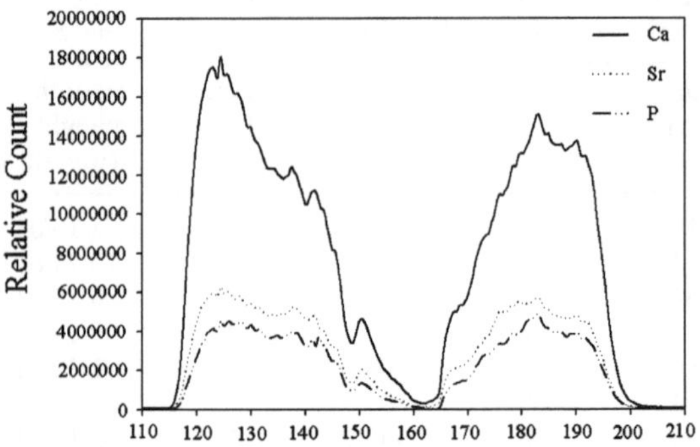

Fig. 2 Representative transect scan for calcium (^{44}Ca; solid line), strontium (^{88}Sr; dotted line), and phosphorus (^{31}P; dashed and dotted line) across a sectioned vertebrae of round stingray sample UH-054

variations in ratios indicative of differential deposition of these elements into the structure (paired t-test, $P = 0.004$; Fig. 3). The spatial periodicity of the changes in Sr:Ca ratio correlated with the growth bands and there was no significant difference in the statistically identified peaks between the ^{44}Ca time scan and the Sr:Ca ratio time scan (paired t-test, $P = 0.86$; Fig. 4). However, no discernable trends were observed between the mean Sr:Ca ion intensity ratio with age or date captured.

To determine whether the observed fluctuations in ^{44}Ca signal intensity within the cross sectioned vertebrae could be used to determine age, a series of analyses were conducted on different vertebrae samples from fish whose age was determined by counting the number of calcified growth bands using standard light microscopy (Hale 2005). Time resolved ion intensity profiles for ^{44}Ca for fish of age classes 0–1, 3–4, 5, and 11–12 showed an increase in the number of peaks ^{44}Ca in older aged fish (Fig. 5 a–d). However, the magnitude of the ^{44}Ca signal differed for samples of similar age (Fig. 5 a–d). Demarcations in the ^{44}Ca signal used to identify the tentative position of a growth band were determined visually in a manner comparable to that used optically to assess the age of the fish (Cailliet and Radtke 1987). There was a strong positive correlation between the number of ^{44}Ca peaks and the assessed age ($P < 0.001$, $r = 0.98$, Fig. 6).

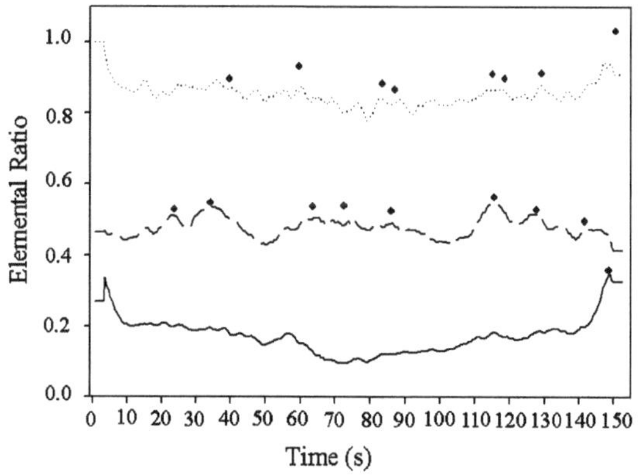

Fig. 3 Smoothed elemental ratios of ^{88}Sr (dashed line), ^{31}P (dotted line) and ^{24}Mg (solid line) normalized to the ^{44}Ca ion intensity across the vertebrae of round stingray sample UH-114. The statistically determined peaks are indicated with a diamond (♦)

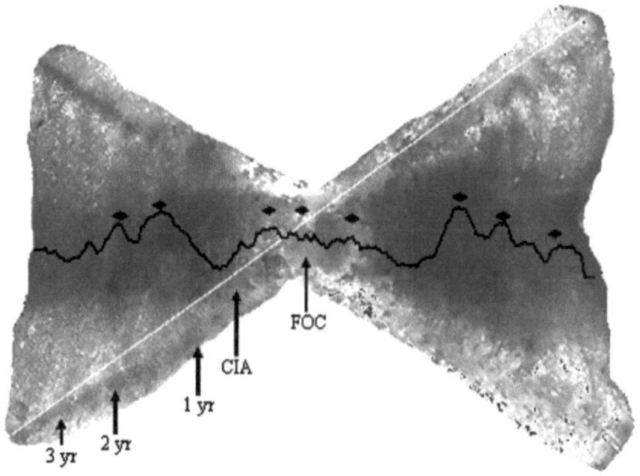

Fig. 4 Laser scan across a sectioned vertebrae of round stingray sample UH-114 showing the relative distribution of ^{88}Sr. The focus (FOC), change in angle (CIA), and year marks are indicated with arrows. The statistically determined peaks are indicated with a diamond (♦)

Discussion

Hale (2005) used centrum edge analysis to demonstrate that round stingrays expressed a pattern of seasonal growth in their vertebrae, with faster growth occurring in the summer months and slower growth occurring in the colder winter months. The observed patterns of peaks and valleys in ^{44}Ca correspond spatially with the seasonal growth bands, with higher peaks of ^{44}Ca corresponding to the wide opaque bands laid down in the summer months of warmer water and longer day lengths and lower valleys of ^{44}Ca corresponding to the narrow translucent bands deposited during the winter months of colder water and shorter day lengths. Cailliet and Radtke (1987) ascertained that the variation in calcium and phosphorus in the vertebrae of the grey reef shark and the common thresher shark was seasonal, with higher levels in the summer and lower levels in the winter. Similar patterns of seasonal deposition have been found using LA-ICP-MS in a bivalve, which had low strontium, magnesium, barium, and manganese levels in the winter and peaks in these elements in the summer (Leng and Pearce 1999).

The seasonal nature of the mineralization process produces a periodicity in the ^{44}Ca enriched bands that correlates closely with the assessed age of the fish. This correlative relationship is particularly strong in younger fish (up to age 5)

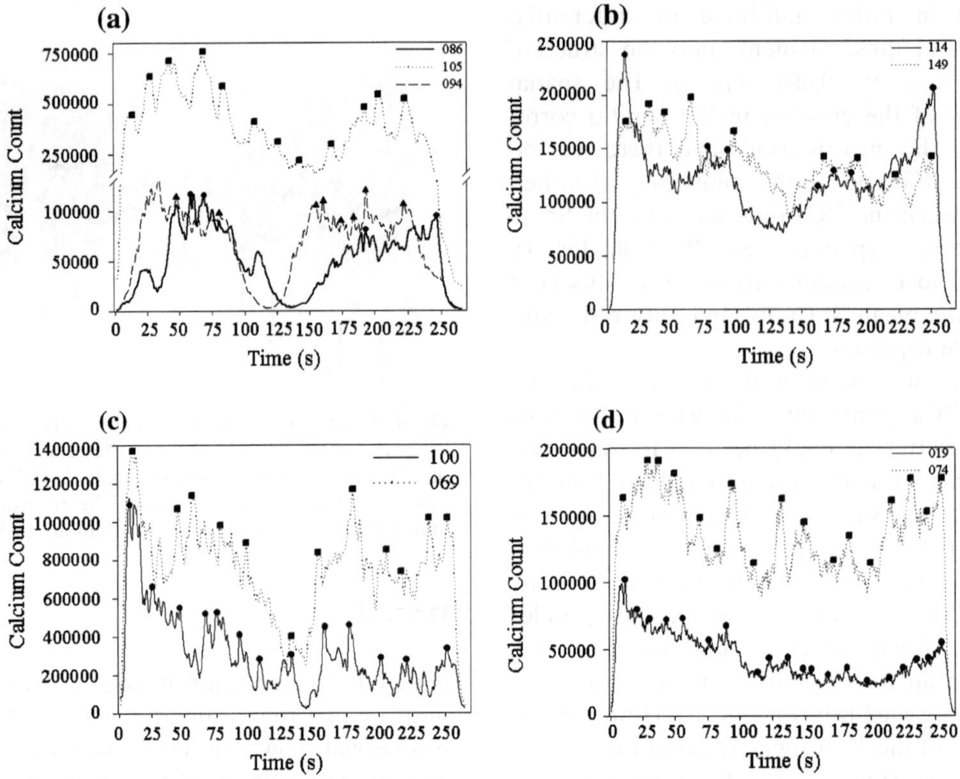

Fig. 5 (**a**) Transect scan for calcium for round stingray samples aged 0–1 year old. Statistically determined peaks are indicated with a symbol. (**b**) Transect scan for calcium for round stingrays aged 3–4 years old. Statistically determined peaks are indicated with a symbol. (**c**) Transect scan for calcium for round stingray samples aged 5 years old. Statistically determined peaks are indicated with a symbol. (**d**) Transect scan for calcium for round stingray samples aged 11–14 years old. Statistically determined peaks are indicated with a symbol

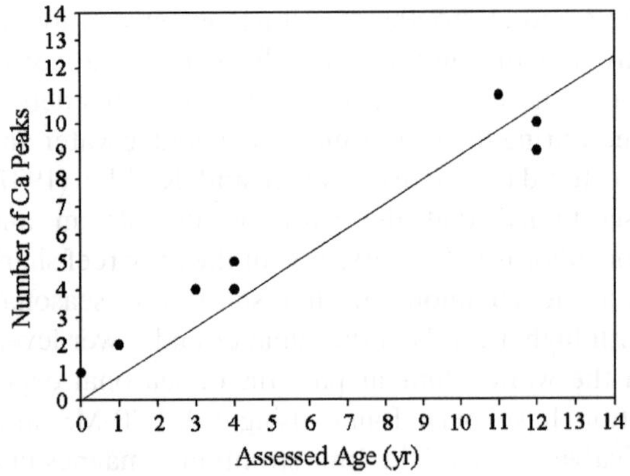

Fig. 6 Relationship between assessed age (yr) and number of calcium peaks in all samples from round stingrays ($P = 0.83$, $r = 0.98$)

where the bands are discrete and are sufficiently spatially resolved to enable the probe to ablate between the bands to give statistically definable depressions in the signals. However, this correlation is less defined in the older round stingrays, where the ICP-MS data show a tendency to underestimate the age of round stingrays over 11 years old. These findings are in accord with those of Panfili and Tomas (2001) who found that peaks in elements tended to underestimate the true age of tilapia in general, and that this underestimation was more pronounced in older fish.

One explanation for the discrepancy between the number of optically and analytically defined bands in older round stingrays is that in elasmobranchs the growth bands towards the perimeter of the vertebrae tend to become more closely opposed (Goldman 2004). Thus, light microscope estimations of the inter-band distances show a progressive decrease in distance with the age of the band with the 1–5 year old round stingray growth bands 95–160 μm apart while growth bands for 5–13 year old round stingrays were

44–66 μm apart (Hale unpublished data). Spatial resolution studies by Sanborn and Telmer (2003) on natural and artificially zoned materials have shown that the spatial resolution of LA-ICP-MS is influenced by a number of factors including spot size, scan rates, concentration gradients, memory effects and instrument acquisition parameters. Based on their data, it would appear that the distances between the outer bands of older stingrays are close to the expected attainable limit of spatial resolution using a LA-ICP-MS (Sanborn and Telmer 2003). If so, then a loss in signal amplitude below the statistical confidence window necessary for peak detection caused by signal overlap and averaging as the laser moves across adjacent bands would result in an underestimation in the numbers of bands in these older fish.

There was no discernable pattern in the relative amount of calcium present within the different annuli within a given vertebrae. Similarly, no trends were observed in the degree of mineralization with the age of the fish, and significant variability was noted in the ^{44}Ca signals between individuals of the same or similar age which could not be accounted for by the inherent imprecision of the methodology. Kalish (1989) attributed this variability in salmon as within-individual differences in growth based on location, health, or somatic growth. Cailliet and Radtke (1987) also observed inter-annual variation in the magnitude and periodicity of the peaks and valleys of calcium and phosphorus with some individuals showing double peaks within a year, which they attributed to perturbations such as stress, changing temperature, or altered food availability. The variability in the calcium patterns in the round stingray may also be due to such perturbations, but further studies are necessary to resolve this issue.

Despite the observed intra- and inter-vertebrae variability in ion signals, reduced ion signals were consistently noted in the center of the structure and elevated ^{44}Ca signals were invariably observed from the edges of the structure when compared to the focus of the vertebrae. These findings are consistent with those of Cailliet and Radtke (1987) who also found a lower level of calcium and phosphorus in the vertebrae of the gray reef shark and the common thresher shark at the time of embryonic growth that was then followed by a peak in calcium and phosphorus at birth and first feeding. The focus of the vertebrae of the round stingray appears to have much lower levels of elements than the other parts of the vertebrae (other than barium, which appears to be enriched), but further analysis of this region is necessary to identify the natal signature of the round stingray. The elevated signals for ^{44}Ca at the outside edges of the vertebrae could be due to active calcification at the vertebrae perimeter (Cailliet and Radtke 1987). However, since all samples had an increase in analyte signal at the vertebrae edge irrespective of month captured, the elevated levels of elements could also be due to a topographical effect and the vertebral edge amplifying the efficiency of the ablation process (Reish and Mason 2003).

It is clear that LA-ICP-MS has great future potential for the elemental fingerprinting of elasmobranch cartilage. In particular, field and laboratory studies should be conducted to determine the degree of fidelity and distinctiveness of the elemental fingerprint to the chemistry of the surrounding environment. Round stingrays are seasonally abundant at Seal Beach (Hoisington and Lowe 2005) and Vaudo and Lowe (2006) found that these fish show seasonal site fidelity to this area. Additionally, the location of pupping for the round stingray is currently unknown, although it is thought to occur in inshore estuaries (Hoisington and Lowe 2005; Vaudo and Lowe 2006). Subtle changes in isotopic elemental signatures may provide a tool to track round stingray movements between estuarine and coastal habitats (Gillanders and Kingsford 2003; Gillanders 2005) and distinguish between genetically distinct populations of the round stingray in southern California (Campana et al. 1994, 2000; Thresher 1999; FitzGerald et al. 2004).

In conclusion, this preliminary study represents the first attempt to conduct spatial elemental analysis on the vertebrae of the round stingray. While the sensitivity of LA-ICP-MS is better than most other forms of microanalysis, the spatial resolution of the technique is inferior to both EDX and WDX. At present, the major limitation of the technique is that it is, at best, a semi-quantitative procedure (Durrant and Ward 2005).

Precision tends to be poor (5–8% under current conditions) even with short-wavelength, high energy lasers having high absorption efficiency and reduced ablation and plasma induced fractionation, and accurate quantification of elements in solid samples requires that the unknowns be carefully matrix-matched to certified standard reference materials (Durrant and Ward 2005). Consequently, accurate estimations of elemental content cannot be readily made by direct comparisons of signal intensities between different samples and this inherent limitation restricts the usefulness of the procedure to qualitatively study the relative distribution and content of elements within the sample (Durrant and Ward 2005).

Nevertheless, this technique appears to have potential for verifying age estimations and also quantify seasonal patterns in calcification. As suggested by Cailliet et al. (1986) and Cailliet and Radtke (1987), elemental microanalyses of growth bands lend credibility to age assessments derived from optical measurements. At the present time, the methods commonly used to assess the age of elasmobranchs are highly subjective (Goldman 2004). Optically-derived age estimates require the personnel be highly trained to ensure precision and decrease ageing errors and biases (Goldman 2004). Age assessment using LA-ICP-MS therefore provides an alternative, objective method to standard light microscopy for the verification of growth bands.

Acknowledgements Funding for field collections of stingrays was provided by University of Southern California Sea Grant Program, part of the National Sea Grant College Program, National Oceanic and Atmospheric Administration, U.S. Department of Commerce (Grant # NA86RG0054). Funding for the acquisition of the New-Wave laser ablation system was provided by CSULB University College Extension. The GBC instrument was purchased by NSF MRI Grant# BCS 0321361. Funding for the analyses in this project was provided by CSUPERB. We would like to thank J. Neer for editorial comments, and J. Carlson and K. Goldman for the opportunity to publish these results with the proceedings of the symposia.

References

Andrews AH, Burton EJ, Kerr LA, Cailliet GM, Coale KH, Lundstrom CC, Brown TA (2005) Bomb radiocarbon and lead-radium disequilibria in otoliths of boccacio rockfish (*Sebastes paucispinis*): a determination of age and longevity for a difficult-to-age fish. Mar Freshwat Res 56:517–528

Cailliet GM, Radtke RL (1987) A progress report on the electron microprobe analysis technique for age determination and verification in elasmobranchs. In: Summerfelt RC, Hall GE (eds) The age and growth of fish. Iowa State University Press, Iowa, pp 359–369

Cailliet GM, Radtke RL, Weldon BA (1986) Elasmobranch age determination and verification: a review. In: Uyeno T, Arai R, Taniuchi T, Matsuura K (eds) Indo-Pacific fish biology: proceedings of the second international conference on Indo-Pacific fishes. Ichyol. Soc. Jpn., Tokyo, pp 385–360

Cailliet GM, Andrews AH, Burton EJ, Watters DL, Kline DE, Ferry-Graham LA (2001) Age determination and validation studies of marine fishes: do deep-dwellers live longer? Exp Gerontol 36:739–764

Campana SE (2005) Otolith science entering the 21st century. Mar Freshwat Res 56:485–495

Campana SE, Fowler AJ, Jones CM (1994) Otolith elemental fingerprinting for stock identification of Atlantic cod (*Gadus morhua*) using laser ablation ICP-MS. Can J Fish Aquat Sci 51:1942–1950

Campana SE, Thorrold SR, Jones CM, Gunther D, Tubrett M, Longerich H, Jackson S, Halden NM, Kalish JM, Piccoli P, de Pontual H, Troadec H, Panfili J, Secor DH, Severin KP, Sie SH, Thresher R, Teesdale WJ, Campbell JL (1997) Comparison of accuracy, precision, and sensitivity in elemental assays of fish otoliths using the electron microprobe, proton-induced X-ray emission, and laser ablation inductively coupled plasma mass spectrometry. Can J Fish Aquat Sci 54:2068–2079

Campana SE, Chouinard GA, Hanson JM, Frechet A, Brattey J (2000) Otolith elemental fingerprints as biological tracers of fish stocks. Fish Res 46:343–357

Campana SE, Natanson LJ, Myklevoll S (2002) Bomb dating and age determination of large pelagic sharks. Can J Fish Aquat Sci 59:450–455

Casselman J (1983) Age and growth assessments of fish from their calcified structures: techniques and tools. In: Prince E, Pulos L (eds) Proceedings of the international workshop on age determination of oceanic pelagic fishes: tunas, billfishes, and sharks. NOAA Tech. Rep. NMFS 8, pp 1–17

Durrant SF, Ward NI (2005) Recent biological and environmental applications of laser ablation inductively coupled plasma mass spectrometry (LA-ICP-MS). J Anal Atom Spectrom 20:821–829

FitzGerald JL, Thorrold SR, Bailey KM, Brown AL, Severin KP (2004) Elemental signatures in otoliths of larval walleye pollock (*Theragra chalcogramma*) from the northeast Pacific Ocean. Fish Bull 102:604–616

Gillanders BM (2005) Using elemental chemistry of fish otoliths to determine connectivity between estuarine and coastal habitats. Estuar Coast Shelf Sci 64:47–57

Gillanders BM, Kingsford MJ (2003) Spatial variation in elemental composition of otoliths of three species of fish (Family Sparidae). Estuar Coast Shelf Sci 57:1049–1064

Goldman KG (2004) Age and growth of elasmobranch fishes. In: Musick JA, Bonfil R (eds) Elasmobranch fisheries management techniques. APEC Fisheries Working Group, pp 97–132

Hale LF (2005) Age and growth of the round stingray, *Urobatis halleri*, at Seal Beach, California. MS Thesis, California State University Long Beach, p 42

Hoisington G, Lowe CG (2005) Abundance and distribution of the round stingray, *Urobatis halleri*, near a heated effluent outfall. Mar Environ Res 60(4):437–453

Jones BC, Geen GH (1977) Age determination of an elasmobranch (*Squalus acanthias*) by X-ray spectrometry. J Fish Res Board Canada 34:44–48

Kalish JM (1989) Otolith microchemistry: validation of the effects of physiology, age and environment on otolith composition. J Exp Mar Biol Ecol 132:151–178

Kastelle CR, Kimura DA, Jay SR (2000) Using $^{210}Pb/^{226}Ra$ disequilibrium to validate conventional ages in Scorpaenids (genera *Sebastes* and *Sebastolobus*). Fish Res 46:299–312

Leng MJ, Pearce NJG (1999) Seasonal variation of trace element and isotopic composition in the shell of a coastal mollusk, *Mactra isabelleana*. J Shellfish Res 18(2):569–574

Panfili J, Tomas J (2001) Validation of age estimation and back-calculation of fish length based on otoliths microstructures in tilapias (Pisces, Cichlidae). Fish Bull 99:139–150

Reish DJ, Mason AZ (2003) Radiocarbon dating and metal analyses of 'fossil' and living tubes of *Protula* (Annelida: Polychaeta). Hydrobiologia 496:371–383

Sanborn M, Telmer K (2003) The spatial resolution of LA-ICP-MS line scans across heterogeneous materials such as fish otoliths and zoned minerals. J Anal Atom Spectrom 18:1231–1237

Sinclair DJ, Kinsley LPJ, McCulloch MT (1998) High resolution analysis of trace elements in corals by laser ablation ICP-MS. Geochem Cosmochem Acta 62:1889–1901

Thresher RE (1999) Elemental composition of otoliths as a stock delineator in fishes. Fish Res 43:165–204

Trejos T, Montero S, Almirall JR (2002) Analysis and comparison of glass fragments by laser ablation inductively coupled plasma mass spectrometry (LA-ICP-MS) and ICP-MS. Anal Bioanal Chem 376(8):1255–1264

Vaudo J, Lowe CG (2006) Movement patterns of the round stingray, *Urobatis halleri* (Cooper), near a thermal outfall. J Fish Biol 68:1756–1766

Welden BA, Cailliet GM, Flegal AR (1987) Comparison of radiometric with vertebral band age estimates in four California elasmobranchs. In: Summerfelt RC, Hall GE (eds) The age and growth of fish. Iowa State University Press, Iowa, pp 301–315

Environ Biol Fish (2006) 77:327–336
DOI 10.1007/s10641-006-9107-3

ORIGINAL PAPER

Bomb dating and age validation using the spines of spiny dogfish (*Squalus acanthias*)

Steven E. Campana · Cynthia Jones · Gordon A. McFarlane · Sigmund Myklevoll

Received: 2 June 2006 / Accepted: 22 June 2006 / Published online: 12 October 2006
© Springer Science+Business Media B.V. 2006

Abstract Bomb radiocarbon has previously been used to validate the age of large pelagic sharks based on incorporation into vertebrae. However, not all sharks produce interpretable vertebral growth bands. Here we report the first application of bomb radiocarbon as an age validation method based on date-specific incorporation into spine enamel. Our results indicate that the dorsal spines of spiny dogfish, *Squalus acanthias*, recorded and preserved a bomb radiocarbon pulse in growth bands formed during the 1960s with a timing which was very similar to that of marine carbonates. Using radiocarbon assays of spine growth bands known to have formed in the 1960s and 1970s as a dated marker, we confirm the validity of spine enamel growth band counts as accurate annual age indicators to an age of at least 45 year. Radiocarbon incorporation into northeast Atlantic dogfish spines occurred in similar years as those in the northwest Atlantic and northeast Pacific, although the amount of radiocarbon differed in keeping with the radiocarbon content of the different water masses. Published reports suggesting that Pacific dogfish are longer lived and slower growing than their Atlantic counterparts appear to be correct, and are not due to errors in interpreting the spine growth bands. Radiocarbon assays of fin spine enamel appears to be well suited to the age validation of sharks with fin spines which inhabit the upper 200 m of the ocean.

Keywords Age determination · Shark · Longevity · Growth rate

S. E. Campana (✉)
Bedford Institute of Oceanography, P.O. Box 1006, Dartmouth, Nova Scotia, Canada B2Y 4A2
e-mail: campanas@mar.dfo-mpo.gc.ca

C. Jones
Center for Quantitative Fisheries Ecology, Old Dominion University, Norfolk, Virginia 23508, USA

G. A. McFarlane
Pacific Biological Station, Nanaimo, British Columbia, Canada V9R 5K6

S. Myklevoll
Institute of Marine Research, P.O. Box 1870 Nordnes, N-5817 Bergen, Norway

Introduction

Spiny dogfish, *Squalus acanthias*, are small squaloid sharks common in the surface mixed layer of coastal temperate oceans around the world. Studies to date suggest that they are both long-lived and slow-growing, making them among the least productive of the shark species (Smith et al. 1998; Cortés 2000). Most of these productivity calculations have been based on the dogfish population in the northeast Pacific Ocean, which appears to grow much more slowly and reach a much greater longevity than does the population in the northwest Atlantic (Ketchen 1975; Nammack et al. 1985; Saunders and McFarlane 1993). While it is possible that the reported differences are real and reflect true differences in growth rate, it is also possible that the differences are due to errors in age determination associated with the interpretation of the growth bands on the fin spines, a process which is subjective and whose accuracy is difficult to confirm. Based on recaptures of oxytetracycline-tagged individuals, the spine growth bands have been demonstrated to form annually in dogfish of the northeast Pacific (Beamish and McFarlane 1985; McFarlane and Beamish 1987). However, age validation studies are not available for the northwest Atlantic spiny dogfish population, nor have any studies been done which demonstrate that the interpretation of the growth bands was comparable between the two populations.

Dogfish lack the otoliths, which are typically used to age teleost fish. Whereas otoliths grow concentrically in all dimensions around a central core, with no subsequent resorption or reworking of deposited material (Campana and Thorrold 2001), dogfish spines grow both from the base of the spine, and internally along the length of the vascularized pulp cavity (Holden and Meadows 1962; Beamish and McFarlane 1985). A non-mineralized dentine layer separates the pulp cavity from the outer, mineralized enamel layer. Although there is some evidence of annual markings in the dentine layer, their presence can be difficult to detect and it is not clear that the entire growth sequence can be observed (McFarlane and Beamish 1987). Only the growth bands in the enamel layer are both readily observed and believed to be metabolically static. However, the composition of the spine enamel (which is unknown, but may be hydroxyapatite) is very different than that of otoliths (calcium carbonate). The carbon source also differs: whereas the primary source is dissolved inorganic carbon in the case of otoliths (Schwarcz et al. 1998), it is likely to be metabolic carbon in the case of spines. These differences make the growth and content characteristics of the spine somewhat more difficult to predict than in the otolith.

Atmospheric testing of atomic bombs in the 1950s and 1960s resulted in a rapid and well-documented increase in radiocarbon (^{14}C) in the world's oceans (Druffel and Linick 1978). The period of radiocarbon increase was almost synchronous in marine carbonates such as corals, bivalves and fish otoliths around the world (Kalish 1993; Weidman and Jones 1993; Campana 1997), allowing the period of increase to be used as a dated marker in calcified structures exhibiting growth bands. A similar pattern of increase, lagged by several years due to the incorporation of dietary carbon, has been documented in shark vertebrae (Campana et al. 2002). Here we report the first radiocarbon assays of the fin spines of spiny dogfish, demonstrating that they too recorded and preserved a bomb radiocarbon pulse in growth bands formed during the 1960s. Through comparison of radiocarbon assays in recently formed spine material collected between 1960 and 2002 with growth bands of older dogfish matched for year of formation, we confirm that growth bands form annually on the dogfish spine. Based on the age-validated spines, we then test for differences in the growth rate and longevity of spiny dogfish between the northwest Atlantic and the northeast Pacific.

Methods

Second dorsal spines were collected from 32 spiny dogfish caught in the northwest Atlantic (n = 8), northeast Atlantic (n = 11) and northeast Pacific (n = 13) between 1959 and 2002 (Table 1). Samples from the northeast Atlantic (which occasionally consisted of first dorsal spines rather than second dorsal) were obtained as part of a tag-recapture program carried out by the Institute of Marine Research in Bergen, Norway between 1958 and 1980, while those from the northwest Atlantic were obtained from the commercial dogfish fishery off of Nova Scotia in 2002. Northeast Pacific dogfish samples were obtained on research vessel cruises carried out by the Pacific Biological Station in 1980–1981, and again in 2003. The total length of samples from all regions ranged from 69 cm to 99 cm. All spines had been stored dry in paper envelopes after removal from the dogfish.

Table 1 Summary of $\delta^{13}C$ (‰) and $\Delta^{14}C$ assay results for annual growth bands microsampled from dogfish spines from three populations. Growth bands sampled are shown as the range of bands sampled, counting distally from the spine base. Mean year of growth band formation is that based on the mean number of growth bands

Area	Dogfish ID	Year collected	Age	Growth bands sampled	Mean year of band formation	$\delta^{13}C$ of band	$\Delta^{14}C$ of band
NE Atlantic	H265	1959	22	13–22	1942.0	–14.1	–69.4
				1–2	1958.0	–13.8	–71.9
	H401	1960	15	7–15	1949.5	–14.1	–81.1
				1–2	1959.0	–13.6	–67.6
	H661	1963	25	17–25	1942.5	–13.7	–61.0
				1–2	1962.0	–14.3	–50.0
	H2257	1964	19	16–19	1947.0	–13.5	–81.7
				1–2	1963.0	–15.1	–23.4
	H1719	1965	14	11–14	1953.0	–13.8	–80.4
				1–2	1964.0	–14.0	–44.8
	H6136	1966	15	12–15	1953.0	–14.2	–72.4
				1–2	1965.0	–14.2	1.5
	HB1294	1976	16	18	1958.5	–13.7	–21.3
				1–6	1973.0	–13.6	98.2
	HB1201	1977	19	18–19	1959.0	–13.6	–40.3
				1–8	1973.0	–14.3	101.7
	HB1170	1978	16	14–16	1963.5	–13.7	61.7
				1–6	1975.0	–15.2	85.3
	HB1263	1978	22	20–22	1957.5	–13.9	–11.8
				1–7	1974.5	–14.2	109.5
	NOR11529	1979	15	12–15	1966.0	–13.7	92.9
				1–8	1975.0	–14.7	89.5
NW Atlantic	85D	1996	10	3–4	1993.0	–14.4	30.5
	1416	2002	21	18–21	1983.0	–15.0	50.8
	1834	2002	21	17–21	1983.5	–14.1	62.7
	254	2002	21	17–21	1983.5	–14.3	50.3
	260	2002	25	22–25	1979.0	–14.2	54.3
	904	2002	21	18–21	1983.0	–14.0	27.8
	1412	2002	17	3–4	1999.0	–15.0	1.8
				14–17	1987.0	–13.1	54.6
	2238	2002	20	18–19	1984.0	–14.0	62.3
NE Pacific	10	1980	39	30–39	1946.0	–13.9	–113.0
				25–28	1954.0	–12.6	–106.0
				8–11	1971.0	–12.8	–12.1
				1–6	1977.0	–12.4	49.7
	17	1980	9	1–6	1977.0	–13.9	–49.2
	2	1980	31	28–30	1951.5	–13.0	–108.0
				8–11	1971.0	–13.2	19.5
				1–6	1977.0	–13.5	3.2
	5	1980	52	44–52	1932.5	–14.0	–133.0
				28–32	1950.5	–12.7	–104.0
				8–11	1971.0	–12.3	29.7
				1–6	1977.0	–11.7	59.6
	DF22220	1980	13	10–13	1969.0	–13.2	36.0
				1–4	1978.0	–14.5	0.8
	DF22170	1980	12	11–12	1969.0	–13.8	–11.8
				1–8	1976.0	–13.8	–4.8
	DF19979	1980	9	9	1971.5	–13.1	–8.7
	DF23042	1981	11	10–11	1971.0	–12.2	–1.2
				1–8	1977.0	–14.4	18.5
	12	1981	7	1–4	1979.0	–14.0	0.5

Table 1 continued

Area	Dogfish ID	Year collected	Age	Growth bands sampled	Mean year of band formation	$\delta^{13}C$ of band	$\Delta^{14}C$ of band
	DF43719	1989	11	9–11	1979.5	−12.8	−8.4
				1–7	1985.5	−14.4	24.1
	DF100	2003	52	26–34	1973.5	−14.0	2.8
				16–24	1983.5	−13.1	11.6
				1–10	1998.0	−13.7	−2.2
	DF15	2003	52	46–52	1954.5	−13.8	−121.8
				27–34	1973.0	−14.0	3.4
				16–25	1983.0	−13.7	16.2
				1–10	1998.0	−13.4	5.7
	DF94	2003	51	40–51	1958.0	−13.8	−6.0
				27–35	1972.5	−13.8	10.8
				16–24	1983.5	−12.9	8.6
				1–10	1998.0	−13.4	4.0

All ages were based on counts of presumed annual growth increments (growth bands) that were visible on the external spine surface after light polishing. The strong white band near the spine base was assumed to represent a check rather than a growth band. Age interpretation was carried out while working at 6× magnification under reflected light with a binocular microscope. Spine growth bands were also digitally photographed at a minimum resolution of 1,280 × 1,024 then enhanced. Transverse and longitudinal sections of the spine did not reveal interpretable growth bands, nor were complete growth sequences visible in whole or sectioned vertebrae.

To insure comparability in age interpretation among regions, all age interpretations used to guide micromilled samples were made by the same experienced age reader of Atlantic dogfish spines, and replicated by a second experienced age reader (SC). To insure that these age interpretations were also consistent with the original interpretations of the region, a subsample of Pacific spines were aged by both the above-mentioned age reader and the primary reader of dogfish spines in the Pacific (GM). Age comparisons were restricted to the no-wear region of the spine, so as to remove all error associated with worn growth bands. Bias between age readers was evaluated with age bias plots, whereas precision was quantified using the coefficient of variation (CV) (Campana 2001).

Multiple samples (n = 64), usually representing groups of 1–3 growth bands (up to 10 bands, in areas of slow growth), were micromilled from the enamel layer of each spine. In most cases, attempts were made to isolate samples from near the base (representing the period just prior to collection), the unworn region nearest the tip (representing the oldest portion of the spine), and the mid-region of each spine. Enamel samples were isolated as solid pieces with a Merchantek computer-controlled micromilling machine using steel cutting bits and burrs. Any dentine which remained attached to the enamel sample after micromilling was removed using a Gesswein high-speed hand tool fitted with a steel burr. The date of sample formation was calculated as the year of dogfish collection minus the growth band count from spine base to the midpoint of the range of growth bands comprising the sample, to which was added 0.5 to adjust for the fact that the spine base was counted as the first growth band. After sonification in Super Q water and drying, the sample was weighed to the nearest 0.1 mg in preparation for ^{14}C assay with accelerator mass spectrometry (AMS). Since the carbon content of the spine enamel was about 10%, the minimum sample weight which could be assayed was about 3 mg. AMS assays also provided $\delta^{13}C$ (‰) values, which were used to correct for isotopic fractionation effects and provide information on the source of the carbon. Radiocarbon values were subsequently reported as $\Delta^{14}C$, which is the per mil (‰) deviation of the sample from the radiocarbon concentration of 19th-century wood, corrected for sample decay prior to 1950 according to methods outlined by Stuiver and Polach (1977).

The onset of nuclear testing in the late 1950s resulted in a marked and widespread increase in $\Delta^{14}C$ in marine dissolved inorganic carbon (DIC) which is easily detected in all marine carbonates growing in surface waters during the 1960s (Druffel 1989; Campana and Jones 1998). To assign dates of formation to an unknown sample, it is necessary that the $\Delta^{14}C$ of the unknown sample be compared with a $\Delta^{14}C$ chronology based on known-age material (a reference chronology). In the case of carbonates, the years corresponding to the onset of radiocarbon increase are synchronous in reference chronologies based on corals, bivalves and otoliths, and are thus interchangeable (Campana 1999). In the case of spine material however, synchrony with carbonates could not be assumed. Therefore, we used newly formed dogfish spine enamel (within two growth bands of the spine base) for our known-age, or reference, chronology. Dogfish spines are known to grow outward from the base of the spine, with the newest material being found at the base (Holden and Meadows 1962; Beamish and McFarlane 1985). Since the date of collection was known for all dogfish, and making the reasonable assumption that 1–2 growth bands from the base represents 1–2 years before the date of collection with a maximum error of no more than ±2 year, the dates of formation of the basal growth bands were considered to be of known age.

The reference $\Delta^{14}C$ carbonate chronology for the northwest Atlantic was derived from assays of known-age fish otoliths formed between 1949 and 2000. The collection and radiocarbon assay of 56 age 1–3 haddock, *Melanogrammus aeglefinus*, and redfish, *Sebastes* spp., otoliths has been described elsewhere (Campana 1997; Campana et al. 2002); the chronology was supplemented by 17 age 1–2 haddock and yellowtail flounder, *Limanda ferruginea*, otoliths collected between 1980 and 2000 and prepared in a similar manner. There was no detectable difference between the haddock, redfish and yellowtail chronologies; therefore they were pooled and used as the reference carbonate chronology. The $\Delta^{14}C$ chronology of aragonitic fish otoliths in the NW Atlantic parallels that of North Atlantic corals and bivalves (Campana 1997), and thus is a good proxy for the $\Delta^{14}C$ DIC history of the spiny dogfish environment. The reference chronologies for the northeast Pacific Ocean were drawn from known-age assays of Pacific halibut and rockfish otoliths (Kerr et al. 2004; Piner and Wischniowski 2004).

Results

Growth band sequences were clearly visible along fin spines, although it was often difficult to distinguish between a broad diffuse band and two or more closely packed growth bands in spines from the Atlantic (Fig. 1). In such cases, similarity in appearance and spacing to adjacent growth bands was used as a guide to interpretation. Spines from the Pacific tended to be more clear, and characterized by narrower growth increments, than were those from the Atlantic (Fig. 1). In most cases, the most medial growth mark appeared as a white check, which we interpreted as approximating the region of current enamel formation. The distal tips of larger dogfish (>60 cm fork length) were

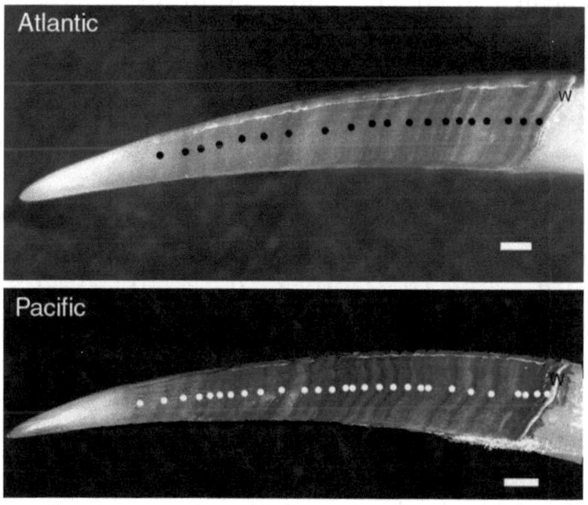

Fig. 1 Second dorsal spine of spiny dogfish collected in the northwest Atlantic (Top) and northeast Pacific (Bottom) annotated to show annual growth bands. The ages of the Atlantic (fork length 74 cm) and Pacific (total length 95 cm) dogfishes were interpreted as 20 and 28 years, respectively, before adjustment for growth bands formed before birth and the expected number of unobserved bands in the worn tip. The white check (W) was assumed to represent the region of active enamel formation. Scale bar is 2 mm

often worn to the extent that enamel was no longer visible. Since the tip of the spine corresponded to the region formed at or around birth, the oldest enamel of the spine (corresponding to the age of the dogfish) was often not available for assay.

A comparison between Atlantic and Pacific age readers indicated that the spines from the two oceans were interpreted consistently. Age bias plots indicated no bias between Atlantic and Pacific age readers interpreting Pacific spines, and no bias between multiple Atlantic age readers interpreting Atlantic spines. Ageing precision was similar in both populations: the CV was 7.2% between Atlantic and Pacific agers of Pacific spines, and 8.0% between multiple Atlantic age readers of NW Atlantic spines.

Although spine $\delta^{13}C$ would not be expected to change very much through the period of bomb testing, it can be a reflection of the carbon source. Spine $\delta^{13}C$ ranged between –11.7‰ and –15.2‰ with an overall mean of –13.7‰ (SE = 0.09) (Table 1). Differences in $\delta^{13}C$ among areas were significant (ANOVA, $P < 0.05$), with the Pacific spines (mean = –13.3‰) significantly less depleted than those from the northeast Atlantic (mean = –14.0‰) and northwest Atlantic (mean = –14.2‰). $\delta^{13}C$ declined weakly but significantly with both year of formation ($P < 0.05$, $r^2 = 0.20$) and age at formation ($P < 0.05$, $r^2 = 0.21$) in the northeast Atlantic, but not in the other areas.

Although the magnitudes differed among areas, the increase in the spine $\Delta^{14}C$ through the 1960s was very similar to the bomb signal expected of $\Delta^{14}C$ in the marine environment (Fig. 2). In the Atlantic, spine $\Delta^{14}C$ increased sharply from about –80 in spine enamel formed before 1958, to values of about 100 after 1970, declining slowly thereafter to just above 0 in 1998 (Fig. 2A). Both the northeast and northwest Atlantic spine samples resembled the reference chronology for the northwest Atlantic (based on known-age otoliths), suggesting that the uptake of ^{14}C into the spine enamel mirrored that of marine DIC. In the northeast Pacific, spine $\Delta^{14}C$ also reflected the otolith-based reference chronologies for the region, albeit with more variance for the post-bomb (1970–1998) section of the curve

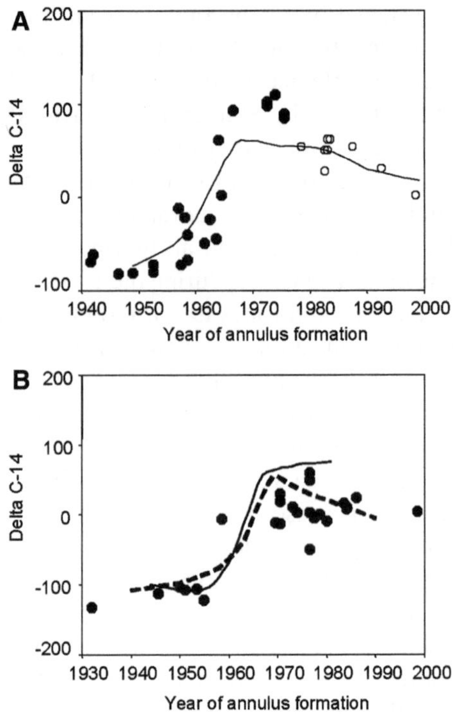

Fig. 2 $\Delta^{14}C$ in individual growth bands of dogfish spines versus year of formation inferred from counts of the growth bands. The $\Delta^{14}C$ chronology of the spines (symbols) was similar to that of a reference carbonate chronology (lines smoothed with a loess curve) from the same area, with the crucial feature being the period of rapid increase. (Top) Fish from the northeast Atlantic (•) and northwest Atlantic (○) have been plotted with the reference chronology from the Atlantic (solid line); (Bottom) Fish from the northeast Pacific (▲) have been plotted with the Pacific reference chronologies from Piner and Wischniowski (2004) (solid line) and Kerr et al. (2004) (dashed line)

(Fig. 2b). Pre-bomb $\Delta^{14}C$ (between 1932 and 1958) measured in Pacific spines averaged about –110, while post-bomb spine $\Delta^{14}C$ tended to be above 0.

Bomb radiocarbon can only serve as a reliable dated marker if its position relative to the growth bands remains static through the subsequent life of the animal. There was little evidence of metabolic reworking of the radiocarbon signal in the spine enamel. Of the 22 dogfish from which multiple growth bands were sampled, all showed the same within-spine bomb signal across year of formation as that demonstrated by the across-spine analysis (Fig. 3). Ontogenetic effects were not evident, indicating that the bomb signal was

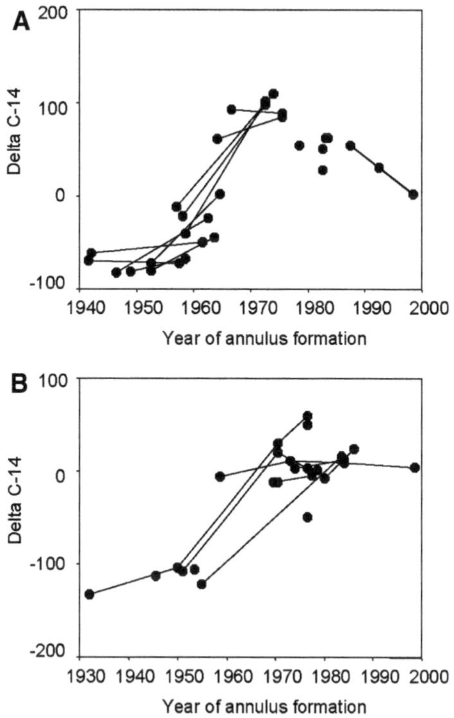

Fig. 3 $\Delta^{14}C$ in individual growth bands of dogfish spines versus year of formation inferred from counts of the growth bands. The $\Delta^{14}C$ chronology within each dogfish spine (each line corresponds to an individual spine) paralleled the across-spine chronology, indicating that the bomb signal of a single growth band remained stable as the dogfish grew older. (Top) Atlantic Ocean; (Bottom) Pacific Ocean

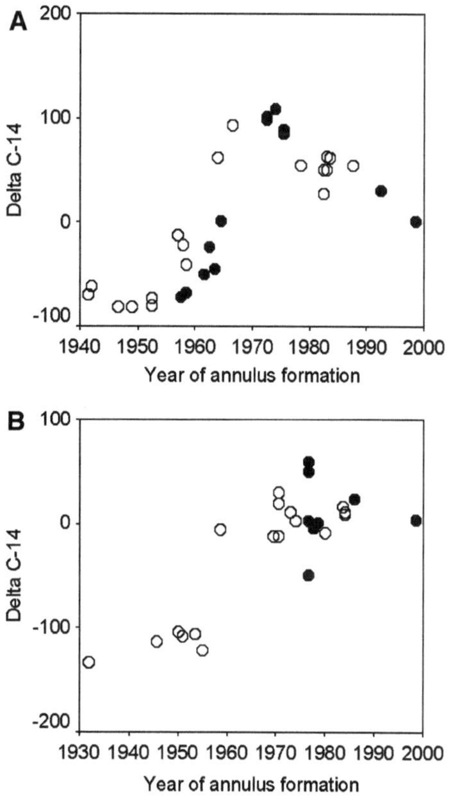

Fig. 4 $\Delta^{14}C$ in individual growth bands of dogfish spines versus year of formation inferred from counts of the growth bands. The radiocarbon chronology apparent in growth bands formed shortly before collection of the dogfish (implying that the date of band formation was known to within 1–2 year) (•) was similar to that of growth bands formed earlier in life (nearer the spine tip) (○). This indicated that the age of the growth bands was relatively accurately interpreted in the older dogfish. (Top) Atlantic Ocean; (Bottom) Pacific Ocean

not diluted by subsequent growth over a period of up to 44 year.

It is safe to assume that samples milled from near the base represent enamel formed within 1–2 years of capture; hence, a year of formation can be assigned to these samples with considerable certainty. The $\Delta^{14}C$ chronology prepared using these known-date samples was very similar to that based on earlier-formed growth bands nearer to the spine tip (Fig. 4). Since any errors in the age interpretation of the older spines would have been evident in $\Delta^{14}C$ values that lay-off the curve defined by the known-date spines, the dogfish aged 7–52 years must have been aged correctly (on average) based on the number of growth bands. No known-date samples from pre-bomb growth bands were available for the Pacific dogfish, which limited the comparisons, which were possible. However, in the case of the Atlantic dogfish, both the known-date and the test samples extended through the period of rapid increase in bomb radiocarbon through the 1960s. In light of the relative stability in $\Delta^{14}C$ after 1970, it would have been difficult to detect any under-ageing in dogfish born after 1970. However, over-ageing would have been easily detected, since a $\Delta^{14}C$ value of more than 0 would have been completely incompatible with a date of formation before about 1963. Thus the $\Delta^{14}C$ values between 60 and 95 in the growth bands of dogfish collected in 1978 and 1979, and presumed to have formed 14 years before collection based on growth band counts from the spine base, cannot have been aged with an error of more than ±1–2 year or the assay values would have been totally inconsistent with both the known-date assay curve and the reference chronology. Similarly, under-ageing of any sharks born before 1958 would have been easily detected based upon the presence of depleted $\Delta^{14}C$ values in apparently post-bomb samples. None of the spines with pre-bomb $\Delta^{14}C$ values were underaged to an extent that would place them into the period of radiocarbon increase. Nevertheless, there were three samples with presumed dates of formation in the late 1950s whose $\Delta^{14}C$ was slightly higher than expected based on the known-date samples. These samples, micromilled 18–21 growth bands from the spine base, may either have been overaged by a few years or may have inadvertently included underlying dentine representing later years of formation.

Discussion

Our results indicated that the amount and timing of bomb radiocarbon incorporated into dogfish spine enamel was very similar to that of DIC in the surrounding water. Comparison of $\Delta^{14}C$ values in spine growth bands formed at different ages within a spine indicated that the bomb signal was not transported across growth bands through time. Similar stability in deposited radiocarbon was reported for vertebral growth bands in large pelagic sharks (Campana et al. 2002). Since the period of increase in radiocarbon in the water is both known and predictable throughout the surface marine waters of the world (Druffel and Linick 1978; Campana 2001), and given the metabolic stability of the radiocarbon once deposited in the spines, radiocarbon in spiny dogfish spines appears to be well suited as a dated marker for age validation. Since ageing errors would have shifted the presumed date of growth band formation, the annual correspondence between spine and DIC radiocarbon chronologies supported the conclusion that our interpretation of spine growth bands as age indicators must have been accurate, at least on average, to an age of 52 year.

Spine $\delta^{13}C$ averaged –13.7, which is markedly different than the values of –1 to –3 typically found in otoliths (Kalish 1993; Campana 1997). Whereas the carbon source for otoliths is largely DIC with a $\delta^{13}C$ close to zero (Schwarcz et al. 1998), strongly depleted values such as those observed in the dogfish spines are more characteristic of metabolic and dietary carbon (Fry 1988). As such, it seems likely that the carbon source for the spine enamel is of dietary origin, similar to that of shark vertebrae (Campana et al. 2002). Campana et al. (2002) documented a phase-shifting between the vertebral radiocarbon of porbeagle sharks and the reference carbonate chronology due to the presence of deepwater, ^{14}C-depleted prey in the diet. The absence of this phase-shifting in dogfish spines is also consistent with a dietary source for spine carbon, since most of the dogfish diet would be expected to be found with the dogfish in shallow coastal waters (<200 m) (Jones and Geen 1977; Alonso et al. 2002), and thus reflect the radiocarbon content of the surface mixed layer DIC.

There was a visible but small tendency for the recently formed growth bands in some of the spines growing during the 1960s to be slightly depleted in $\Delta^{14}C$ compared to growth bands formed earlier in life. A similar phenomenon in porbeagle vertebrae was attributed to diet switching to larger, older prey by bigger sharks (Campana et al. 2002). It is possible that a similar explanation applies to dogfish. The tendency for spine $\Delta^{14}C$ to decrease with dogfish age would be consistent with this view, since larger, older prey would provide more depleted $\Delta^{14}C$ if tissue accumulated during the period of bomb ^{14}C increase. Alternatively, some of the spines may have been over-aged, or underlying dentine may not have been completely removed from the enamel samples.

The pre-bomb spine $\Delta^{14}C$ was very similar to the otolith carbonate chronologies from the same waters, as would be expected given the surface mixed layer depths occupied by dogfish and their diet. The post-bomb spine $\Delta^{14}C$ was also similar to that of the otolith carbonate chronology in each region, but the correspondence in the northeast Pacific was much more variable than in the Atlantic; since the absolute value of post-bomb $\Delta^{14}C$ in the water column is sensitive to water mass mixing rates and residence times (Druffel 1989; Weidman and Jones 1993), the area- and time-specific upwelling of the eastern Pacific would be expected to introduce more variability into radiocarbon levels. This variability in water-borne Pacific $\Delta^{14}C$ was also evident in the difference between the two North Pacific reference chronologies shown in Fig. 2. Of more importance to age validation is the timing of the initial post-bomb increase in spine radiocarbon relative to that in the DIC, since it is the year of initial increase that is most invariant throughout the world, and thus serves as the most reliable dated marker (Piner and Wischniowski 2004). The correspondence between the Atlantic spine chronology and the reference chronology was quite precise, both appearing to increase around 1958. With the data available, it was more difficult to pinpoint the initial increase in the Pacific spine chronology. However, with the Pacific spine $\Delta^{14}C$ showing peak values around 1970 and minima around 1954, the period of increase must have been within a couple of years of that evident in the reference chronologies.

Do Pacific dogfish grow more slowly and live longer than northwest Atlantic dogfish? This study indicates that our interpretation of the growth bands on the spines is accurate in both oceans. Moreover, our age comparisons in which spines from both oceans were aged by the same primary age reader indicates that our age interpretations for both populations are consistent with those that have been published. Therefore, the published work suggesting that Pacific spiny dogfish are both longer lived and slower growing than their northwest Atlantic counterparts appears to be accurate, and is not due to interpretational errors in one or both populations (Ketchen 1975; Nammack et al. 1985; Saunders and McFarlane 1993). The cause of the difference in growth rates between the two oceans remains unknown, but may be linked to population-level differences in reproductive output (Campana, unpublished).

The use of spine enamel bomb radiocarbon to determine age and confirm the periodicity of growth bands should be appropriate for all surface mixed layer shark species with fin spines where at least some of the growth bands were formed prior to 1965. If fin spines in teleost fishes form similarly to those in sharks, the method should also be appropriate for age validation in teleost fishes with fin spines. It is unlikely that deepwater shark species could be used, due to the delayed arrival of the bomb radiocarbon signal at depth. Whereas bomb signals in fish otoliths are limited by the small amounts of material available for assay (Campana 1999), sequential sampling of multiple growth bands from a single spine can be used to prepare a complete $\Delta^{14}C$ chronology and thus confirm the age of a single shark. Adjustments for phase shifting of the bomb signal in vertebrae is sometimes required depending on location and depth; although this was not an issue with spiny dogfish, phase shifting due to diet might be expected in the case of other shark species.

Acknowledgements We thank W. Joyce and L. Marks for expert technical assistance. Two anonymous referees provided helpful comments on the manuscript.

References

Alonso MK, Crespo EA, Garcia NA, Pedraza SN, Mariotti PA, Mora NJ (2002) Fishery and ontogenetic driven changes in the diet of the spiny dogfish, *Squalus acanthias*, in Patagonian waters, Argentina. Environ Biol Fishes 63:193–202

Beamish RJ, McFarlane GA (1985) Annulus development on the second dorsal spine of the spiny dogfish (*Squalus acanthias*) and its validity for age determination. Can J Fish Aquat Sci 42:1799–1805

Campana SE (1997) Use of radiocarbon from nuclear fallout as a dated marker in the otoliths of haddock, *Melanogrammus aeglefinus*. Mar Ecol Prog Series 150:49–56

Campana SE (1999) Chemistry and composition of fish otoliths: pathways, mechanisms and applications. Mar Ecol Prog Series 188:263–297

Campana SE (2001) Accuracy, precision and quality control in age determination, including a review of the use and abuse of age validation methods. J Fish Biol 59:197–242

Campana SE, Jones CM (1998) Radiocarbon from nuclear testing applied to age validation of black drum, *Pogonias cromis*. US Fishery Bull 96:185–192

Campana SE, Natanson LJ, Myklevoll S (2002) Bomb dating and age determination of large pelagic sharks. Can J Fish Aquat Sci 59:450–455

Campana SE, Thorrold SR (2001) Otoliths, increments and elements: keys to a comprehensive understanding of fish populations? Can J Fish Aquat Sci 58:30–38

Cortés E (2000) Life history patterns and correlations in sharks. Rev Fish Sci 8:299–344

Druffel EM (1989) Decadal time scale variability of ventilation in the North Atlantic: high-precision measurements of bomb radiocarbon in banded corals. J Geophys Res 94:3271–3285

Druffel EM, Linick TW (1978) Radiocarbon in annual coral rings of Florida. Geophys Res Lett 5:913–916

Fry B (1988) Food web structure on Georges Bank from stable C, N and S isotopic compositions. Limnol Oceanogr 33:1182–1190

Holden MJ, Meadows PS (1962) The structure of the spine of the spur dogfish (*Squalus acanthias*) and its use for age determination. J Mar Biol Assoc UK 42:179–197

Jones BC, Geen GH (1977) Food and feeding of spiny dogfish (*Squalus acanthias*) in British Columbia waters. J Fish Res Board Canada 34:2067–2078

Kalish JM (1993) Pre- and post-bomb radiocarbon in fish otoliths. Earth Planet Sci Lett 114:549–554

Kerr LA, Andrews AH, Frantz BR, Coale KH, Brown TA, Cailliet GM (2004) Radiocarbon in otoliths of yelloweye rockfish (*Sebastes ruberrimus*): a reference time series for the coastal waters of southeast Alaska. Can J Fish Aquat Sci 61:443–451

Ketchen KS (1975) Age and growth of dogfish *Squalus acanthias* in British Columbia waters. J Fish Res Board Canada 32:43–59

McFarlane GA, Beamish RJ (1987) Validation of the dorsal spine method of age determination for spiny dogfish. In: Summerfelt RC, Hall GE (eds) Age and growth of fish. Iowa State University Press, Ames, Iowa, pp 287–300

Nammack MF, Musick JA, Colvocoresses JA (1985) Life history of spiny dogfish off the northeastern United States. Transac Am Fish Soc 114:367–376

Piner KR, Wischniowski SG (2004) Pacific halibut chronology of bomb radiocarbon in otoliths from 1944 to 1981 and a validation of ageing methods. J Fish Biol 64:1060–1071

Saunders MW, McFarlane GA (1993) Age and length at maturity of the female spiny dogfish, *Squalus acanthias*, in the Strait of Georgia, British Columbia, Canada. Environ Biol Fish 38:49–57

Schwarcz HP, Gao Y, Campana S, Browne D, Knyf M, Brand U (1998) Stable carbon isotope variations in otoliths of Atlantic cod (*Gadus morhua*). Can J Fish Aquat Sci 55:1798–1806

Smith SE, Au DW, Show C (1998) Intrinsic rebound potentials of 26 species of Pacific sharks. Mar Freshwater Res 49:663–678

Stuiver M, Polach HA (1977) Reporting of C-14 data. Radiocarbon 19:355–363

Weidman CR, Jones GA (1993) A shell-derived time history of bomb C-14 on Georges Bank and its Labrador Sea implications. J Geophys Res 98:14577–14588

Environ Biol Fish (2006) 77:337–353
DOI 10.1007/s10641-006-9125-1

ELASMOBRANCHS

Investigations of $\Delta^{14}C$, $\delta^{13}C$, and $\delta^{15}N$ in vertebrae of white shark (*Carcharodon carcharias*) from the eastern North Pacific Ocean

Lisa A. Kerr · Allen H. Andrews · Gregor M. Cailliet · Thomas A. Brown · Kenneth H. Coale

Received: 8 June 2006 / Accepted: 30 June 2006 / Published online: 27 September 2006
© Springer Science+Business Media B.V. 2006

Abstract The white shark, *Carcharodon carcharias*, has a complex life history that is characterized by large scale movements and a highly variable diet. Estimates of age and growth for the white shark from the eastern North Pacific Ocean indicate they have a slow growth rate and a relatively high longevity. Age, growth, and longevity estimates useful for stock assessment and fishery models, however, require some form of validation. By counting vertebral growth band pairs, ages can be estimated, but because not all sharks deposit annual growth bands and many are not easily discernable, it is necessary to validate growth band periodicity with an independent method. Radiocarbon (^{14}C) age validation uses the discrete ^{14}C signal produced from thermonuclear testing in the 1950s and 1960s that is retained in skeletal structures as a time-specific marker. Growth band pairs in vertebrae, estimated as annual and spanning the 1930s to 1990s, were analyzed for $\Delta^{14}C$ and stable carbon and nitrogen isotopes ($\delta^{13}C$ and $\delta^{15}N$). The aim of this study was to evaluate the utility of ^{14}C age validation for a wide-ranging species with a complex life history and to use stable isotope measurements in vertebrae as a means of resolving complexity introduced into the ^{14}C chronology by ontogenetic shifts in diet and habitat. Stable isotopes provided useful trophic position information; however, validation of age estimates was confounded by what may have been some combination of the dietary source of carbon to the vertebrae, large-scale movement patterns, and steep ^{14}C gradients with depth in the eastern North Pacific Ocean.

Keywords Radiocarbon · Stable isotope ratios · Age validation · Shark vertebrae

L. A. Kerr (✉)
Chesapeake Biological Laboratory, University of Maryland Center of Environmental Science, P.O Box 38, Solomons, MD 20688, USA
e-mail: kerr@cbl.umces.edu

A. H. Andrews · G. M. Cailliet · K. H. Coale
Moss Landing Marine Laboratories, California State University, 8272 Moss Landing Road, Moss Landing, CA 95039, USA

T. A. Brown
Center for Accelerator Mass Spectrometry, Lawrence Livermore National Laboratory, 700 East Avenue, Livermore, CA 94551, USA

Introduction

The life history of the white shark, *Carcharodon carcharias*, varies seasonally, geographically, and ontogenetically, making it difficult to fully characterize the lifestyle of this animal. The white shark is globally distributed, ranging in habitat from temperate coastal and shelf to pelagic waters (Compagno 2001; Boustany et al. 2002). Recent satellite tagging data revealed that juvenile white sharks occur in nearshore waters, whereas adults are wide-ranging, with extensive periods of oceanic travel and what appears to be distinct oceanic and coastal phases (Boustany et al. 2002; Dewar et al. 2004; Bonfil et al. 2005). Both juvenile and adult white sharks exhibit deep-diving behavior below the ocean's mixed layer, with juveniles documented to dive to depths of 100 m (Dewar et al. 2004) and adults to 980 m (Bonfil et al. 2005). The diet of the white shark is variable and it is often described as a scavenger, feeding upon a wide range of prey taxa that includes marine mammals, teleost fishes, and invertebrates (Compagno 2001). In addition, an ontogenetic shift in diet has been documented for the white shark, with diet mainly composed of fishes (for sharks less than 2 m) to a diet of marine mammals (for sharks >3 m; Tricas and McCosker 1984; Klimley 1985; Compagno 2001). Although our understanding of the life history of the white shark is advancing, there still exists relatively limited knowledge of the habitat and diet of this animal over its lifetime. In addition, basic demographic information, including reliable age, growth, and longevity estimates, useful for stock assessment and fishery models, are not well defined.

Age determination of sharks is most commonly performed by counting growth band pairs in the vertebral centra that comprise the vertebral column (Ridewood 1921; Cailliet 1990). Vertebral centra are calcified cartilage, composed primarily of the mineral hydroxyapatite $[Ca_{10}(PO_4)_6(OH)_2]$ deposited within an organic matrix, of which collagen is the primary component (Urist 1961). In most elasmobranchs, this mineralization process occurs incrementally, resulting in vertebral growth bands (Cailliet 1990). One translucent and one opaque band comprise a band pair that is often assumed to represent one year of growth (Cailliet et al. 1983; Cailliet and Goldman 2004). The only age and growth study for the Pacific coast white shark population revealed they may have a slow growth rate and be long-lived based on growth band counts (Cailliet et al. 1985). However, no age validation has been successfully undertaken to date. Results from the Cailliet et al. (1985) study indicated that white shark may mature at an age of 9–10 years and live up to 27 years (assuming a maximum size of 7.6 m). Sharks were estimated to grow at a rate of 25–30 cm year^{-1} for young animals and 22 cm year^{-1} for older animals. In addition to a relatively late age at maturity and a slow growth rate, life history characteristics such as low fecundity and hypothesized infrequent reproduction frequency (Mollet et al. 2000), may make the white shark particularly vulnerable to exploitation (Cailliet et al. 1985). Validation of age and age estimation procedures is essential because proper management strategies rely heavily on accurate growth rates, age, and longevity.

Age can be estimated by counting vertebral growth band pairs; however, not all sharks deposit annual growth bands and many are not easily discernable (Cailliet and Goldman 2004). Thus, it is necessary to validate growth band periodicity with an independent method. Traditional age validation techniques, such as captive rearing, mark-recapture, and tag-recapture, can be difficult or impractical for these long-lived, pelagic fishes (Cailliet 1990). Marginal increment analysis and oxytetracycline injection of white sharks was attempted off the coast of South Africa, but results were inconclusive with regard to determining periodicity of growth band formation (Wintner and Cliff 1999). Radiometric age validation using lead-210 dating was also explored for this species, but results were inconclusive (Welden et al. 1987), due to failed assumptions. The irregular radiometric results were attributed to possible metabolic reworking of the vertebrae (likely the inorganic component, hydroxyapatite), or the result of an ontogenetic shift in habitat and diet of individual animals. A recent captive juvenile white shark held at the Monterey Bay Aquarium established the longest time span and record of growth in captivity. Prior to its release, the white shark was shown to have increased from 5 feet (1.52 m) TL and 62 lbs (28.1 kg) to 6 feet 4.5 inches (1.94 m) TL and 162 lbs (73.5 kg) during the 198 days of captivity[1]. This shark had a growth rate more than double that estimated by Cailliet et al. (1985); however, this high growth

[1] Monterey Bay Aquarium News Release, 31 March 2005

rate is not reflective of growth rates in the wild due to the sizeable feeding regime of this animal in captivity.

Measurement of the change in radiocarbon levels ($\Delta^{14}C$) produced by atmospheric testing of thermonuclear devices in the 1950s and 1960s has been established as an effective method for validating age estimates in calcified skeletal structures (Campana 2001). The discrete $\Delta^{14}C$ signal created by this testing was incorporated into the oceans of the world and has been used as a time-specific marker. Recent studies have correlated the changes in marine ^{14}C over time and used this temporal information to either: (1) make estimates of age and growth where no reliable age estimations were possible (e.g. calcareous algae, invertebrates, and some fishes; Frantz et al. 2000; Ebert and Southon 2003; Andrews et al. 2005; Frantz et al. 2005); or (2) validate estimates of age and growth (e.g. fishes; Kalish 1995; Kerr et al. 2005). For some $\Delta^{14}C$ records, age and growth of an organism was validated with another method and used to establish a $\Delta^{14}C$ reference time-series (e.g. hermatypic corals and fishes; Guilderson et al. 1998; Kerr et al. 2004; Piner and Wischniowski 2004). The utility of the method is dependent upon the ^{14}C signal retained in the skeletal structure of marine organisms as a permanent record of the ^{14}C present in ambient seawater at the time of formation. In fishes, this application is most commonly applied to the calcified ear bones, or otoliths. Recently, the use of this technique was expanded to shark vertebrae, including the school shark, *Galeorhinus galeus*, (Kalish and Johnston 2001) and two lamnoid sharks (Campana et al. 2002). Results of $\Delta^{14}C$ analyses for porbeagle, *Lamna nasus*, vertebrae indicated the $\Delta^{14}C$ signal was conserved across growth bands through time. Therefore, it was concluded that metabolic reworking of the organic or cartilaginous component of the vertebrae was minimal (Campana et al. 2002). This notion was further supported by a preliminary determination of the $\Delta^{14}C$ values for four growth bands from a single shortfin mako, *Isurus oxyrinchus*, vertebrae (Campana et al. 2002).

Radiocarbon age validation has been successfully applied to marine teleosts and sharks inhabiting surface waters during the period of life sampled from the growth structure for $\Delta^{14}C$ analysis. It has been noted, however, that the application of this technique can be problematic for species inhabiting waters below the mixed layer; due to the dependence of the ^{14}C signal in deeper waters on oceanic circulation and mixing rates (Kalish 1995). Therefore, knowledge of both seasonal and ontogenetic movements of a study species is important for interpreting $\Delta^{14}C$ values, especially in the case of elasmobranchs for which we are able to serially sample growth bands from vertebrae over the lifetime of the individual.

Interpretation of vertebral $\Delta^{14}C$ values is further complicated by the source of ^{14}C to the shark vertebrae. Unlike fish otoliths, which primarily obtain ^{14}C from ambient seawater (70–90% derived from dissolved inorganic carbon (DIC) in seawater and 10–30% is dietary; Kalish 1991; Farrell and Campana 1996), shark vertebrae reflect the ^{14}C composition of their diet (Kalish and Johnston 2001; Campana et al. 2002). Thus, an understanding of a species' diet is also necessary for interpreting shark vertebral $\Delta^{14}C$ values. Documented large-scale movements (Boustany et al. 2002; Bonfil et al. 2005) and a variable diet (Tricas and McCosker 1984; Klimley 1985) present complications in the application of ^{14}C age validation to the white shark. We proposed to approach these complications using stable isotopes in concert with ^{14}C measurements in vertebrae to discern possible changes in life history and diet over the lifetime of animals.

The objectives of this study were to: (1) measure $\Delta^{14}C$ in aged vertebrae as a means of determining the validity of white shark age estimates and the periodicity of growth band formation, and (2) use $\delta^{13}C$ and $\delta^{15}N$ measurements in white shark vertebrae to better understand the trophic position and carbon source to the vertebrae and to aid in the interpretation of $\Delta^{14}C$ values.

Materials and methods

Archived white shark vertebrae collected off the coast of central and southern California (capture years ranging from 1936 to 1994) were obtained from various collections for this study (sources included vertebrae from collections at Moss Landing Marine Laboratories, Los Angeles County Museum, California Academy of Sciences, Sea World San Diego, and from Leonard J.V. Compagno at the Shark Research Center, Iziko-Museums of Cape Town South African Museum, Cape Town, South Africa (Table 1).

Age estimation

One whole vertebra from each specimen was transversely sectioned, using the thin-section technique, for age estimation purposes. Sectioning was performed on a low-speed saw with two diamond blades separated by a spacer (2–3 mm) and polished with a Buehler® Ecomet III lapping wheel using 600 and 800 grit silicon-carbide wet/dry sandpaper for optimal viewing thickness. Sections were examined and images captured using a Leica dissecting microscope with an attached Spot RT® video camera. Transmitted light was used to make growth bands visible for counting. A growth band pair was defined as one translucent and one opaque growth band. Thin section age estimates were estimated by one reader (3 independent reads) and compared to whole vertebra age estimates and/or calculated age (for those vertebrae that were not used in the original age and growth study by Cailliet et al. 1985). Calculated age was estimated with the von Bertalanffy growth function (VBGF) determined by Cailliet et al. (1985). Coefficient of Variation (CV) was calculated for thin section age estimates as a measure of ageing precision (Chang 1982).

Radiocarbon analysis

Vertebrae from nine individual white sharks were sectioned (described previously) for $\Delta^{14}C$ analysis. In total, 22 white shark vertebra growth band pairs, with estimated growth years ranging from the pre-bomb 1930s to the post-bomb mid-1980s, were extracted for $\Delta^{14}C$ analyses. The limited availability of archived vertebrae with known capture years restricted the selection of vertebra and growth band years to be analyzed for $\Delta^{14}C$. Individual growth band pairs were sampled from thin-sections of the corpus calcareum using a New Wave® micro-milling machine with a small-scale end mill. The width of growth band pairs was used to guide extraction and minimize the amount of older or younger material incorporated in the sample. Hence the amount of material extracted decreased as growth slowed and band width decreased. The first growth band pair past the birth band (estimated as the first year of growth after birth) and one to three subsequent growth band pairs further up the corpus calcareum were extracted for analysis from each vertebrae. The last band pair, corresponding to the last year of

Table 1 Summary of data from white shark collected off the coast of California. Sample number identifies the individual white shark with corresponding sex, year of capture, capture location and total length (TL). Application indicates if vertebrae from the individual was used for ageing (Age), stable isotope analysis (SI), and radiocarbon analysis (^{14}C). The number of samples analyzed for stable isotopes or ^{14}C for each individual is noted in parentheses

Sample #	Sex	Year of Capture	Capture location	TL (cm)	Application
WH 1	M	1978	Moss Landing, CA	393	Age, SI (5), ^{14}C (3)
WH 3	M	1968	Half Moon Bay, CA	234	Age, SI (2), ^{14}C (3)
WH 6	F	1959	Tomales Bay, CA	277.5	Age, SI (3), ^{14}C (4)
WH 7	F	1936	Malibu, CA	167.6	Age, ^{14}C (1)
WH 8	M	1981	southern CA	147.3	Age, ^{14}C (1)
WH 9	M	1981	southern CA	159	Age
WH 12	?	1977	Ventura, CA	210	Age, ^{14}C (1)
WH 17	M	1982	southern CA	460.9	Age, SI (4), ^{14}C (4)
WH 25	?	1984	Half Moon Bay, CA	168[a]	Age, ^{14}C (1)
WH 90	M	Unknown	California	471	SI (4)
WH 128	F	1994	California	534.4	SI (5)
WH 26694	?	1959	NE Pacific, CA	225.4	Age, ^{14}C (1)

[a] TL not recorded and calculated from a total length to centrum diameter regression

growth, was targeted to provide a sample where time of formation was constrained by the collection date. In some cases, the penultimate growth band pair was targeted because the width of the last growth band pair provided insufficient sample size. Extraction of the sample with the micro-mill resulted in a solid sample of material that was weighed to the nearest 0.1 mg.

Pre-treatment of vertebral samples was performed to isolate the organic portion (collagen) and maximize the carbon yield from vertebrae (Brown et al. 1988). Inorganic carbon was removed through the process of demineralization by soaking vertebral samples in 0.25 N HCl at refrigerator temperatures (slows reaction). Treated samples were dried in an oven and placed in clean quartz tubes. CuO (copper oxide, oxidizing agent) and Ag (silver, used to remove impurities: SOx and NOx) were added to the treated organic sample. Three samples from individual vertebrae were replicated and analyzed for $\Delta^{14}C$ without demineralization to evaluate the effect of demineralization on $\Delta^{14}C$ measurements. Quartz tubes were evacuated, sealed, and heated for 2 h at 900°C to convert the organic carbon to CO_2. Sample CO_2 was converted to graphite (Vogel et al. 1984; Vogel et al. 1987) and measured for ^{14}C content using an accelerator mass spectrometer (AMS) at the Center for Accelerator Mass Spectrometry, Lawrence Livermore National Laboratory. The ^{14}C values were reported as $\Delta^{14}C$ (Stuivier and Polach 1977).

A qualitative comparison of the white shark $\Delta^{14}C$ record was made with existing marine records including two otolith-based records, the yelloweye rockfish, *Sebastes ruberrimus*, (Kerr et al. 2004) and Pacific halibut, *Hypoglossus stenolepis*, (Piner and Wischniowski 2004), and three vertebra-based shark records, the western North Atlantic porbeagle (Campana et al. 2002), western North Atlantic shortfin mako (Campana et al. 2002; Ardizzone et al. 2006), and the western South Pacific school shark (Kalish and Johnston 2001).

Stable isotope analysis

Vertebrae from six individual white sharks were sectioned (described previously) for isotopic analysis ($\delta^{15}N$ and $\delta^{13}C$). Four of the six vertebra samples were also analyzed for $\Delta^{14}C$ and many of the growth band pairs analyzed for stable isotopes were targeted to coincide with those analyzed for $\Delta^{14}C$. In total, 23 white shark vertebra growth band pairs were extracted from the corpus calcareum using a New Wave micro-milling machine and analyzed for $\delta^{13}C$ and $\delta^{15}N$. Vertebral samples were demineralized to isolate collagen from the vertebrae (Brown et al. 1988). In addition, lipids were extracted from samples analyzed for $\delta^{13}C$ and $\delta^{15}N$ using a 2:1:0.8 methanol/chloroform/water mixture. Samples were analyzed by continuous flow isotope ratio mass spectrometer (IRMS) at University of California, Davis. Values are reported as $\delta^{13}C$ and $\delta^{15}N$ relative to standards of Pee Dee Belemnite limestone (^{13}C) and atmospheric N_2 (^{15}N).

Trophic position (TP) of the white shark was calculated using the equation:

$$TP = \frac{\lambda + (\delta^{15}N_{\text{consumer}} - \delta^{15}N_{\text{base}})}{\Delta n}$$

Where λ is the trophic position of the organism employed to estimate the baseline $\delta^{15}N$ for the region, $\delta^{15}N_{\text{consumer}}$ is the average value of the consumer, $\delta^{15}N_{\text{base}}$ is the average value of the base organism, and Δn is the average enrichment per trophic level (Post 2002). The northern anchovy, *Engraulis mordax*, was chosen as a representative secondary consumer (assigned a trophic position (λ) of 3.0) for estimating $\delta^{15}N_{\text{base}}$ of the region and an average trophic enrichment ($\Delta^{15}N$) of 3.4 was assumed following Estrada et al. (2003).

Results

Age estimation

The vertebrae from white sharks were relatively difficult to age. A comparison of thin section age estimates with corresponding whole vertebral-based estimates revealed there was reasonable agreement (H_0: no significant difference between techniques; paired *t*-test: $P = 0.65$, $\alpha = 0.05$; Table 2). Thin section age estimates were, on

Table 2 Age estimate comparison between values obtained from thin sections in this study, and whole vertebral reads and calculated age based on von Bertalanffy growth function (VGBF) from Cailliet et al. 1985. Age estimation error was the uncertainty associated with the thin section age estimates (individual coefficient of variation (CV); rounded to the nearest whole number). Mean CV = 16%

Sample #	Estimated age		
	Thin section (±CV)	Whole vertebrae (Cailliet et al. 1985)	Calculated age (VBGF) (Cailliet et al. 1985)
WH 1	7 (±1)	9	9
WH 3	3 (±0)	2	3
WH 6	4 (±1)	2	4
WH 7	1 (±0)	1	1
WH 9	0 (±0)	1	0
WH 12	1 (±1)	2	2
WH 17	18 (±1)	13	12

average, higher than whole vertebral age estimates. The mean CV for thin section age estimates was 16.0% of the individual age. Age estimates from thin sections deviated by as much as 2 years.

Radiocarbon analysis

Radiocarbon values measured in growth band pairs from white sharks varied considerably over time (Table 3). When $\Delta^{14}C$ values were plotted against estimated growth year, the values produced a $\Delta^{14}C$ time series from 1936 to 1984 (Fig. 1). Overall, $\Delta^{14}C$ values increased as growth year progresses into the post-bomb era. However, this increase was not synchronous with the characteristic bomb ^{14}C rise and had considerably more scatter in the post-bomb period than could be explained by $\Delta^{14}C$ measurement uncertainty. Replicate samples from 1956 had very depleted $\Delta^{14}C$ values that were atypical of $\Delta^{14}C$ measured in surface waters of the Pacific Ocean. These replicate samples differed from each other by 87.8‰. Pre-bomb $\Delta^{14}C$ values (1936 to approximately 1959, excluding the 1956 outliers) had a trend similar to the pre-bomb record from the reference time-series, averaging –90.5 ± 5.1‰ (mean ± SD). The first evidence of an increase due to atmospheric testing of thermonuclear devices was the elevated $\Delta^{14}C$ value measured for the 1966 growth band pair (–72.24‰). This value was the first to have a $\Delta^{14}C$ value that was significantly above pre-bomb $\Delta^{14}C$ levels using a +2 SD criteria (+10.2‰). The rise in $\Delta^{14}C$ continued until 1984 (the last growth year sampled) with a maximum observed $\Delta^{14}C$ value of 79.8‰ and no indication of a post-bomb decline.

Radiocarbon values for the first year of growth from five age-1 (±1 year) white sharks created a $\Delta^{14}C$ reference time-series for this species composed of essentially known-age specimens (Fig. 1). The $\Delta^{14}C$ values in these vertebrae showed a response to the bomb pulse that is similar in magnitude to the two otolith reference time-series from the Pacific Ocean, but differed by what appears to be a significant phase lag.

Radiocarbon values from vertebral material near the known year of capture (up to minus 2 years from date of collection) revealed a very different pattern relative to the known-age juvenile series (Fig. 1). Radiocarbon values in these vertebrae showed a depleted and/or delayed bomb ^{14}C signal based on knowledge of the collection year. All other samples, for which age was estimated from growth band pair counts, had $\Delta^{14}C$ values distributed between these two sets of time-constrained samples.

Radiocarbon values for demineralized and untreated vertebral samples (these samples were as close to replicates as we could get based on the position in the corpus calcareum) indicated a difference in the $\Delta^{14}C$ between treated and untreated samples; hence, the inorganic portion of the vertebrae had different $\Delta^{14}C$ levels (Table 3). All untreated samples had elevated $\Delta^{14}C$ values compared to their demineralized counterparts.

Stable isotope analysis

Twenty-three growth band pairs, extracted from the vertebrae of six individual white sharks, were analyzed for $\delta^{13}C$ and $\delta^{15}N$. A decreasing trend with increased age was exhibited for $\delta^{13}C$ values within individual vertebrae (Fig. 2). Growth band pairs ≥ age 6 in the vertebrae were used to discriminate between what we have termed juvenile and adult growth, a criterion based on a reported

Table 3 Summary of vertebral and $\Delta^{14}C$ data from white shark collected off the coast of California. Resolved age is the final age estimate. Birth year is collection year minus the resolved age. Ageing error is the uncertainty associated with the age estimate (CV). Radiocarbon values for the extracted samples were expressed as $\Delta^{14}C$ ± AMS analytical uncertainty

Sample #	Resolved age (years ± age error)	Capture Year	Birth Year	Year Sampled for ^{14}C analysis (years)	$\Delta^{14}C$ (‰)
WH 7	1 ± 0	1936	1935	1936	–94.2 ± 3.5
WH 26694	2 ± 1	1959	1957	1956[a]	–92.7 ± 3.5
WH6	4 ± 1	1959	1955	1956	–318.2 ± 2.9
WH 6	4 ± 1	1959	1955	1956	–230.4 ± 3.0
WH 6[b]	4 ± 1	1959	1955	1956	–63.1 ± 3.6
WH 6	4 ± 1	1959	1955	1957	–83.0 ± 3.5
WH 6	4 ± 1	1959	1955	1959	–92.1 ± 4.0
WH 3	3 ± 0	1968	1965	1966	–72.2 ± 4.0
WH 3	3 ± 0	1968	1965	1967	–21.8 ± 3.7
WH 3	3 ± 0	1968	1965	1968	–98.5 ± 3.7
WH 3 [b]	3 ± 0	1968	1965	1968	–72.9 ± 3.9
WH 17	18 ± 1	1982	1964	1966	–74.1 ± 5.0
WH 17	18 ± 1	1982	1964	1971	–65.6 ± 4.9
WH 17	18 ± 1	1982	1964	1976	–29.2 ± 4.2
WH 17	18 ± 1	1982	1964	1981	34.7 ± 4.5
WH 1	7 ± 1	1978	1971	1972	–59.7 ± 4.2
WH 1	7 ± 0	1978	1971	1975	–58.3 ± 3.5
WH 1	7 ± 1	1978	1971	1978	–55.8 ± 3.6
WH 1[b]	7 ± 1	1978	1971	1978	1.8 ± 3.8
WH 12	1 ± 1	1977	1975	1975	58.2 ± 4.9
WH 8	1 ± 0	1981	1980	1981	75.0 ± 4.2
WH 25	1 ± 0	1984	1983	1984	79.8 ± 4.1

[a] indicates sample was composed of pre-birth material

[b] indicates samples that were not demineralized

size at transition in diet of 3 meters documented by Compagno (2001) and estimated age at transition (6 years) based on the VBGF estimated from Cailliet et al. (1985). Mean stable isotope values associated with juvenile growth (1–5 years: $\delta^{13}C$: –11.81 ± 0.60‰, $\delta^{15}N$: 19.16 ± 1.01‰) and adult growth (6–18 years: $\delta^{13}C$: –12.73 ± 0.62‰, $\delta^{15}N$: 19.34 ± 0.93‰) showed no significant difference in $\delta^{15}N$ (paired *t*-test, $P = 0.70$) or calculated trophic position (paired *t*-test, $P = 0.70$), but did exhibit a significant difference in $\delta^{13}C$ values between juvenile and adult growth (paired *t*-test $P = 0.01$).

Overall, $\delta^{15}N$ values ranged from 17.68 to 20.84‰ (Mean: 19.24‰ ± 0.95‰) and $\delta^{13}C$ values ranged from –13.31 to –11.10‰ (Mean: –12.24 ± 0.76; Table 4). Across longitudinal sections of individual vertebra, nitrogen values differed by a maximum of 0.51–2.85‰ and carbon values differed by a maximum of 0.39–2.05‰ (Table 4). An approximate increase of 1‰ $\delta^{13}C$ and 3–4‰ $\delta^{15}N$ is associated with an increase in one trophic level (Michener and Schell 1994). Four individual vertebrae exhibited differences in $\delta^{13}C$ across longitudinal sections of vertebra >1‰, however no vertebrae exhibited differences in $\delta^{15}N$ values >3‰ across longitudinal sections. White shark stable isotope ratios indicated feeding at an upper trophic level relative to fish and marine mammal stable isotope values from the eastern North Pacific Ocean (Fig. 3). Trophic position calculation of the white shark based on mean $\delta^{15}N$ value resulted in an estimated trophic position of 4.57 (range 4.11–5.04).

Discussion

Age estimation

The observation that thin section age estimates were, on average, higher than those from whole

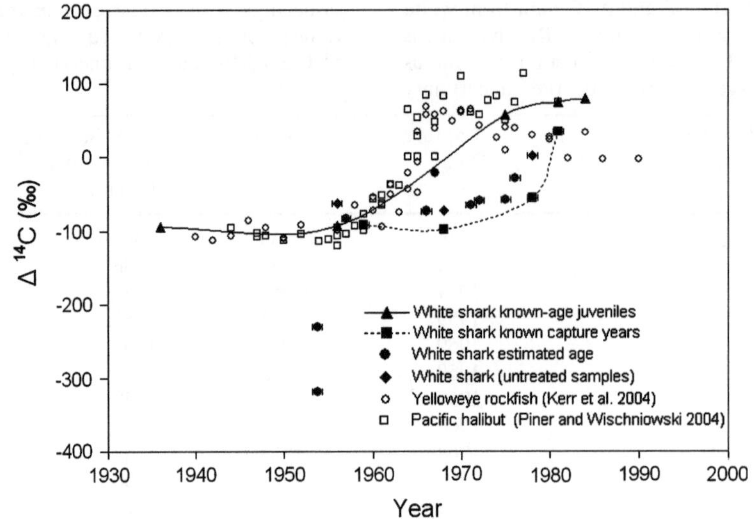

Fig. 1 Radiocarbon ($\Delta^{14}C$) values for white shark, *Carcharodon carcharias*, (solid symbols) vertebral cores ($n = 22$) in relation to year. The solid line connects values from known-age juvenile white sharks, the dashed line connects values from known-collection years of the sharks, the unconnected black circles are $\Delta^{14}C$ values plotted in relation to years based on estimated age, and black diamonds are replicate untreated samples that were analyzed for $\Delta^{14}C$. Horizontal error bars represent the age estimate uncertainty, rounded to the nearest whole number, and vertical error bars represent the 1σ AMS analytical uncertainty. The white shark ^{14}C time series is plotted with two regional $\Delta^{14}C$ chronologies for the yelloweye rockfish (open circles, Kerr et al. 2004) and Pacific halibut otoliths (open squares, Piner and Wischniowski 2004)

vertebrae was expected based on what we know about these two techniques. Currently, thin sectioning is accepted as the more accurate technique (Cailliet and Goldman 2004); and because growth band pairs were isolated from thin sections for $\Delta^{14}C$ analyses, we relied on these age estimates in this study.

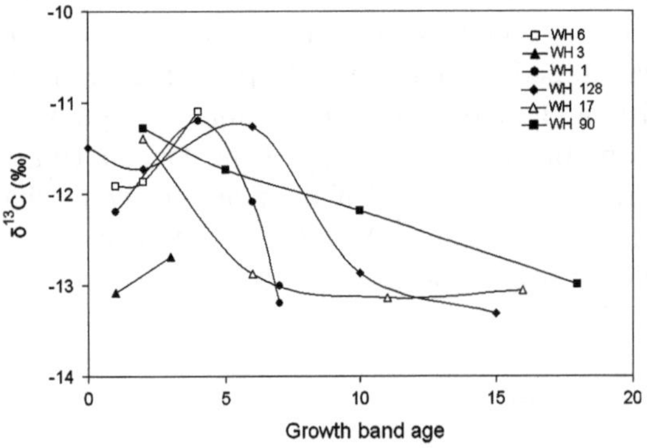

Fig. 2 Stable carbon ratio ($\delta^{13}C$) for samples ($n = 23$) taken across longitudinal sections of six individual white shark, *Carcharodon carcharias*, vertebrae. Unique symbols represent $\delta^{13}C$ results for samples from an individual white shark vertebra

Radiocarbon analysis

It is well established that the initial rise in $\Delta^{14}C$ is nearly synchronous in all marine carbonate records, but the white shark record exhibited an asynchronous $\Delta^{14}C$ time series. The uncharacteristic timing of $\Delta^{14}C$ values differed from the nearest chronologies in the region (Kerr et al. 2004; Piner and Wischchniowski 2004). There are three plausible mechanisms for explaining this uncharacteristic trend: (1) an apparent delay of the $\Delta^{14}C$ signal due to age underestimation; (2) metabolic reworking of the vertebral collagen; and (3) a depletion of the $\Delta^{14}C$ levels due to depleted dietary sources of carbon. Each of these factors could explain the observed trend individually, but a combination of some or all of them cannot be ruled out.

Age underestimation of white shark vertebrae would be manifested as a delayed $\Delta^{14}C$ time series. The degree of age underestimation was estimated by tracing estimated growth years back to their appropriate place on the white shark reference curve (known-age juvenile series). Potential age underestimation ranged from 6–11 years

Table 4 Summary of vertebral and stable isotope data from white shark collected off the coast of California. Sample number identifies the individual white shark. Isotope values are reported as $\delta^{13}C$ and $\delta^{15}N$ relative to standards of Pee Dee Belemnite limestone ($\delta^{13}C$) and atmospheric N_2 ($\delta^{15}N$). Overall difference in stable isotope ratios is the maximum difference observed across longitudinal sections of an individual vertebra

Sample #	$\delta^{13}C$ (‰)	$\delta^{15}N$ (‰)	Estimated Age (year)	Overall difference in $\delta^{13}C$ (‰)	Overall difference in $\delta^{15}N$ (‰)
WH 17	–11.39	19.51	2	1.75	0.51
WH 17	–12.88	20.03	6		
WH 17	–13.14	20.00	11		
WH 17	–13.06	19.81	16		
WH 90	–11.28	17.87	2	1.71	2.16
WH 90	–11.74	19.51	5		
WH 90	–12.19	20.03	10		
WH 90	–12.99	18.67	18		
WH 128	–11.49	19.87	0	2.05	1.69
WH 128	–11.73	19.48	2		
WH 128	–11.26	19.13	6		
WH 128	–12.87	18.18	10		
WH 128	–13.31	19.70	15		
WH 6	–11.92	18.31	1	0.82	0.91
WH 6	–11.87	19.21	2		
WH 6	–11.10	18.86	4		
WH 3	–13.08	17.68	1	0.39	0.53
WH 3	–12.69	18.21	3		
WH 1	–12.20	20.61	1	2.00	2.85
WH 1	–11.20	20.78	4		
WH 1	–12.09	20.84	6		
WH 1	–13.01	18.41	7		
WH 1	–13.20	17.93	7		

(mean 9 ± 3 years). Examination of the residuals (estimated age versus expected age based on the reference time series) for potential age underestimation indicated there was no systematic ageing bias (no significant trend observed, $r^2 = 0.02$). If age underestimation were the sole source of the apparent delay, it would not explain the delay in the time constrained samples taken near the collection year. These samples alone strongly suggest there are other factors responsible for the observed trend.

A more plausible explanation is that there is a metabolic reworking of the vertebrae through the life of the fish. The question of metabolic reworking of vertebrae has been hypothesized for the white shark in relation to radiometric age determination. Analysis of vertebrae for lead-210 indicated this may not be a static or conserved structure with respect to the inorganic portion of the vertebrae (Welden et al. 1987). Unlike cancellous or "true" bone which undergoes remodeling over the lifetime of the animal, hydroxyapatite, the primary component of shark cartilage, is thought to grow by accretion with little remodeling; therefore, recording conserved information for different periods of the animal's life (Koch et al. 1994).

Because there is no empirical evidence to support or refute the temporal stability of the organic portion of vertebral cartilage, we looked to isotopes for indirect evidence. Differences in $\Delta^{14}C$, $\delta^{13}C$, and $\delta^{15}N$ values preserved across longitudinal sections of vertebrae, with differences in $\Delta^{14}C$ as great as 235‰ (sample WH 6), support the conclusion that vertebrae are not completely reworked. This is also supported by $\Delta^{14}C$ results for the porbeagle that indicated metabolic reworking was minimal and the ^{14}C signal was not transported across growth bands through time (Campana et al. 2002). These findings suggest that carbon in the vertebral material is metabolically and temporally stable (Campana et al. 2002).

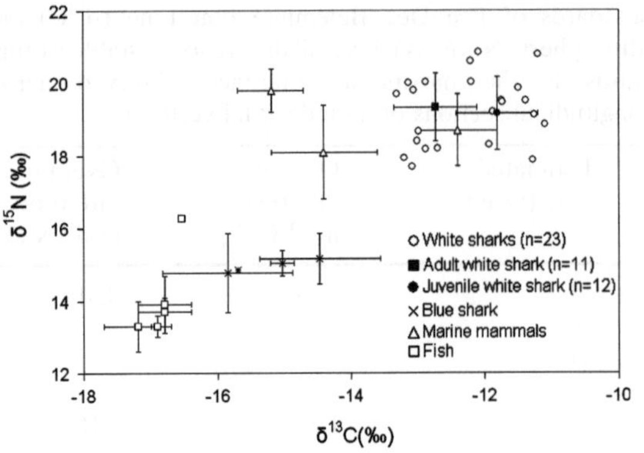

Fig. 3 Stable carbon isotope ratios ($\delta^{13}C$) for individual white shark, *Carcharodon carcharias*, vertebral cores (open circles; $n = 23$) in relation to stable nitrogen ratios ($\delta^{15}N$) plotted with mean marine mammal (open triangles; harbor seal, *Phoca vitulina*, elephant seal *Mirounga angustirostris*, California sea lion, *Zalophus californianus*, Stellar sea lion, *Eumetopias jubata*; Burton et al. 2001; Jamon et al. 1996) and fish stable isotope values (open squares; rockfishes, *Sebastes* spp., white croaker, *Genyonemus lineatus*, shortraker rockfish, *Sebastes brevispinis*, northern anchovy, *Engraulis mordax*, and California market squid, *Loligo opalescens* (Toperoff 1997; Jarman et al. 1996)), and blue shark (crosses, *Prionace glauca*; R. Leaf personal communication). Mean (±SD) isotope ratio for juvenile growth band pairs (closed circle; $n = 12$, growth band pairs 1–5 years) and adult growth band pairs (closed square; $n = 11$, growth band pairs 6–18 years) in white shark vertebrae are shown. This distinction based on ontogenetic shift in diet at size 3 m

However, further research is needed to provide direct evidence to validate the assumption of shark vertebrae functioning as a closed-system with respect to carbon and nitrogen.

A third possibility for the uncharacteristic trend in the white shark $\Delta^{14}C$ signal is that values were depleted due to the dietary source of carbon to the vertebrae. Because the composition of vertebral collagen reflects dietary sources, low $\Delta^{14}C$ values of vertebrae could be the result of prey consumption that has integrated a ^{14}C depleted signal, possibly from a deep-sea source (Campana 1999). Incorporation of ^{14}C in the ocean is related to ocean circulation and mixing, with different water masses reflecting different ^{14}C values and values becoming attenuated with depth (Broeker and Peng 1982). Radiocarbon enters the food web through photosynthetic organisms and is transferred to higher trophic levels (Pearcy and Stuvier 1983). Lower mesopelagic, bathypelagic and abyssopelagic animals in the northeastern Pacific Ocean, however, were found to have depleted $\Delta^{14}C$ values relative to surface dwellers (1973–1976), indicating that surface-derived POC was not the major source of organic carbon for deep-sea fish (Pearcy and Stuvier 1983).

A steep gradient in bomb $\Delta^{14}C$ values with depth was documented in the North Pacific Ocean, with $\Delta^{14}C$ values of DIC in the North Central Pacific reaching approximately –100‰ at about 500 m (1986), and values of –150‰ in the Santa Monica Basin at about 500 m (1986–1987; Druffel and Williams 1990). Similarly, low DOC $\Delta^{14}C$ values have been documented in the North Pacific Ocean (~ –375‰ at 500 m in the North Central Pacific and ~ –300‰ at 100 m in Santa Monica Basin; Druffel and Williams 1990). Druffel and Williams (1990) proposed three possible mechanisms for the incorporation of ^{14}C depleted carbon into the oceanic food chain: (1) uptake of DOC by bacteria in the water column, or adsorption of depleted DOC through webs and mucous of filter feeders (e.g. larvaceans and salps), (2) chemosynthetic production of organic matter from DIC, and (3) resuspension of depleted sedimentary organic carbon. Druffel and Williams (1990) identified this depleted deep-sea source of carbon as responsible for depletion of surface water POC $\Delta^{14}C$ values by an average of 93‰ in these regions. This magnitude of depletion was strikingly similar to the average depletion of white shark vertebral values (93‰ ± 53‰).

Until recently, white sharks in the Pacific Ocean were thought to reside in waters of the upper continental shelf and were not known to dive below 100 m. Satellite tagging results revealed the white shark are deep diving, with a bimodal depth of occurrence at 0–5 and 300–500 m (Boustany et al. 2002). This tagging study, along with Bonfil et al. (2005), revealed that white sharks are wider ranging than previously thought, with extensive periods of oceanic travel and what appears to be distinctive oceanic and coastal phases. Based on what is now known about its depth of occurrence and extensive movements, consumption of deep-dwelling prey or prey that feed in deep waters is a strong possibility. This depleted source offers a reasonable explanation for the unusually low $\Delta^{14}C$ values for replicate measures of the 1956 white shark samples and attenuated values measured for known collection years during the post-bomb era. In addition, the degree of variation in depletion could be related to the variable diet of this pelagic species that inhabits both inshore and offshore habitats and consumes a wide spectrum of prey over the course of their lifetime.

The reference chronology comprised of known-age juveniles reflected a more rapid response to the bomb pulse, most likely due to the habitat utilized during the juvenile stage. Satellite tagging results of a juvenile (young of the year) white shark off the coast of California supported by evidence of capture locations indicate that juvenile white shark remain in coastal waters (Klimley 1985; Dewar et al. 2004). Juveniles inhabiting and feeding nearshore on coastal fish would reflect a $\Delta^{14}C$ signal typical of coastal, surface waters and would be more synchronous with the reference time series.

Employing a simple mass balance equation for $\Delta^{14}C$ allowed us to examine the likelihood the observed values in white shark vertebrae were attributable to the influence of a depleted deep-water carbon source. The equation describing this deep-water influence is:

$$\Delta^{14}C_{\text{observed}} = \Delta^{14}C_{\text{expected}}(x) - \Delta^{14}C_{\text{deep-water}}(1-x)$$

where $\Delta^{14}C_{\text{observed}}$ is the value measured in the vertebrae, $\Delta^{14}C_{\text{expected}}$ is the value expected based on the white shark reference curve, x is the fraction contributed from the "expected" surface water source, $\Delta^{14}C_{\text{deep-water}}$ is a value of a possible deep-water source based on ^{14}C gradients off the coast of California (Druffel and Williams 1990), and $1–x$ is the fraction contributed from the "deep-water" source. As a result of this modeling exercise, most observed $\Delta^{14}C$ values for white shark vertebrae were found to be similar in value to contributions ranging from a small percent of the diet composed of a highly depleted ^{14}C source (25% contribution of a –300‰ source) to a large percentage of the diet composed of a slightly depleted ^{14}C source (75% contribution of a –100‰ source). Based on the highly variable diet of white sharks and this modeling, it is likely that depleted values are attributable to a small contribution of highly depleted ^{14}C sources in the diet.

Factors associated with the sampling method should also be considered. There are two factors that may have influenced the outcome of $\Delta^{14}C$ levels measured in white shark vertebrae: (1) the location of sampling within the vertebrae and (2) the preparation of vertebrae for $\Delta^{14}C$ analysis (demineralized versus untreated vertebrae). The variation in replicated demineralized samples (WH 6: 1956 samples) from the same location within the vertebra indicated that small deviations in coring location can result in large deviations in $\Delta^{14}C$ (an observed difference of 87.8‰). This can be explained by large variations of ^{14}C in food sources that were variably accreted during the period of formation. In addition, relatively little is known about the growth and accretion of white shark vertebrae. If growth was episodic and accretion was not uniform, but rather focused where material is needed from a structural perspective, this could explain regional variations of the $\Delta^{14}C$ signal of the vertebrae.

Results from untreated samples (the inorganic and organic portions) indicated the potential presence of carbonates in the inorganic portion of the vertebrae in two ways. First, we observed bubbling during the acid dissolution of the vertebrae indicating CO_2 was liberated from the inorganic portion of the vertebral matrix. Second, the elevated levels of $\Delta^{14}C$ in replicate untreated samples compared to demineralized (the organic portion) samples indicated a ^{14}C source in the inorganic portion of the vertebrae. The depletion of $\Delta^{14}C$ values in the organic portion of the vertebrae likely reflects the different sources of ^{14}C to the organic and inorganic portions of the vertebrae. The dominant source of carbon to the organic portion of the vertebrae is known to be metabolic (Kalish and Johnston 2001, Campana et al. 2001). The difference between the demineralized and untreated vertebrae represents the addition of elevated ^{14}C values, most likely from a DIC source. Evidence of a similar trend was identified in the treatment of organic (collagen) and inorganic (carbonate) portions of swordfish, *Xiphias gladius*, vertebrae (Kalish and Demartini 2001). Results from the study revealed the organic component of the swordfish vertebrae was more ^{14}C depleted than the inorganic component (Kalish and Demartini 2001). This difference was attributed to the dominant source of carbon to the organic portion of the vertebrae being metabolically derived and the source to inorganic portion derived from DIC. Our results indicate this same trend is present in the white shark vertebrae.

A comparison of the white shark $\Delta^{14}C$ record with three other shark records indicated the white shark record was most temporally similar to the school shark record from the South Pacific Ocean (Fig. 4; Kalish and Johnston 2001). The school shark $\Delta^{14}C$ time series was delayed in relation to the nearest marine carbonate chronology and in this case the delay was attributed to age underestimation. A 3-year lag observed in the porbeagle $\Delta^{14}C$ time series, based on comparison to a juvenile porbeagle reference time series, was attributed to both the mean age and depth of occupied by prey in the diet of porbeagle (Campana et al. 2002). All of these shark records show, to differing degrees, a nonconforming $\Delta^{14}C$ signal compared to their nearest $\Delta^{14}C$ chronology. In the case of the white shark, we hypothesize that there is depletion of the $\Delta^{14}C$ time series due to the influence of depleted ^{14}C sources to the vertebrae. This nonconformity indicates the increased complexity of interpreting $\Delta^{14}C$ values in shark vertebrae for the purpose of validating age and the necessity to interpret values in the context of diet, movement, and regional ^{14}C gradient information.

Stable isotope analysis

Stable isotope ratios can provide valuable information relative to trophic level, carbon flow to a consumer, and location of feeding (Vander Zanden and Rasmussen 2001). While gut content analysis provides a snapshot of what was most recently consumed by an animal, stable isotopes ($\delta^{13}C$ and $\delta^{15}N$) can provide time-integrated diet information (Vander Zanden and Rasmussen 2001). This application is based on predictable enrichment factors of both ^{13}C (approximately 1‰ enrichment) and ^{15}N (approximately 3–4‰ enrichment) in consumer bone collagen relative to prey with increasing trophic level, termed trophic fractionation (Deniro and Epstein 1981; Schoeniger and Deniro 1984; Vander Zanden and Rasmussen 1999). Additionally, $\delta^{13}C$ can provide information relative to feeding habitat (inshore versus offshore; France 1995). A caveat to this approach is that spatial and temporal variability in stable isotope ratios must be considered when interpreting stable isotope values.

In general, adults are reported to feed primarily on marine mammals and juveniles on fishes (Le Boeuf et al. 1982; Tricas and McCosker 1984; Klimley 1985; Compagno 2001). Current

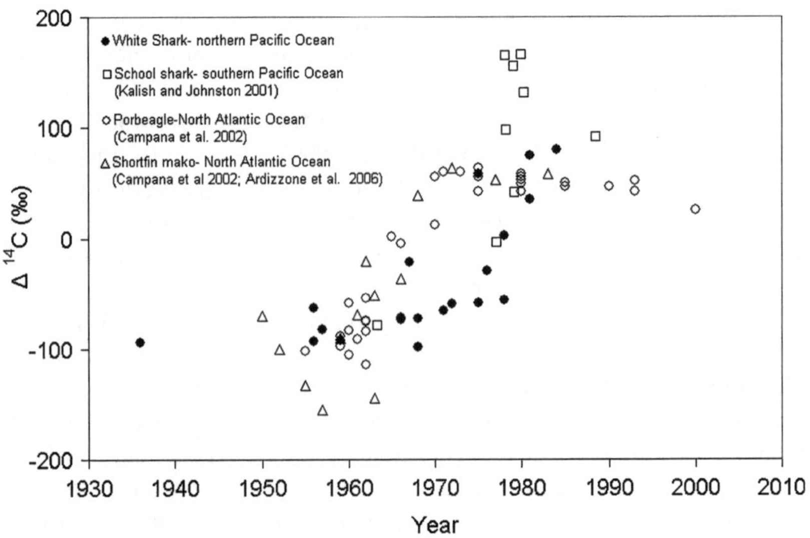

Fig. 4 Radiocarbon ($\Delta^{14}C$) values from white shark, *Carcharodon carcharias* vertebral cores (closed circles) and three vertebral-based shark records, the northwest Atlantic porbeagle (open diamonds, Campana 2002), northwest Atlantic mako shark (solid triangles; Campana 2002; Ardizzone et al. 2006), and the southern Pacific school shark (open squares; Kalish and Johnston 2001)

knowledge of the white shark diet is based on stomach contents from opportunistic nearshore landings and observed feeding events. The main prey items of the white shark include marine cephalopods, crustaceans, ray-finned bony fishes, cartilaginous fishes, mammals, and birds (Klimley 1985; Compagno 2001). In an examination of white sharks captured off the coast of northern and central California, Tricas and McCosker (1984) found stomach contents to be comprised solely of elasmobranchs and teleost fishes that inhabit both pelagic and inshore habitats.

Stable isotope values obtained in this study for white shark vertebrae, when compared with representative prey item values for bony fishes and mammals off the coast of California, indicate that the white shark is an upper trophic level consumer. $\delta^{13}C$ values indicate enrichment relative to fish and most marine mammals and $\delta^{15}N$ values indicate isotopic enrichment relative to fishes and similar values to mammals in the region. Differences in $\delta^{13}C$ and $\delta^{15}N$ values across longitudinal sections of individual vertebra were variable over the lifetime of individuals (Table 4).

Calculated trophic position (TP) for the white shark was greater in value than trophic position calculated for the basking shark *Cetorhinus maximus* = 3.1 (primary prey: zooplankton), blue shark *Prionace glauca* = 3.8 (primary prey: fish), and shortfin mako *Isurus oxyrinchus* = 4.0 (primary prey: fish) and similar to the TP calculated for the common thresher shark *Alopias vulpinus* = 4.5 (primary prey: fish and squid). The white shark had similar variability in TP (TP ranged from 4.11 to 5.04) as calculated for the shortfin mako (TP ranged from 3.6 to 4.5), this was attributed to migration and feeding in both inshore and offshore regions (Estrada et al. 2003). Calculated white shark TP for this study agrees with the determination of trophic level for the family Lamnidae by Cortes (1999) based on diet composition (mean = 4.3, range = 4.22–4.5) and specifically the trophic level determined for the white shark (4.5) based on diet composition. In this analysis the mean TP of marine mammals was estimated at 4.0, placing the trophic level of white sharks somewhat higher than marine mammals (Cortes 1999).

Estrada et al. (2003) contend that TP values obtained for sharks in their study do not support the idea put forth by Fisk et al. (2002), that retention of urea in elasmobranchs may lower $\delta^{15}N$ values and therefore lead to underestimation of trophic position. Based on our limited data, we cannot determine whether urea retention may have affected white shark $\delta^{15}N$ values. However, the similarity of calculated TP in this study to the diet analysis TP supports the idea that urea retention was not a problem.

Values of $\delta^{13}C$ measured in juvenile growth band pairs were significantly enriched compared to adult growth bands; which may reflect variation of ^{13}C due to feeding location rather than trophic level. Isotopic signatures in mobile animals may reflect feeding in different environments or an actual change in diet (Michener and Schell 1994). Gradients of ^{13}C in the northeast Pacific Ocean demonstrate an enriched nearshore signal (benthic-based food web) compared to the offshore, oceanic signal (phytoplankton-based food web) and a ~2‰ enrichment in $\delta^{13}C$ has been associated with marine mammals foraging in nearshore versus offshore waters of California (Burton and Koch 1999). In addition, gradients of ^{13}C with depth have been observed in the northeast Pacific Ocean (off the coast of British Columbia, Canada) where the slope/deep ocean food web was determined to be depleted relative to the pelagic food web (4.3‰ depletion in $\delta^{13}C$ based on measurements of fish larvae; Perry et al. 1999). Therefore, feeding on prey from offshore, deeper waters will reflect a depleted $\delta^{13}C$ signal relative to prey inhabiting nearshore regions.

The trend in $\delta^{13}C$ across individual longitudinal sections of vertebrae tended to decrease with increased age of white shark, thus perhaps reflecting an ontogenetic shift in habitat and foraging location. Evidence suggests that juveniles are residing and feeding in shallow inshore waters and adults spending increased time offshore feeding in deep oceanic waters, incorporating prey reflecting a depleted $\delta^{13}C$ signal relative to nearshore surface waters. However, the decrease observed in $\delta^{13}C$ is a small one (~1.5‰ difference) with respect to potential spatial and temporal variability of $\delta^{13}C$ and cannot be considered as evidence alone for this behavior, but is supported by the tagging literature.

Although an ontogenetic shift in trophic level has been documented in the diet literature and identified through stable isotope analysis of white shark vertebrae from the western North Atlantic (Estrada et al. 2006), we found no significant difference between mean $\delta^{15}N$ or calculated trophic position of juvenile growth and adult growth band pairs (juvenile versus adult classification is based solely on an ontogenetic shift in diet documented at 3 m length). The lack of evidence for an ontogenetic trophic level shift based on $\delta^{15}N$ values may reflect environmental (spatial and temporal) variation in $\delta^{15}N$ (Vander Zanden and Rasmussen 1999) and/or a variable adult white shark diet with lower trophic items remaining an important component in the adult diet (Estrada et al. 2006). Because white shark migratory movements consist of coastal and pelagic periods (Boustany et al. 2002), lower trophic items, such as pelagic fish, could dominate the diet during these pelagic phases. Although most diet studies stress the importance of pinnipeds in adult white shark diets, Tricas and McCosker (1984) documented the greater numerical importance of teleosts and elasmobranchs over marine mammals in white shark diet. If no change in $\delta^{15}N$ is attributable to diet, this finding would support an increased diversity and size spectrum of prey consumed as white shark increase in size, as opposed a strict shift from a fish to a marine mammal diet (Compagno 2001). Alternatively, the time period over which growth bands were sampled from white shark vertebrae coincided with the decline in abundance of pinnipeds in the North Pacific Ocean over the last several decades (Springer et al. 2003). Therefore, the relatively low trophic position observed in adult white sharks, relative to juveniles and marine mammals, during this period may reflect the low abundance of higher trophic level prey (Sora Kim, personal communication).

Conclusions

The uncharacteristic trend in $\Delta^{14}C$ in white shark vertebrae is most likely attributable to the influence of a deep-water depleted carbon source to the vertebrae. Conclusive age validation of vertebral age estimates of white shark was confounded by what may have been some combination of the dietary source of carbon to the vertebrae, large-scale movement patterns, and steep $\Delta^{14}C$ gradients with depth in the eastern North Pacific Ocean. Stable isotopes provided us with time-integrated trophic information, reflecting the upper trophic level status of the white shark. No evidence of an ontogenetic shift in diet was detected, although increased movement into offshore, deeper waters with age was supported by a decrease in $\delta^{13}C$ values.

Future application of this technique is best suited to sharks with well-characterized diet and movement patterns, inhabiting surface waters in an area for which there is knowledge of regional $\Delta^{14}C$ gradients. Sampling strategies should include both known age individuals to construct a reference chronology and older individuals with samples taken from growth bands over the lifetime of the individual. This approach enables indirect confirmation of the stability of the ^{14}C signal within vertebrae. To gain a better understanding of the chemical composition of shark vertebrae, the pathway of elemental uptake, and the metabolic stability of tissue more studies need to be done. Captive rearing studies, feeding small shark species prey with known ^{14}C content and measuring the $\Delta^{14}C$ values in blood, tissue, and vertebrae over time would improve our understanding of carbon pathways and turnover rates. This could be used to provide direct evidence for the assumption of metabolic stability of carbon in shark vertebrae, one of the main assumptions in the application of this technique in elasmobranchs.

Acknowledgments We thank the Los Angeles County Museum, California Academy of Sciences, Sea World San Diego, and Leonard Compagno (Shark Research Center, Iziko-Museums of Cape Town South African Museum, Cape Town, South Africa) for providing vertebral samples. We acknowledge Steve Campana, John Kalish, Sora Kim, Rob Leaf, and Rob Burton for their insights regarding these results and for improving the manuscript. We thank Lisa Natanson for her assistance in white shark vertebrae ageing. This paper was supported in part by the Pacific Shark Research Center and the National Sea Grant College Program of the U.S. Department of

Commerce's National Oceanic and Atmospheric Administration under NOAA Grant #NA06RG0142, project number R/F-190, through the California Sea Grant College Program, and in part by the California State Resources Agency. The views expressed herein do not necessarily reflect the views of any of those organizations. This work was performed, in part, under the auspices of the U.S. Department of Energy by University of California, Lawrence Livermore National Laboratory under Contract No. W-7405-Eng-48.

References

Andrews AH, Burton EJ, Kerr LA, Cailliet GM, Coale KH, Lundstrom CC, Brown TA (2005) Bomb radiocarbon and lead-radium disequilibria in otoliths of bocaccio rockfish (*Sebastes paucispinis*): a determination of age and longevity for a difficult-to-age fish. Mar Freshwater Res 56:517–528

Ardizzone D, Cailliet GM, Natanson LJ, Andrews AH, Kerr LA, Brown TA (2006) Application of Bomb Radiocarbon Chronologies to Shortfin Mako *(Isurus oxyrinchus)* Age Validation. Environ Biol Fishes (this volume)

Bonfil R, Meyer M, Scholl MC, Johnson R, O'Brien S, Oosthuizen H, Swanson S, Kotze D, Paterson M (2005) Transoceanic migration, spatial dynamics, and population linkages of white sharks. Science 310:100–103

Boustany AM, Davis SF, Pyle P, Anderson SD, Le Boef BJ, Block BA (2002) Satellite tagging: expanded niche for white sharks. Nature 415:35–36

Broeker WS, Peng T-H (1982) The anthropogenic invasion: the movement of water through the oceanic thermocline. In: Broeker WS, Peng T-H (eds), Tracers in the sea. Lamont–Doherty Geological Observatory, New York, pp 383–444

Brown TA, Nelson DE, Vogel JS, Southon JR (1988) Improved collagen extraction by modified Longin method. Radiocarbon 30(2):171–177

Burton RK, Koch PL (1999) Isotopic tracking of foraging and long distance migration in northeastern Pacific pinnipeds. Oecologia 119:578–585

Burton RK, Snodgrass JJ, Gifford-Gonzalez D, Guilderson T, Brown T, Koch PL (2001) Holocene changes in the ecology of northern fur seals: insights from stable isotopes and archaeofauna. Oecologia 128:107–115

Cailliet GM (1990) Elasmobranch age determination and verification: an updated review. In: Pratt HL Jr, Gruber SH and Taniuchi T (eds) Elasmobranchs as living resources: advances in the biology, ecology, systematics and the status of the fisheries, U.S. Department of Commerce, NOAA Tech. Rep. NMFS 90, pp 157–165

Cailliet GM, Goldman KJ (2004) Age determination and validation in chondrichthyan fishes. In: Carrier J, Musick JA, Heithaus MR (eds), Biology of sharks and their relatives. CRC Press LLC, Boca Raton, Florida, pp 399–447

Cailliet, GM, Martin LK, Kusher D, Wolf P, Welden BA (1983) Techniques for enhancing vertebral bands in age estimation of California elasmobranchs. In: Prince ED, Pulos LM (eds) Proceedings international workshop on age determination of oceanic pelagic fishes, Tunas, Billfishes, Sharks, NOAA Tech. Rep. NMFS 8, pp 157–165

Cailliet GM, Natanson LJ, Welden BA, Ebert DA (1985) Preliminary studies on the age and growth of the white shark, *Carcharodon carcharias*, using vertebral bands. Memoirs Southern Calif Acad Sci 9:49–60

Campana SE (1999) Chemistry and composition of fish otoliths: pathways, mechanisms and applications. Mar Ecol Progr Ser 188:263–297

Campana SE (2001) Accuracy, precision and quality control in age determination, including a review of the use and abuse of age validation methods. J Fish Biol 59:197–242

Campana SE, Natanson LJ, Myklevoll S (2002) Bomb dating and age determination of large pelagic sharks. Can J Fish Aquat Sci 59:450–455

Chang WYB (1982) A statistical method for evaluating reproducibility of age determination. Can J Fish Aquat Sci 39:1208–1210

Compagno LJV (2001) Sharks of the world. An annotated and illustrated catalogue of shark species known to date, vol 2. Bullhead, mackerel and carpet sharks (Heterodontiformes, Lamniformes and Orectolobiformes). FAO Species Catalogue for Fishery Purposes. No. 1, vol. 2. Rome, FAO, 269 pp

Cortes E (1999) Standardized diet compositions and trophic levels of sharks. ICES J Mar Sci 56:707–717

DeNiro MJ, Epstein S (1981) Influence of diet on the distribution of nitrogen isotopes in animals. Geochimica et Cosmochimica Acta 45:341–351

Dewar H, Domeier M, Nasby-Lucas N (2004) Insights into young of the year white shark, *Carcharodon carcharias*, behavior in the southern California Bight. Environ Biol Fish 70(2):133–143

Druffel ERM, Williams PM (1990) Identification of a deep marine source of particulate organic carbon using bomb ^{14}C. Nature 347:172–174

Ebert TA, Southon JR (2003) Red sea urchins (*Strongylocentus franciscanus*) can live over 100 years: confirmation with A-bomb ^{14}carbon. Fish Bull 101:915–922

Estrada JA, Rice AN, Lutcavage ME, Skomal GB (2003) Predicting trophic position in sharks of the north-west Atlantic Ocean using stable isotope analysis. J Mar Biol Assoc UK 83:1347–1350

Estrada JA, Rice AN, Natanson LJ, Skomal GB (2006) Use of isotopic analysis of vertebrae in reconstructing ontogenetic feeding ecology in white sharks. Ecology 87(4):829–834

Farrell J, Campana SE (1996) Regulation of calcium and strontium deposition on the otoliths of juvenile tilapia, *Oreochromis niloticus*. Comp Biochem Physiol 115A:103–109

Fisk AT, Tittlemier SA, Pranschke JL, Norstrom RJ (2002) Using anthropogenic contaminants and stable isotopes to assess the feeding ecology of greenland sharks. Ecology 83:2162–2172

France RL (1995) Carbon-13 enrichment in benthic compared to planktonic algae: foodweb implications. Mar Ecol Progr Ser 124:307–312

Frantz BR, Foster MS, Riosmena-Rodriguez R (2005) *Clathromorphum nereostratum* (Corallinales, Rhodophyta): the oldest alga? J Phycol 41(4):770–773

Frantz BR, Kashgarian M, Coale KH, Foster MH (2000) Growth rate and potential climate record from a rhodolith using ^{14}C accelerator mass spectrometry. Limnol Oceanogr 45(8):1773–1777

Guilderson TP, Schrag DP, Kashgarian M, Southon J (1998) Radiocarbon variability in the western equatorial Pacific inferred from a high-resolution coral record from Nauru Island. J Geophys Res 103(C11):24641–24650

Jarman WM, Hobson KA, Sydeman WJ, Bacon CE, McLaren EB (1996) Influence of trophic position and feeding location on contaminant levels in the Gulf of the Farallones food web revealed by stable isotope analysis. Environ Sci Technol 30(2):654–660

Kalish JM (1991) ^{13}C and ^{18}O isotopic disequilibria in fish otoliths: metabolic and kinetic effects. Mar Ecol Progr Ser 75:191–203

Kalish JM (1995) Radiocarbon and fish biology. In: Secor DH, Dean JM, Campana SE (eds) Recent developments in fish otolith research. University of South Carolina Press, Columbia, South Carolina, pp 637–653

Kalish JM, DeMartini E (2001) Determination of swordfish (*Xiphias gladius*) age based on analysis of radiocarbon in vertebral carbonate and collagen. In: Kalish JM (ed), Use of the bomb radiocarbon chronometer to validate fish age. Final Report FRDC Project 93/109. Fisheries Research and Development Corporation, Canberra, Australia, pp 340–351

Kalish JM, Johnston J (2001) Determination of school shark age based on analysis of radiocarbon in vertebral collagen. In: Kalish JM (ed), Use of the bomb radiocarbon chronometer to validate fish age. Final Report. FDRC Project 93/109. Fisheries Research and Development Corporation, Canberra, Australia, pp 116–122

Kerr LA, Andrews AH, Frantz BR, Coale KH, Brown TA, Cailliet GM (2004) Radiocarbon in otoliths of yelloweye rockfish (*Sebastes ruberrimus*): a reference time series for the waters of southeast Alaska. Can J Fish Aquat Sci 61:443–451

Kerr LA, Andrews AH, Frantz BR, Coale KH, Brown TA, Munk K, Cailliet GM (2005) Age validation of quillback rockfish (*Sebastes maliger*) using bomb radiocarbon. Fish Bull 103(1):97–107

Klimley AP (1985) The areal distribution and autoecology of the white shark, *Carcharodon carcharias*, off the west coast of North America. Southern Calif Acad Sci 9:15–40

Koch PL, Fogel ML, Tuross N (1994) Tracing the diet of fossil animals using stable isotopes. In: Lajtha K, Michener RH (eds), Stable isotopes in ecology and environmental science. Blackwell Scientific Publications, New York, New York, USA, pp 63–92

Le Boeuf BJ, Riedman ML, Keyes RS (1982) White shark predation on pinnipeds in California coastal waters. Fish Bull 80:891–895

Michener RH, Schell DM (1994) Stable isotope rations as tracers in marine aquatic food webs. In: Michener RH, Schell DM (eds), Stable isotopes in ecology and environmental science. Blackwell Scientific Publications, New York, New York, USA, pp 138–157

Mollet HF, Cliff G, Pratt HL Jr, Stevens JD (2000) Reproductive biology of the female shortfin mako *Isurus oxyrinchus* Rafinesque 1810, with comments on the embryonic development of lamnoids. Fish Bull 98(2):299–318

Pearcy, Stuvier (1983) Vertical transport of carbon-14 into deep-sea food webs. Deep-Sea Res 30(4A):427–440

Perry RI, Thompson PA, Mackas DL, Harrison PJ, Yelland DR (1999) Stable carbon isotopes as pelagic food web tracers in adjacent shelf and slope regions off British Columbia, Canada. Can J Fish Aquat Sci 56:2477–2486

Piner K, Wischniowski SG (2004) Pacific halibut chronology of bomb radiocarbon in otoliths from 1944 to 1981 and a validation of ageing methods. J Fish Biol 64:1060–1071

Post DM (2002) Using stable isotopes to estimate trophic position: models, methods, and assumptions. Ecology 83:703–718

Ridewood WG (1921) On the calcification of the vertebral central in sharks and rays. Philadelphia Trans Roy Soc Lond Ser B 210:311–407

Schoeninger GJ, Deniro GJ (1984) Nitrogen and carbon isotopic composition of bone collagen from marine and terrestrial animals. Geochimica Cosmochimica Acta 48:625–639

Springer AM, Estes JA, van Vliet GB, Williams TM, Doak DF, Danner EM, Forney KA, Pfister B (2003) Sequential megafaunal collapse in the North Pacific Ocean: an ongoing legacy of industrial whaling? Proc Nat Acad Sci 100(21):12223–12228

Stuvier M, Polach HA (1977) Discussion reporting of ^{14}C data. Radiocarbon 19(3):355–363

Toperoff AK (1997) Diet of harbor porpoise (*P. phocoena*) using stomach contents and stable isotope analyses. A thesis presented to the faculty of California State University, San Jose through Moss Landing Marine laboratories, 103 p

Tricas TC, McCosker JE (1984) Predatory behavior of the white shark (*Carcharodon carcharias*), with notes on its biology. Proc Calif Acad Sci 43:221–238

Urist MR (1961) Calcium and phosphorus in the blood and skeleton of the elasmobranchii. Endocrinology 69:778–801

Vander Zanden MJ, Rasmussen JB (1999) Primary consumer $\delta^{15}N$ and $\delta\ ^{13}C$ and the trophic position of aquatic consumers. Ecology 80(4):1395–1404

Vander Zanden MJ, Rasmussen JB (2001) Variation in $\delta\ ^{15}N$ and $\delta\ ^{13}C$ trophic fractionation: implications for aquatic food web studies. Limnol Oceanogr 46(8):2061–2066

Vogel JS, Nelson DE, Southon JR (1987) ^{14}C background levels in an accelerator mass spectrometry system. Radiocarbon 29(3):323–333

Vogel JS, Southon JR, Nelson DE, Brown TA (1984) Performance of catalytically condensed carbon for use in accelerator mass spectrometry. Nuclear Instrument Methods Phys Res B5:289–293

Welden BA, Cailliet GM, Flegal AR (1987) Comparison of radiometric with vertebral band age estimates in four California elasmobranchs. In: Summerfelt RC, Hall GE (eds), Age and growth of fish. Iowa State University Press, Ames, Iowa, pp 301–315

Wintner SP, Cliff G (1999) Age and growth determination of the white shark, *Carcharodon carcharias*, from the east coast of South Africa. Fish Bull 97:153–169

Environ Biol Fish (2006) 77:355–366
DOI 10.1007/s10641-006-9106-4

ORIGINAL PAPER

Application of bomb radiocarbon chronologies to shortfin mako (*Isurus oxyrinchus*) age validation

Daniele Ardizzone · Gregor M. Cailliet · Lisa J. Natanson · Allen H. Andrews · Lisa A. Kerr · Thomas A. Brown

Received: 2 June 2006 / Accepted: 22 June 2006 / Published online: 9 August 2006
© Springer Science+Business Media B.V. 2006

Abstract Age estimation is an issue for the shortfin mako, *Isurus oxyrinchus*, because of disagreement on vertebral band-pair deposition periodicity. In the 1950s–1960s, thermonuclear testing released large amounts of radiocarbon into the atmosphere, which diffused into the ocean through gas exchange. This influx created a time-specific marker that can be used in age validation. Annual band-pair deposition in the porbeagle, *Lamna nasus*, was validated in a previous study and indicated preliminary annual deposition in the shortfin mako, using four samples from one vertebra. In the present study, age estimates from 54 shortfin mako vertebrae collected in 1950–1984 ranged 1–31 years. Ageing error between readers was consistent, with 76% of the estimates ranging within 2 years. Twenty-one $\Delta^{14}C$ values from eight shortfin mako vertebrae (collected in the western North Atlantic in 1963–1984) ranged –154.8‰ to 86.8‰. The resulting conformity with the $\Delta^{14}C$ timeline for the porbeagle supported annual band-pair deposition in vertebrae of the shortfin mako.

Keywords Shark vertebrae · Lamnidae · Band pairs · Growth

D. Ardizzone (✉) · G. M. Cailliet · A. H. Andrews
Moss Landing Marine Laboratories, California State University, 8272 Moss Landing Road, Moss Landing, CA 95039, USA
e-mail: dardizzone@mlml.calstate.edu

L. J. Natanson
Narragansett Laboratory, NOAA Fisheries Northeast Fisheries Science Center, 28 Tarzwell Drive, Narragansett, RI 02882, USA

L. A. Kerr
Chesapeake Biological Laboratory, University of Maryland Center of Environmental Science, 1 Williams Street, P.O. Box 38, Solomons, MD 20688, USA

T. A. Brown
Lawrence Livermore National Laboratory, Center for Accelerator Mass Spectrometry, 7000 East Avenue, Livermore, CA 94550, USA

Introduction

There is an ongoing disagreement regarding the ageing of the shortfin mako due to a difference of interpretation in the periodic deposition of vertebral growth band pairs, especially for the larger size classes. Pratt and Casey (1983), using analysis of length-month information, tagging data, and length-frequency analysis, concluded that two band pairs were formed in the vertebral centrum every year (biannual band-pair interpretation). Cailliet et al. (1983), however, presented growth parameters based on the common assumption that one band pair forms annually (annual band-pair interpretation). Therefore, growth rates obtained by Pratt and Casey (1983) were twice that of Cailliet et al. (1983) and could lead to age discrepancies of about 15 years for maximum estimated ages on the order of 30 from the annual band-pair interpretation.

Serious consequences in the population dynamics could occur for this species if inputs are based on an invalid age interpretation. The latest Fishery Management Plan (FMP) for Highly Migratory Species (HMS), for example, adopted the biannual band-pair deposition hypothesis because it apparently fit the observed growth patterns best (Pacific Fishery Management Council 2003). However, the ongoing uncertainty about the ageing of the shortfin mako was acknowledged and it was recommended that an endeavor to resolve this issue be made.

Since 1983, five additional studies on the age and growth of the shortfin mako have been conducted (Chan 2001; Campana et al. 2002; Hsu 2003; Ribot-Carballal et al. 2005; Bishop et al. 2006). Using Marginal Increment Ratio (MIR), Hsu (2003) indicated the formation of annual translucent bands from July to September in western North Pacific Ocean shortfin makos. Using Marginal Increment Analysis (MIA) Ribot-Carballal et al. (2005) supported the annual band-pair interpretation for 109 shortfin makos collected in the eastern Pacific Ocean. Although the study provided support for annual band-pair deposition, no statistical test was performed and the number of samples for MIA analysis was insufficient for some months. Hence, unequivocal validation of shortfin mako age estimates has yet to be accomplished.

Atmospheric testing of thermonuclear devices in the 1950s and 1960s effectively doubled the natural atmospheric radiocarbon (^{14}C; Druffel and Suess 1983; Broeker et al. 1985). The elevated ^{14}C levels were first recorded in 1957–1958, with a peak around 1963 (Nydal and Lovseth 1983). As a consequence, ^{14}C entered the ocean through gas exchange with the atmosphere at the ocean surface and in terrestrial runoff (Broeker et al. 1985). Despite variable oceanographic conditions, a worldwide rise of the bomb ^{14}C signal entered the ocean mixed layer as dissolved inorganic carbon (DIC) in 1957–1958 (Druffel 1980). The large amounts of ^{14}C released from the bomb tests produced a signature that can be followed through time, throughout the marine food web, and into deeper waters (Pearcy and Stuiver 1983). The marked increase of radiocarbon levels was first measured in the DIC of seawater and in biogenic marine carbonates of hermatypic corals in Florida (Druffel and Linick 1978). Subsequently, this record was documented in corals from other regions (e.g. Nozaki et al. 1978; Druffel 1980; Druffel and Suess 1983; Druffel 1989; Guilderson et al. 1998) and in the thallus of rhodoliths (Frantz et al. 2000, 2005). The accumulation of radiocarbon in the hard parts of most marine organisms in the mixed layer (such as fish otoliths and bivalves) was synchronous with the coral time-series (Kalish 1993; Weidman and Jones 1993).

This technique has been used to validate age estimates and longevity of numerous bony fishes to date, as well as to establish bomb radiocarbon chronologies from different oceans (e.g. Kalish 1993; Campana 1997; Campana and Jones 1998; Piner and Wischniowski 2004; Andrews et al. 2005; Kerr et al. 2005). In the first application of this technique to lamnoid sharks, Campana et al. (2002) validated annual band-pair deposition in vertebral growth bands for the porbeagle, *Lamna nasus*, aged up to 26 years. Radiocarbon values from samples obtained from 15 porbeagle caught in the western North Atlantic Ocean (some of which were known-age) produced a chronology similar in magnitude to the reference carbonate chronology for that region. The observed phase shift of about 3 years was attributed to different sources of carbon between vertebrae and those for otoliths, bivalves and corals. In the same study by Campana et al. (2002), a single vertebra from a shortfin mako caught in 1977 was aged at 21 and 10 years, using the annual versus the biannual deposition hypotheses, respectively. Vertebral samples were extracted from the first, last, and two intermediate bands and were assayed for radiocarbon. The results indicated the ageing interpretation for the vertebra from this fish best fit the timing of the porbeagle time-series by adopting the annual band-pair interpretation.

To provide a more comprehensive basis for valid ageing criteria and a definitive growth function for the shortfin mako, more radiocarbon assays were required (Campana et al. 2002). The goal of our research was to take heed of this suggestion and continue the use of bomb radiocarbon to validate the ageing of the shortfin mako, and specifically to resolve the validity of either annual or biannual band-pair age interpretations.

Materials and methods

Age assessment

Vertebral samples were collected from shortfin makos caught in the western North Atlantic, eastern North Pacific, western South Pacific, and western South Indian Oceans between 1950 and 1984 to obtain sufficient sample size across size classes. These centra were sectioned for age estimation using the growth bands visible in the corpus calcareum, an approach commonly used in shark age estimation. Each vertebral centrum was sectioned along the sagittal plane using a Ray Tech gem saw with two diamond blades separated by a 0.6 mm spacer at the NOAA Fisheries Northeast Fisheries Science Center, Narragansett Laboratory, Rhode Island (NMFS). The resulting sections were placed under water with a black background and digitally photographed using an MTI® CCD 72 video camera mounted on an Olympus® SZX9 stereomicroscope under reflected light. Magnification ranged between 4× and 12× according to the size of the vertebra. For larger vertebrae, an additional photograph at higher magnification was taken to resolve the finer bands toward the edge. The sections were stored in 70% EtOH.

Ages were estimated by counting the number of band pairs in the images using the annual band-pair interpretation. Brightness and contrast of the images was adjusted to enhance the banding pattern of the section using Adobe Photoshop 7.0. Each band pair consisted of one opaque and one translucent band, corresponding to the light-dark pattern observed against a dark background under reflected light.

Vertebrae were independently aged by a second reader at the NMFS Narragansett lab to assess reading precision and for cross-calibration between readers. Age discrepancies were plotted to show the age difference frequency between readers. An age-bias plot was produced to detect bias among the size range and to suggest trends in this type of error. In addition, Average Percent Error (APE), Percent Error (*D*) and Coefficient of Variation (CV) were calculated to measure precision of the readings (Beamish and Fournier 1981; Chang 1982; Campana 2001). The resulting CV was used as a measure of the potential age range in the assigned year of formation for the samples assayed for radiocarbon. Because the uncertainty in assigning the year of formation decreased from the first-formed band to the last-formed band (where the collection date was known), the closer the sample to the band pair corresponding to the year of collection, the smaller the error and vice versa. Hence, the effect of this uncertainty increased as the samples approached the birth year. Individual CV values for each sample were calculated multiplying the difference between the collection year and the year of band-pair formation by CV.

Sample preparation

To address the validity of age interpretations, samples for radiocarbon analysis were extracted from growth bands of specimens collected from the western North Atlantic Ocean with estimated collection years between 1963 and 1984. The sharks were from the same collection as those used by Campana et al. (2002). To avoid the use of vertebrae sectioned for ageing, adjacent vertebrae were sectioned along the sagittal plane to a thickness of 1.3 mm using a Buehler Isomet® 1000 low-speed saw with a diamond blade. Samples corresponding to specific years of formation were extracted from the center of growth bands using a New Wave Research® Micromill with Brasseler® Carbide tips (0.3 mm). The drill bits were rinsed in 10% HCl and MilliQ water between sampling events to prevent cross-contamination. Vertebral sections were mounted for milling using double-sided tape and samples were extracted by drilling a series of holes along the boundary of the two growth bands. The extracted samples were solid pieces composed of one opaque and one translucent band, corresponding to what would be interpreted as one year of growth, based on the annual band-pair interpretation, or half a year's growth, based on the biannual band-pair interpretation. Samples were weighed to the nearest 0.1 mg and placed in glass culture tubes (13 × 100 mm) for processing prior to radiocarbon analysis. For all but one specimen, the first-formed band pair and a series of other growth band pairs ranging to the most recently formed were extracted from the corpus calcareum of each vertebral section. Four of these samples were cored from vertebrae belonging to the same shortfin mako assayed by Campana et al. (2002) to provide a basis for direct comparisons.

Prior to radiocarbon analysis, samples were demineralized to remove any carbon bound into what could be more transient carbonate forms. This step was necessary to attribute all carbon, and measured ^{14}C, to the organic component of the cartilage. To demineralize the sample, culture tubes containing the samples were filled half way with 0.25 N HCl (GFS Chemicals, double distilled acid); acidified samples were capped with foil and placed in a refrigerator for 2 h to slow the reaction (Brown et al. 1988). Hydrochloric acid was drawn off with a pipette from the remaining solid sample, leaving the extracted piece of cartilage at the bottom of the tube; the tubes were re-filled with fresh acid and placed in a refrigerator overnight to complete the dissolution of the inorganic component.

Demineralization was considered complete once the solid material appeared translucent. Samples were rinsed, centrifuged, and transferred into pre-weighed 6 mm quartz tubes. Samples were dried overnight at 95°C and weighed. Dried samples were prepared for combustion by adding the equivalent of 20 times the sample weight in Alpha Aesar® CuO (99.9999%) and 2 mg of Alpha Aesar® Ag (99.999% metals basis). Tubes were evacuated and sealed, placed in an oven at 900°C for 3.5 h and analyzed for $\Delta^{14}C$ by Accelerator Mass Spectrometry (AMS).

Bomb radiocarbon dating

The $\Delta^{14}C$ values from each sample were plotted against the year of band formation based on estimated age and compared to reference time-series obtained from the western North Atlantic for the porbeagle by Campana et al. (2002), the haddock *Melanogrammus aeglefinus* (Campana 1997), and the bivalve *Arctica islandica* (Weidman and Jones 1993). Results from the shortfin mako bomb carbon analysis were expected to produce a sigmoidal curve similar to the three $\Delta^{14}C$ time-series from the western North Atlantic Ocean. A phase lag was defined as a shift in time of onset and increase between chronologies. $\Delta^{14}C$ values were subjectively considered elevated when above 30‰.

To verify the synchrony of the onset and increase of the bomb radiocarbon signal within individual sharks, sequential samples obtained from each specimen were plotted with those from the porbeagle chronology. Differences in timing of the bomb radiocarbon signal observed when the two different ageing interpretations were applied, radiocarbon values from samples of the two longest-living specimens available (life span from pre- to the post-bomb era) were also plotted with the reference time-series. To provide the best analysis using known-age samples for the shortfin mako, samples from the edge of the vertebrae were plotted with the porbeagle reference time-series. The resulting time-series was used to verify the presence of the bomb radiocarbon signal. These values served also as a reference time-series for the shortfin mako because the year of formation was known or had little uncertainty.

Results

Age estimates for the 54 shortfin mako vertebrae ranged from 1 years to 31 years, using the annual band-pair interpretation. Estimated birth years for those samples spanned from 1948 to 1972, covering the period of the bomb radiocarbon increase. If the biannual band-pair interpretation was used, the age range would be reduced to 1–15 years with corresponding birth years between 1957 and 1978. Vertebral growth bands in the early life stages of shortfin mako sharks appeared broad and clear, as shown in transverse section; however, the bands became narrower and less defined, often blending with other bands, as age increased (Fig. 1).

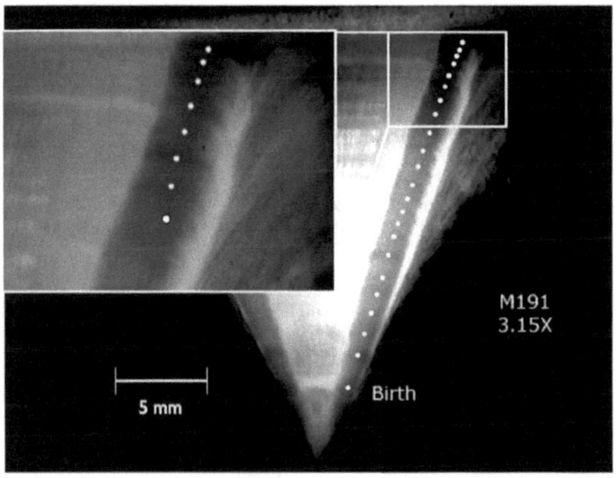

Fig. 1 Transverse section of a vertebra (M191) from a 350-cm (TL) shortfin mako shark collected in 1979 and aged at 22 years. Each dot is centered on the opaque region of a band pair, but represents one whole year of growth. Age 0+ was assigned to the first-formed band (marked as Birth)

Band-pair counts remained consistent between two readers and showed no ageing bias throughout the available size range (Fig. 2). Comparison of age estimates with the second reader yielded an APE value of 11.5% and D value of 7.7%. The CV of 10.8% was used as a measure of the potential age range in the assigned year of formation for the samples assayed for radiocarbon.

The vertebrae of two male and six female shortfin makos, (133–366 cm TL), were analyzed for ^{14}C, producing a total of 21 samples (Table 1). Collection dates ranged from 1963 to 1984. Ages of the selected specimens, adopting the annual band-pair criteria, were estimated at 2–31 years with corresponding birth years of 1948–1972. The error associated with assigned year of band-pair formation for each sample (CV) ranged between 0.1 years and 3.3 years.

For the samples assayed, radiocarbon values for 21 samples ranged between –154.8‰ (±3.2 SD) and 86.8‰ (±4.6) $\Delta^{14}C$ and were estimated to have formed between 1948 and 1983, using the annual band-pair interpretation. The lowest value (–154.8‰ ± 3.2) corresponded to the estimated year 1957; the highest values (86.5‰ ± 4.8 and 86.8‰ ± 4.6) corresponded to the estimated years 1968 and 1978, respectively. Radiocarbon values of shortfin mako vertebrae, relative to their estimated year of band-pair formation, revealed the presence of a bomb radiocarbon signal and followed the general predicted trend (Fig. 3). The resulting chronology was comparable to existing reference time-series of the North Atlantic Ocean. The phase lag of the rise in $\Delta^{14}C$ levels toward more recent dates, however, was most similar to the porbeagle and the increase in started as early as 1960, whereas the first elevated values for the shortfin mako were recorded between 1962 and 1968. The most elevated radiocarbon values for the shortfin mako were 86.5‰ ± 4.8 (1968) and 86.8‰ ± 4.6 (1978), whereas those for the haddock (Campana 1997) and the porbeagle (Campana et al. 2002) were 75‰ (1970) and 63‰ (1975), respectively.

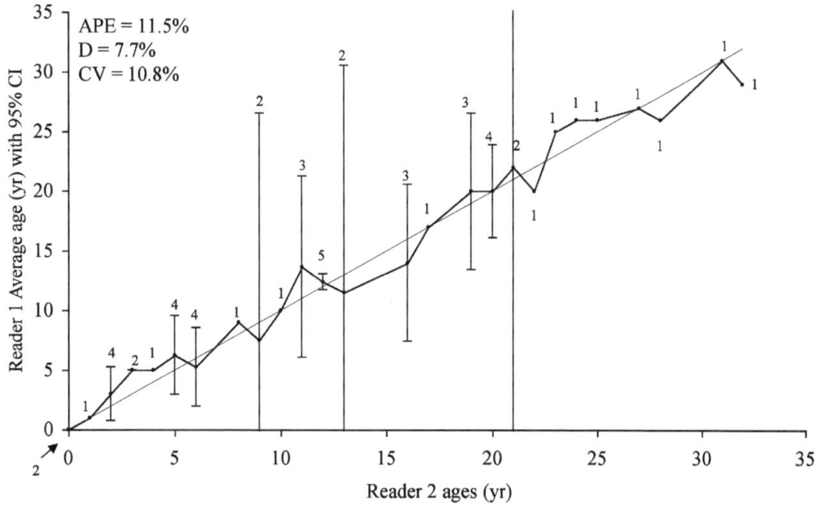

Fig. 2 Age-bias plot showing inter-reader comparison, Average Percent Error (APE), Percent Error (D), and Coefficient of Variation (CV) (n = 53). Vertical error bars represent 95% Confidence Interval (CI). The unusually long bars for years 9, 13 and 21 are due to small sample size of the pair-wise comparisons for those year classes

Table 1 Summary of fish and sample data for vertebral sections that were analyzed for $\Delta^{14}C$

Sample	Sex	TL (cm)	Collection date	Estimated age (years)	Estimated birth date	Estimated year of band-pair formation	Estimated age at band-pair formation	CV of assigned year of band-pair formation[a]	Sample weight (mg)[b]	$\Delta^{14}C$(‰)	±SD	$\partial^{13}C$ (‰)[c]
M001	F	133	1965	2	1963	1963	0+	0.2	17.1	–144.0	3.3	–15
M011	M	232	1963	13	1950	1950	0+	1.4	22.2	–69.8	4.4	–15
						1955	5	0.9	21.1	–132.7	4	–15
						1962	12	0.1	18	–50.6	3.7	–14
M016	M	183	1964	5	1959	1961	2	0.3	20	–69.4	3.6	–15
						1963	4	0.1	19	–51.9	3.6	–15
M151	F	355	1977	21	1956	1956	0+	2.3	20.2	–73.3	4.0	–15
						1956	**0+**			**–100.2**		
						1960	**4**			**–90.1**		
						1965	9	1.3	21.7	–64.2	3.6	–15
						1968	14	0.8	21.3	–45.8	4.6	–15
						1970	**16**			**38.8**		
						1976	20	0.1	20	64.0	5.2	–15
						1977	**21**			**52.6**		
M181	F	240	1979	11	1968	1968	0+	1.2	16	86.5	4.8	–14
						1972	4	0.8	20	35.6	4.1	–14
M186	F	366	1979	31	1948	1948	0+	3.3	16.1	–63.1	3.8	–15
						1955	7	2.6	18	–75.3	3.9	–15
						1966	18	1.4	18	–65.9	4.5	–14
						1978	30	0.1	20	86.8	4.6	–15
M191	F	350	1979	22	1957	1957	0+	2.4	21	–154.8	3.2	–15
						1962	5	1.8	15	–20.8	5.4	–15
						1966	9	1.4	21	–36.9	3.7	–15
M316	F	309	1984	12	1972	1972	0+	1.3	20	62.8	4.7	–15
						1983	11	0.1	20.5	58.0	4.3	–15

Age estimates were based on the annual deposition of one band pair per year. Values in bold represent the four vertebral samples assayed by Campana et al. (2002)

[a] CV = 10.8%

[b] Pre-demineralization

[c] $\partial^{13}C$ values are the assumed values according to Stuiver and Polach (1997)

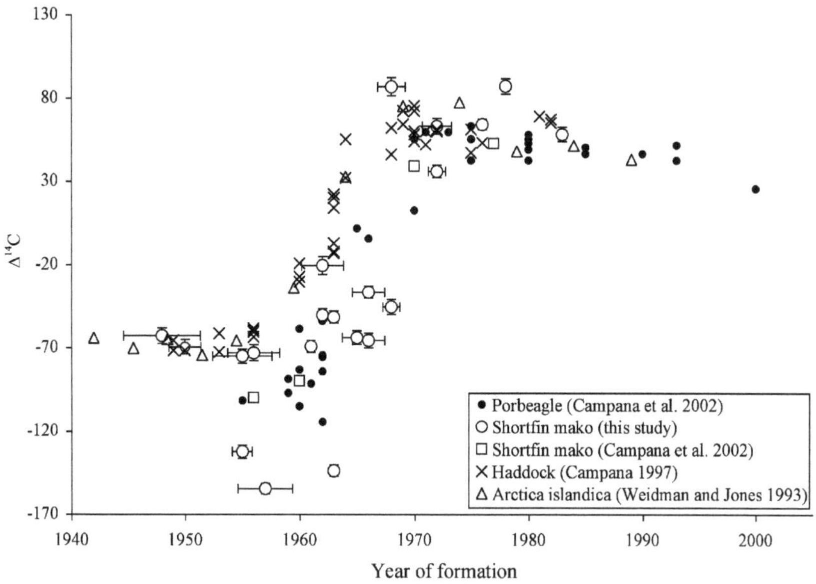

Fig. 3 Values of $\Delta^{14}C$ from individual growth band pairs of shortfin mako vertebrae collected from the western North Atlantic and plotted with the reference time series available from the same region, with the single shortfin mako previously assayed by Campana et al. (2002). The estimated years of formation were based on the annual band-pair interpretation. Horizontal error bars represent uncertainty associated with age estimation from growth bands (CV = 10.8%). Vertical bars represent the 1 SD AMS analytical uncertainty

Some $\Delta^{14}C$ values were lower than other pre-bomb values for the shortfin mako (>50‰) and the reference time-series. Depleted radiocarbon levels were found in the first-formed band pair of two specimens (M001 and M191) and in an intermediate band pair of another (M011). These values ranged between –132.7‰ ± 4 SD and –154.8‰ ± 3.2 SD and were estimated to have formed in 1955 (±0.9 years), 1957 (± 2.4years), and 1963 (±0.2 years).

Within-shark vertebral analysis suggested that radiocarbon levels within individual sharks followed the same trend (Fig. 4). Bomb radiocarbon values obtained from sequential samples of the oldest sharks used in this study (M151 and M186) indicated the sharks began life in the pre-bomb phase, continued to live through the sharp increase in radiocarbon and into the post-bomb phase. All samples from those specimens tended to agree with the expected bomb radiocarbon trend. Some samples showed a phase lag of about 3–5 years relative to the porbeagle chronology for a total lag of 6–8 years relative to more synchronous records. The remainder was in phase with the porbeagle time-series. The most evident shift from the expected trend appeared in four individual samples from three different specimens (M151-1965, M151-1968, M186-1966, and M191-1966) assumed to have formed between 1965 and 1968, during the rapid increase of radiocarbon levels. These samples were collected from growth bands formed at ages estimated between 9 years and 18 years, based on annual deposition.

There was a measurable discrepancy in radiocarbon values for one growth band between the present study and that reported by Campana et al. (2002) (circled samples, Fig. 4). A difference of 26.9‰ was recorded between values from samples from the first-formed band of a specimen (M151) sampled in Campana et al. (2002) and the present study. These samples were similar by location on the vertebral column, location in the centrum and estimated age, but were not replicates. Application of the two age interpretations to the two largest specimens, which were estimated to span throughout the bomb ^{14}C era, revealed two different scenarios (Fig. 5). Using data from M151 and M186, the initial rise and subsequent peak of the bomb radiocarbon levels differed markedly when applying the annual

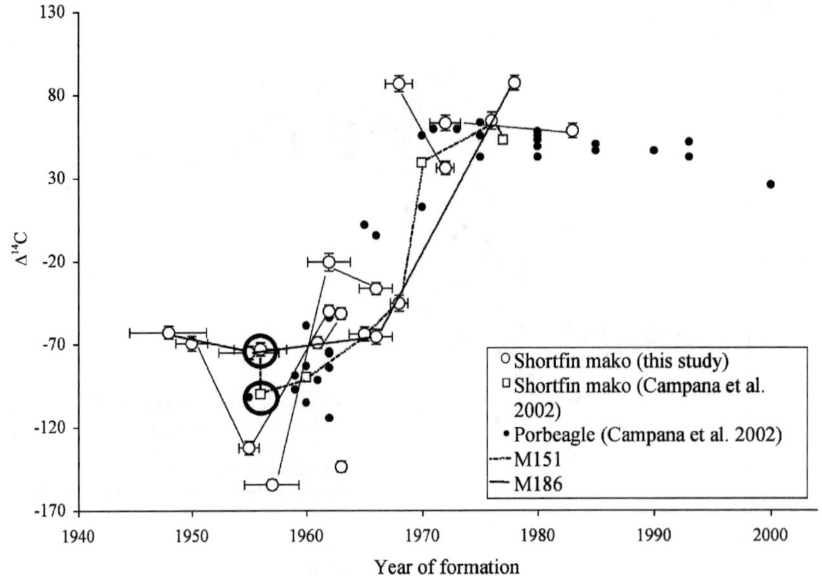

Fig. 4 Radiocarbon values from sequential samples (of vertebral growth bands) in individual sharks. Each line connects growth bands sampled from the same specimen and is plotted with the porbeagle time-series (Campana et al. 2002). The estimated years of formation of the sampled bands used the annual band-pair interpretation. Circles highlight values from the first-formed band pair assayed both in this study and that by Campana et al. (2002) for M151. Horizontal error bars represent uncertainty associated with age determination from growth zones (CV = 10.8%). Vertical bars represent the 1 SD AMS analytical uncertainty

Fig. 5 Radiocarbon values for two individual shortfin makos versus the estimated year of formation as the result of two growth band interpretations. Solid lines connect sequential sampling from specimens aged using the annual band-pair interpretation. Dotted lines connect sequential growth bands of specimens aged using the biannual band-pair interpretation

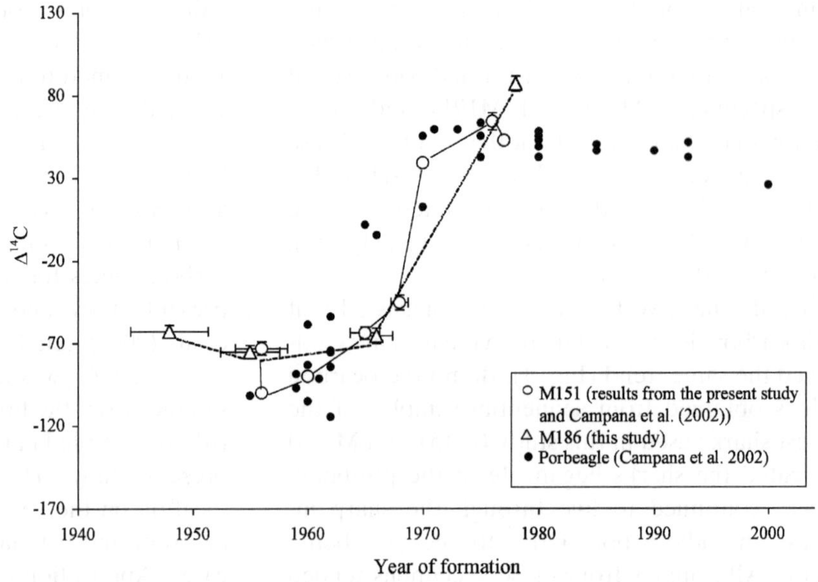

versus biannual band-pair deposition hypotheses. The outcome was an apparent 3–5 years of lag relative to the porbeagle chronology with the annual band-pair interpretation. The phase lag would be 10–13 years relative to the porbeagle for this interpretation to be valid. The biannual band-pair interpretation resulted in time series that indicated the first effects of the bomb radiocarbon signal were not recorded before 1970 and a much sharper rise.

Discussion

Application of the bomb radiocarbon technique to the vertebrae of shortfin mako has demonstrated its utility as an age validation tool for this species. The bomb radiocarbon signal measured in the shortfin mako was temporally similar to the records observed in western North Atlantic Ocean bivalves and fish (Weidman and Jones 1993; Campana 1997), although more comparable to the slightly out-of-phase record determined for the porbeagle (Campana et al. 2002). Records from water samples (Nydal et al. 1998) and corals (Druffel and Linick 1978; Druffel 1980, 1995; Guilderson et al. 1998) show a synchronous rise in radiocarbon levels in the late 1950s to the early 1960s, though the years of peak differ depending on the oceanography and depth of sampling. For fishes, most records have been from teleosts and followed the same trend (Kalish 1993; Campana 1997; Campana and Jones 1998; Kalish et al. 2001; Kerr et al. 2004). For elasmobranchs, one possible exception is with the school shark, *Galeorhinus galeus*, from the western South Pacific Ocean (Kalish and Johnston 2001); the first effects of the bomb radiocarbon signal may have appeared in 1979, more than 15 years after the onset recorded in otoliths from the same region (Kalish 1993). This was attributed to underestimation of age.

The dietary source of carbon for the shortfin mako, combined with slight underageing, could be responsible for the phase shift in the bomb radiocarbon signal and some of the unusually low levels (Campana et al. 2002; Kerr et al. this issue). The source of carbon in elasmobranch non-calcified tissue comes from the diet (Fry 1988; Campana et al. 2002). The incorporation of this carbon in the vertebrae carries the signature of the prey and may be generally older than the surrounding water because of a trophic level delay. This effect may be more evident in larger prey because radiocarbon is carried through more trophic levels. Therefore, large sharks may have incorporated a more delayed signal because they often feed on larger prey (Stillwell and Kohler 1982). The four samples showing more of a phase-lag were taken from growth bands formed at an older age (age 9–18), hence a slight depletion in radiocarbon could explain the discrepancy.

The $\Delta^{14}C$ levels from three samples are among the lowest recorded for the pre-bomb period in the western North Atlantic Ocean. These values are about twice as low as the pre-bomb $\Delta^{14}C$ levels from other regional marine organisms and water samples, which range between –40‰ and –80‰ (Druffel 1989; Campana 1997; Campana and Jones 1998; Nydal et al. 1998; Baker and Wilson 2001; Kalish et al. 2001). Only the porbeagle has pre-bomb values more depleted than corals and fishes in the western North Atlantic Ocean (as low as –114‰; Campana et al. 2002). These depleted levels can be explained based on a diet of prey that live in waters deeper than the mixed layer, depths where the radiocarbon is depleted, relative to surface waters (Broeker et al. 1985; Kalish 2001).

Movement and feeding studies support the notion that an increase in the variability of the source of radiocarbon could partially explain the phase lag observed in some samples. Analyses of feeding habits and tracking studies indicated deep-water squid were a component of the shortfin mako diet in the western North Atlantic Ocean (Carey et al. 1978; Stillwell and Kohler 1982). Tracking studies in the eastern North Pacific Ocean indicated young shortfin mako (up to 1.7–1.8 m TL) frequent the upper 50 m of the water column (corresponding to the 18–20°C range), but make excursions to deeper waters, presumably to feed (Holts and Bedford 1993; Klimley et al. 2002; Sepulveda et al. 2004).

Although dietary effects can explain some of the variation, there is no empirical evidence to rule out some degree of metabolic reworking. Based on ^{210}Pb distributions within vertebrae, Welden et al. (1987) determined that the inorganic component in shark vertebrae may be metabolically unstable. The within-shark $\Delta^{14}C$ signal for the adult shortfin makos that lived through the pre-bomb to post bomb era, however, indicated there is strong evidence that the organic carbon is conserved to a significant extent. Otherwise, no recorded increase with time would be evident and the distribution of pre- and post-bomb values would likely be randomly distributed. The observed variability within this general trend, however, could be due to either minor differences in sample extraction or differences in localized growth and deposition. This notion is supported by the different measured $\Delta^{14}C$ levels in the sample that was similar in vertebral location to one analyzed by Campana et al. (2002) and similar findings were recorded for the white shark (Kerr et al. this issue). However, the samples from Campana et al. (2002) were not demineralized prior to radiocarbon analysis.

The geographic location of an individual shark during a particular period of its life can also introduce further variability in the radiocarbon signal. Follows and Marshall's (1996) model, consistent with the observed patterns, indicated there was an excess of bomb ^{14}C in the western part of the North Atlantic Ocean basin, due to enhanced eddy stirring in the western region. Results of a study on movements of 231 shortfin mako sharks in the western North Atlantic Ocean indicated that 31 (13.4%) sharks recaptured were caught after traveling distances exceeding 1000 nautical miles (Casey and Kohler 1992). One of the sharks tagged off the northeastern United States was recaptured off the coast of Spain, after a straight-line journey of 2452 nautical miles. Because shortfin mako are highly migratory, exposure to prey in different regions might play a role in the variable radiocarbon levels seen here.

The results of this work support the annual band-pair interpretation for western North Atlantic shortfin mako aged up to 31 years, but does not rule out the early age findings for biannual band-pair deposition. The biannual band-pair interpretation could be valid for the first few years of growth, as was indicated by length-frequency and tag/recapture data (Pratt and Casey 1983). However, extrapolation of the early growth pattern to older ages is not supported by the findings of this study.

Acknowledgements This work was supported in part by funds from NOAA/NMFS to the National Shark Research Consortium, in part by the National Sea Grant College Program of the U.S. Department of Commerce's National Oceanic and Atmospheric Administration under NOAA Grant no. NA06RG0142, project number R/F-190, through the California Sea Grant College Program, and in part by the California State Resources Agency. This work was performed, in part, under the auspices of the U.S. Department of Energy by University of California, Lawrence Livermore National Laboratory under Contract W-7405-Eng-48. This research was also funded in part by the Pacific States Marine Fisheries Commission, Earl H. and Ethel M. Myers Oceanographic and Marine Biological Trust, Packard Foundation, PADI Project AWARE, Lerner-Gray Memorial Fund for Marine Research, CSU Fresno Research Merit Award, CSU Fresno Travel Grant. I would also like to thank Steven Campana and the anonymous reviewers for their comments on the manuscript.

References

Andrews AH, Burton EJ, Kerr LA, Cailliet GM, Coale KH, Lundstrom CC, Brown TA (2005) Bomb radiocarbon and lead-radium disequilibria in otoliths of bocaccio rockfish (*Sebastes paucispinis*): a determination of age and longevity for a difficult-to-age fish. Mar Freshwater Res 56:517–528

Baker MS, Wilson CA (2001) Use of bomb radiocarbon to validate otolith section ages of red snapper *Lutjanus campechanus* from the northern Gulf of Mexico. Limnol Oceanogr 46(7):1819–1824

Beamish RJ, Fournier DA (1981) A method for comparing the precision of a set of age determinations. Can J Fish Aquat Sci 38:982–983

Bishop SDH, Francis MP, Duffy C, Montgomery JC (2006) Age, growth, maturity, longevity and natural mortality of the shortfin mako (*Isurus oxyrinchus*) in New Zealand waters. Mar Freshwater Res 57:143–154

Broeker WS, Peng T, Ostlund G, Stuiver M (1985) The distribution of bomb radiocarbon in the ocean. J Geophys Res 90(C4):6953–6970

Brown TA, Nelson DE, Vogel JS, Southon JR (1988) Improved collagen extraction by modified Longin method. Radiocarbon 30(2):171–177

Cailliet GM, Martin LK, Harvey JT, Kusher D, Welden BA (1983) Preliminary studies on the age and growth of blue (*Prionace glauca*), common thresher (*Alopias vulpinus*), and shortfin mako (*Isurus oxyrinchus*) sharks from California waters. pp 179–188. In: Prince ED, Pulos M (eds) Proceedings, International Workshop on Age Determination of Oceanic Pelagic Fishes-Tunas, Billfishes, Sharks. NOAA Technical Report NMFS 8

Campana SE (1997) Use of radiocarbon from nuclear fallout as a dated marker in the otoliths of haddock *Melanogrammus aeglefinus*. Mar Ecol Prog Ser 150:49–56

Campana SE (2001) Accuracy, precision and quality control in age determination, including a review of the use and abuse of age validation methods. J Fish Biol 59:197–242

Campana SE, Jones CM (1998) Radiocarbon from nuclear testing applied to age validation of Black drum *Pogonia chromis*. Fish Bull 96:185–192

Campana SE, Natanson LJ, Myklevoll S (2002) Bomb dating and age determination of large pelagic sharks. Can J Fish Aquat Sci 59:450–455

Carey FG, Middleton L, Stillwell CE, Pratt HL, Kohler NE, Cavin C (1978) Mako shark experiment on Wieczno, March 1978. National Marine Fisheries Service, Northeast Fisheries Center, Woods Hole Laboratory, Document no. 80-7. 12 pp

Casey JG, Kohler NE (1992) Tagging studies on the shortfin mako shark (*Isurus oxyrinchus*) in the Western North Atlantic. Aust J Mar Freshwater Res 43:45–60

Chan RWK (2001) Biological studies on sharks caught off the coast of New South Wales. Ph.D. thesis, University of New South Wales, Sydney, Australia, 323 pp

Chang WYB (1982) A statistical method for evaluating reproducibility of age determination. Can J Fish Aquat Sci 39:1208–1210

Druffel EM (1980) Radiocarbon in annual coral rings of Belize and Florida. Radiocarbon 22(2):363–371

Druffel EM (1989) Decade time scale variability of ventilation in the North Atlantic: high-precision measurements of bomb radiocarbon in banded corals. J Geophys Res 94(C3):3271–3285

Druffel EM, Linick TW (1978) Radiocarbon in annual coral rings of Florida. Geophys Res Lett 5:913–916

Druffel EM (1995) Pacific bomb radiocarbon coral data, IGBP PAGES/World Data Center-A for Paleoclimatology Data Contribution Series # 1995-011. NOAA/NGDC Paleoclimatology Program, Boulder CO, USA

Druffel EM, Suess HE (1983) On the radiocarbon record in banded corals: exchange parameters and net transport of $^{14}CO_2$ between atmosphere and surface ocean. J Geophys Res 88(C2):1271–1280

Follows MJ, Marshall JC (1996) On models of bomb ^{14}C in the North Atlantic. J Geophys Res 101(C10):22577–22582

Frantz BR, Kashgarian M, Coale KH, Foster MS (2000) Growth rate and potential climate record from a rodolith using ^{14}C accelerator mass spectrometry. Limnol Oceanogr 45(8):1773–1777

Frantz BR, Foster MS, Riosmena-Rodriguez R (2005) *Clathromorphum nereostratum* (Corallines, Rhodophita): the oldest alga? J Phycol 41:770–773

Fry B (1988) Food web structure on Georges Bank from stable C, N and S isotopic composition. Limnol Oceanogr 33:1182–1190

Guilderson TP, Schrag DP, Kashgarian M, Southon J (1998) Radiocarbon variability in the western equatorial Pacific inferred from a high-resolution coral record from Nauru Island. J Geophys Res 103(C11):24641–24650

Holts DB, Bedford DW (1993) Horizontal and vertical movements of the shortfin mako shark, *Isurus oxyrinchus*, in the Southern California Bight. Aust J Mar Freshwater Res 44:901–909

Hsu H-H (2003) Age, growth, and reproduction of shortfin mako, Isurus oxyrinchus in the northwestern Pacific. MS thesis, National Taiwan Ocean University, Keelung, Taiwan, 107 pp

Kalish JM (1993) Pre- and post-bomb radiocarbon in fish otoliths. Earth Planet Sci Lett 114:549–554

Kalish JM (2001) Use of the bomb radiocarbon chronometer to validate fish age. Final Report FRDC Project 93/109. Fisheries Research and Development Corporation, Canberra, Australia, 384 pp

Kalish JM, Johnston JM (2001) Determination of school shark age based on analysis of radiocarbon in vertebral collagen. In: Kalish JM (ed) Use of the bomb radiocarbon chronometer to validate fish age. Final Report FRDC Project 93/109. Fisheries Research and Development Corporation, Canberra, Australia, 384 pp

Kalish JM, Nydal R, Nedreaas KH, Burr GS, Eine GL (2001) A time history of pre- & post-bomb radiocarbon in the Barents Sea derived from Arcto-Norwegian cod otoliths. Radiocarbon 43(2B):843–855

Kerr LA, Andrews AH, Frantz BR, Coale KH, Cailliet GM (2004) Radiocarbon in otoliths of the yelloweye rockfish *Sebastes ruberrimus*, a unique chronometer for the waters of southeast Alaska. Can J Fish Aquat Sci 61(3):443–451

Kerr LA, Andrews AH, Munk K, Coale KH, Frantz BR, Cailliet GM, Brown TA (2005) Age validation of quillback (*Sebastes maliger*) using bomb radiocarbon. Fishery Bulletin 103(1):97–107

Kerr LA, Andrews AH, Cailliet GM, Brown TA, Coale KH (This Issue) Investigations of $\Delta^{14}C$ and $\delta^{13}C$ and $\delta^{15}N$ in vertebrae of white shark (*Carcharodon carcharias*) from the eastern North Pacific Ocean. In: Carlson JK, Goldman KJ (eds) Age and growth of chondrichthyan fishes: new methods, techniques, and analyses. Environmental Biology of Fishes

Klimley PA, Beaver SC, Curtis TJ, Jorgensen SJ (2002) Movements and swimming behavior of three species of sharks in La Jolla Canyon, California. Environmental Biology of Fishes 63(2):117–135

Nozaki Y, Rye DM, Turekian KK, Dodge RE (1978) A 200-year record of carbon-13 and carbon-14 variations in Bermuda coral. Geophys Res Lett 5:825–828

Nydal R, Lovseth K (1983) Tracing bomb ^{14}C in the atmosphere 1962–1980. J Geophys Res 88(C6):3621–3642

Nydal R, Brenkert AL, Boden TA (eds) (1998) Carbon-14 Measurements in Surface Water CO_2 from the Atlantic, Indian and Pacific Oceans, 1965–1994. ORNL/CDIAC-104, NDP-057A. Carbon Dioxide Information Analysis Center, Oak Ridge National Laboratory, Oak Ridge, TN, 131 pp

Pacific Fishery Management Council (2003) Fishery Management Plan and Environmental Impact Statement for U. S. West Coast Fisheries for Highly Migratory Species. Pacific Fisheries Management Council, 7700 NE Ambassador Place, Suite 200, Portland, OR

Pearcy WG, Stuiver M (1983) Vertical transport of carbon-14 into deep-sea food webs. Deep-Sea Res 30(4A):427–440

Piner K, Wischniowski SG (2004) Pacific halibut chronology of bomb radiocarbon in otoliths from 1944 to 1981 and a validation of aging methods. J Fish Biol 64:1060–1071

Pratt HL, Casey JG (1983) Age and growth of the shortfin mako *Isurus oxyrinchus*, using four methods. Can J Fish Aquat Sci 40:1944–1957

Ribot-Carballal MC, Galvan-Magana F, Quinonez-Velazquez C (2005) Age and growth of the shortfin mako, *Isurus oxyrinchus*, from the western coast of Baja California Sur, Mexico. Fish Res 76:14–21

Sepulveda CA, Kohin S, Chan C, Vetter R, Graham JB (2004) Movement patterns, depth preferences, and stomach temperatures of free-swimming juvenile mako sharks, *Isurus oxyrinchus*, in the Southern California Bight. Mar Biol 145:191–199

Stillwell CE, Kohler NE (1982) Food, feeding habits, and estimates of daily ration of the shortfin mako (*Isurus oxyrinchus*) in the Northwest Atlantic. Can J Fish Aquat Sci 39:407–414

Stuiver M, Polach HA (1977) Discussion: reporting of ^{14}C data. Radiocarbon 19(3):355–363

Weidman CR, Jones GA (1993) A shell-derived time history of bomb ^{14}C on Georges Bank and its Labrador Sea implications. J Geophys Res 98(C8):14577–14588

Welden BA, Cailliet GM, Flegal AR (1987) Comparison of radiometric with vertebral band age estimates in four California elasmobranchs. In: Summerfelt RC, Hall GE (eds) Age and growth of fish. Iowa State University Press, Ames, Iowa, pp 301–315

Environ Biol Fish (2006) 77:367–383
DOI 10.1007/s10641-006-9127-z

SI- ELASMOBRANCHS

Validated age and growth estimates for the shortfin mako, *Isurus oxyrinchus*, in the North Atlantic Ocean

Lisa J. Natanson · Nancy E. Kohler · Daniele Ardizzone · Gregor M. Cailliet · Sabine P. Wintner · Henry F. Mollet

Received: 12 June 2006 / Accepted: 3 July 2006 / Published online: 29 August 2006
© Springer Science+Business Media B.V. 2006

Abstract Age and growth estimates for the shortfin mako, *Isurus oxyrinchus*, derived from vertebral centra of 258 specimens (118 males, 140 females), ranging in size from 64 to 340 cm fork length (FL) were compared with data from 22 tag–recaptured individuals (74–193 cm FL) and length–frequency data from 1822 individuals (1035 males, 787 females; 65–215 cm FL). Annual band-pair deposition, confirmed by a concurrent bomb radiocarbon validation study, was used as the basis for band interpretation. Validation was further confirmed with a tetracycline-injected male shortfin mako recaptured after being at liberty off South Africa for 1 year and aged at 18 years. Growth rates from tag–recapture analysis (GROTAG) were higher than those derived from vertebral annuli and were only available from sharks up to 193 cm FL at recapture. Modal length–frequency data were used to verify the first four age classes. Growth curves were fit using both von Bertalanffy and Gompertz models. The 3-parameter version of the von Bertalanffy growth function produced the most biologically reasonable values for males, based on observed data (L_∞ = 253 cm FL, K = 0.125 year^{-1} (estimated longevity = 21 year), and L_0 = 72 cm). The 3-parameter version of the Gompertz growth function produced the most biologically reasonable estimates, for females (L_∞ = 366 cm FL, K = 0.087 year^{-1} (estimated longevity = 38 year) and L_0 = 88 cm. Males and females were aged to 29 (260 cm FL) and 32 years (335 cm FL), respectively. Both sexes grew similarly to age 11 (207 cm FL, 212 cm FL for males and females, respectively) when the curve leveled in males and continued to rise in females. Age at 50% maturity was estimated at 8 years for males (185 cm FL) and 18 years for females (275 cm FL). The species grows slower, matures later and has a longer life span than previously reported in North Atlantic waters.

Keywords tag/recapture · length–frequency · mako shark · validation · bomb carbon · tetracycline

L. J. Natanson (✉) · N. E. Kohler
USDOC/NOAA/NMFS, 28 Tarzwell Drive, Narragansett, RI 02882, USA
e-mail: Lisa.Natanson@noaa.gov

D. Ardizzone · G. M. Cailliet · H. F. Mollet
Moss Landing Marine Laboratories, 8272 Moss Landing Road, Moss Landing, CA 95039-9647, USA

S. P. Wintner
Natal Sharks Board, Private Bag 2, Umhlanga Rocks 4320, South Africa

Introduction

Age and growth studies of lamnoid sharks have often been confounded by debate over the periodicity of band-pair formation. Parker and Stott (1965) first suggested that two growth band pairs formed each year (biannual band-pair deposition) in their study of the basking shark, *Cetorhinus maximus*. Pratt and Casey (1983) assumed the same band deposition pattern for the shortfin mako based on consistency with length–frequency and tag/recapture data. Branstetter and Musick (1994) also suggested biannual band-pair deposition for the sand tiger shark, *Carcharias taurus*, based on marginal increment analysis (MIA) and examination of aquarium-reared sharks. Additionally, Chen et al. (1990) proposed biannual band-pair periodicity for the scalloped hammerhead, *Sphyrna lewini*, which was further extended by Anislado-Tolentino and Robinson-Mendoza (2001).

For each species where biannual periodicity has been proposed, an alternate study was performed where annual periodicity was assumed or biannual periodicity refuted. Wintner (unpublished data)[1] is re-aging the basking shark assuming annual periodicity. Goldman et al. (this volume) re-evaluated the age of the sand tiger shark. Using relative MIA and results from two oxytetracycline (OTC)-injected captive sharks they determined that the band pairs are deposited annually. Schwartz (1983) used marginal increments to support annual periodicity in the scalloped hammerhead and Branstetter's (1987) marginal increment data on this species, though limited, also supported annual band-pair deposition. Additionally, while Cailliet et al. (1985) assumed annual periodicity for the white shark, *Carcharodon carcharias* in the Pacific, Wintner and Cliff (1999) could not conclusively validate band periodicity using MIA in this species off the coast of South Africa, but one OTC-injected recapture suggested annual deposition.

Data from Natanson (2001) indicate that Pratt and Casey (1983) overestimated the number of band pairs on vertebral centra of small sharks. Since this was the portion of the growth curve used for comparison with the other methods, the decision to divide their counts in half to make the vertebral growth coincide with the other methods was unfounded and resulted in vastly overestimating the growth rate in larger fish. Bomb radiocarbon analysis was used to examine the periodicity of the bands on the shortfin mako from the North Atlantic Ocean (Ardizzone et al. this volume; Campana et al. 2002). These data clearly demonstrate that a single band pair per year is formed in sharks with between two and 31 band pairs and that the two band-pair per-year-hypothesis for this species is incorrect.

Annual band periodicity has been validated for the porbeagle, *Lamna nasus*, using OTC, known-age fish and bomb radiocarbon dating (Campana et al. 2002; Natanson et al. 2002). Recently, four studies have been conducted on the age of the shortfin mako in the Pacific, Bishop (2004) assumed annual periodicity based on Cailliet et al. (1983b) and Campana et al. (2002). Using whole vertebrae, annual band-pair deposition was supported by MIA for shortfin makos off Mexico up to approximately 7 years of age (Ribot-Carballal et al. 2005), and in the northwestern Pacific by Hsu (2003), though the results of this latter study were not statistically evaluated. Chan (2001) assumed biannual deposition following Pratt and Casey (1983). Three of the four studies showed distinct differences in growth between males and females (Bishop 2004; Chan 2001; Hsu 2003); the one study that did not (Ribot-Carballal et al. 2005), had very few samples from animals that were larger than the size at male maturity, the point at which the growth begins to diverge between the sexes.

In view of these recent validation studies that contradict the biannual band-pair deposition hypothesis, and the necessity of age information for management, a revision of the age estimates for the shortfin mako from the North Atlantic Ocean using updated techniques and increased sample sizes was undertaken.

Materials and methods

Vertebrae were obtained from shortfin makos caught on research cruises, commercial and recreational fishing vessels, and at sport fishing tournaments between 1962 and 2004. Primary sampling took place between Cape Hatteras, N.C.

[1] Wintner SP (2005 unpublished data) Natal Sharks Board, Private Bag 2, Umhlanga Rocks, 4320 South Africa

and the Gulf of Maine (NE coast of the US), although sampling extended south into the Gulf of Mexico. Vertebrae between the number 15 and 20 were excised from each specimen, except on fish that were commercially dressed, in which samples were obtained closer to the head. The vertebrae were trimmed of excess tissue and stored either frozen or preserved in 10% buffered Formalin or 70% ethanol (ETOH). To determine if the number of growth bands differed along the vertebral column, whole columns from four animals (166–208 cm FL) were removed.

Samples that had measured fork length (FL—tip of the snout to the fork in the tail, over the body-OTB), straight line FL (FL_{SL}), total length (TL—tip of the snout to a point on the horizontal axis intersecting a perpendicular line extending downward from the tip of the upper caudal lobe to form a right angle—OTB), or total weight (WT) were used. All lengths reported are in OTB FL unless otherwise noted. Conversions used in this study were:

$$FL = 0.9286\ (TL) - 1.7101\, N = 199 r^2 = 0.99\ (\text{Kohler et al. 1996}) \tag{1}$$

$$WT = 5.2432 * 10^{-6} * FL^{3.1407}\, N = 2081 r^2 = 0.96\ (\text{Kohler et al. 1996}) \tag{2}$$

$$OTBFL = 1.03\ (FL_{SL}) - 0.79\, N = 30 r^2 = 0.99\ (\text{Bishop et al. 2006}) \tag{3}$$

in which weights and lengths are expressed in metric units (kg and cm).

One vertebra from each sample and every fifth vertebrae from the whole columns were removed for processing. The centra were sectioned using a Ray Tech Gem Saw[2] with two diamond blades separated by a 0.6 mm spacer. Each centrum was sectioned through the middle along the sagittal plane, and the resulting bow-tie sections were stored in individual capsules in 70% ETOH. Each section was digitally photographed with an MTI CCD 72 video camera attached to a SZX9 Olympus stereomicroscope using reflected light. Band pairs (consisting of one opaque and one translucent band) were counted and measured on the images using Image Pro 4 software. Measurements were made from the midpoint of the notochordal remnant of the full bow-tie to the opaque growth bands at points along the internal corpus calcareum. The radius of each centrum (VR) was measured from the midpoint of the notochordal remnant to the distal margin of the intermedialia along the same diagonal as the band measurements.

The relationship between FL and VR was used to determine if the vertebrae grew relative to FL and were therefore suitable as an aging structure. Potential differences in vertebral growth between the sexes were tested using the linear interaction model of Neter and Wasserman (1974) to test for statistically significant differences between the sexes.

Vertebral centrum interpretation

Entire vertebral columns were collected from sharks of various lengths to examine the band-pair counts along the column. Band-pair count was plotted against location along the vertebral column for every fifth vertebrae to determine if the counts changed based on location along the vertebral column. Presuming the counts remained the same any vertebrae obtained could then be used for aging.

A band-pair consisted of one opaque and one translucent band. The criteria for designating a band pair were based on broad opaque and translucent bands, each of which was composed of layers of distinct thinner rings (sensu Cailliet et al. 1983a; Martin and Cailliet 1988). A solid broad opaque band through the intermedialia and continuing to the corpus calcareum as a translucent band constituted a growth band.

Twenty-one vertebrae from this study had previously been used in bomb carbon age validation studies (Campana et al. 2002; Ardizzone et al. this volume). Thus, these samples were considered to be of known age and the criteria for the band pairs were based on the bands identified in sections from these samples.

Validation of annual band periodicity was also obtained through recapture of an OTC-injected

[2] Reference to trade names does not imply endorsement by the National Marine Fisheries Service.

and tagged individual. Shortfin makos of various lengths have been tagged and injected with 25 mg OTC/kg of OTC in both the North Atlantic Ocean and off the east coast of South Africa. A returned vertebra from a recaptured shark from South Africa was examined with reflected UV light for the OTC mark. The number of band pairs distal to the mark was then compared with the number of years at liberty.

The first opaque band distal to the focus was defined as the birth band (BB). A slight angle change in the corpus calcareum coincided with this band. Additionally, the identity of the birth band was confirmed with back-calculation and comparison of the birth band with the vertebral radius from young of the year (YOY).

Data analysis

To ensure that vertebral counts were consistent with those of researchers aging this species in other regions, a three-laboratory intercalibration study was done among researchers at the NMFS Narragansett, RI Laboratory, Moss Landing Marine Laboratories (MLML), Moss Landing, CA and The University of Auckland (UA), Auckland, New Zealand. Digital images of 53 vertebrae were exchanged with MLML and 50 with UA; criteria were discussed and readers counted the bands without prior knowledge the FL of the samples. All counts were made using digital images although the actual samples were available if necessary. Aging bias and precision of bands counts were examined using age-bias plots and the coefficient of variation (Campana et al. 1995).

Once the criteria for the bands were determined using the intercalibration, the first author counted the entire sample twice. Pairwise comparisons of precision and bias were conducted on the two counts. Samples that still did not agree were recounted and then sent to the other laboratories for confirmation of the band number, if a consensus was not reached at that time the sample was discarded.

Von Bertalanffy growth functions (VBGF) were fit to length-at-age data using the original equation of von Bertalanffy (1938) with size at birth L_0 rather than t_0:

$$L(t) = L_\infty - (L_\infty - L_0)\exp^{-kt} \quad (4)$$

where $L(t)$ = predicted length at time t;
L_∞ = mean asymptotic fork length;
k = a rate parameter (year^{-1}); and
L_0 = fork length at birth.

Three variations of the model were used: 3-parameter calculation estimated L_∞, k and L_0, 2-parameter method estimated L_∞ and k and incorporated a set L_0 = 70 cm FL (H. Mollet unpublished data[3]), and a 1-parameter method estimated k with observed values for both L_0 = 70 cm and L_∞ = 338.85 and 267.7 cm females and males, respectively. L_∞ was estimated by taking the mean FL of the three largest specimens from each sex in our sample. These values were considered justified because our sample included all of the largest confirmed measured shortfin makos measured in the sampling area in the past 40 years (N. Kohler unpublished data[4]).

As an alternative to the VBGF analysis, we also used the Gompertz growth function (GGF) as described in Ricker (1979):

$$L(t) = L_0 e^{G(1-e^{(-kt)})} \quad (5)$$

where:
$L_\infty = L_0 e^G$ is the mean maximum FL ($t = \infty$);
k (=g in Ricker 1979) is a rate constant (year^{-1}), and
L_0 = fork length at birth.

Three variations of this model were also fit to the data as above with either unconstrained parameters and with certain set parameters using the same values as with the VBGF.

All of the growth equations were fit to the length and vertebral band count data using non-linear regression in Statgraphics (Manuguistics)®.[2] Counts of vertebral band pairs were adjusted for the date of capture assuming a theoretical birthday of 1 March based on the beginning of the estimated period of parturition from Mollet et al. (2000). Thus a specimen with three complete bands caught

[3] Henry F. Mollet Moss Landing Marine Laboratories, 8272 Moss Landing Road, Moss Landing, CA 95039-9647

[4] Nancy E. Kohler National Marine Fisheries Service, 28 Tarzwell Dr., Narragansett RI 02879

6 months after 1 March would be assigned an age of 3.5 years.

Variable band-pair deposition

Additional growth curves were generated using the assumption that individuals less than or equal to 160 cm FL deposited two band pairs per year and greater than 160 cm FL deposited one band pair per year. Data were sorted based on size and an average band count was calculated for sharks between 150 and 160 cm FL, which was 5 for males and 6 for females. For specimens less than or equal to 160 cm FL, the band count was divided by two to obtain an age estimate. In sharks greater than 160 cm FL, the average band count calculated for the 150–160 cm grouping was subtracted from the total count and the average age of the 150–160 grouping was added to the band count. Thus a 241 cm FL male with a band count of 15 would have an age of 12.5. These data were then used to calculate growth curves using the methods detailed above.

Length–frequency analysis

Length–frequency data were obtained from sharks caught by commercial and recreational fishermen and by biologists operating along the US Atlantic Coast primarily between the Gulf of Maine and Florida Keys between 1961 and 2004. The data set was examined in two ways: combined for all years; and for 20 year groupings based on approximate 20 year generation times: 1961–1980 and 1981–2004, to determine if there were changes over time. Monthly length–frequency histograms separated by sex were initially produced in addition to one with sexes combined. Histograms were plotted in 5-cm intervals using the above three scenarios. The months of June, July and August had the largest samples and were used as comparison to follow the first four age classes.

Tagging data

From 1963 through 2003, members of the NMFS Cooperative Shark Tagging Program tagged 6334 and recaptured 757 shortfin makos. Only those sharks reliably measured by biologists or fishermen trained by NMFS biologists at both tagging and recapture were used in the analyses.

The Gulland and Holt (1959), Fabens (1965), and Francis (1988a) models were used to generate VBGFs from the tag–recapture data. The Gulland and Holt (1959) model uses graphical interpretation of the recapture data to produce estimates of L_∞ and k. Specifically, annualized growth rate (cm/year) was plotted against average FL (cm) between tagging and recapture to calculate linear regression coefficients.

$$\mathrm{d[FL]}(t)/\mathrm{d}t = a + b[\mathrm{FL}] = k[\mathrm{FL}]_\infty - k[\mathrm{FL}] \quad (6)$$

is the Gulland and Holt (1959) equation where: $\mathrm{d[FL]}(t)/\mathrm{d}t$ is the first derivative of FL as a function of time, i.e. annualized growth rate, $b = -k$ is the slope, $a = k[\mathrm{FL}]_\infty$ is the growth rate at size 0 (y-axis intercept), and $[\mathrm{FL}]_\infty = -a/b$ (x-axis intercept) = a/k is the mean maximum size.

$$[\mathrm{FL}]_{\mathrm{END}} = [\mathrm{FL}]_{\mathrm{INI}} + ([\mathrm{FL}]_\infty - [\mathrm{FL}]_{\mathrm{INI}})(1 - e^{-kT}) \quad (7)$$

is the Fabens (1965) equation where: T is the time between two consecutive measurements $[\mathrm{FL}]_{\mathrm{INI}}$ and $[\mathrm{FL}]_{\mathrm{END}}$, and the parameters $[\mathrm{FL}]_\infty$ and k are the same as in the Gulland and Holt (1959) equation. Only sharks at liberty for at least 0.9 years were included in these analyses.

The Francis (1988a) model (GROTAG) uses maximum likelihood techniques to estimate growth parameters and variability from tagging data. A coefficient of variation of growth variability (v), measurement errors (m and s) and outlier contamination (p) are estimated as well as growth rates at two user selected lengths (α and β). The reference lengths, α and β, were chosen to lie within the range of tagged individuals. The form of the von Bertalanffy equation becomes:

$$\Delta L = \left[\beta g_\alpha - \alpha g_\beta g_\alpha - g_\beta - L_1\right] \left[1 - \left(1 + g_\alpha - g_\beta \alpha - \beta\right)^{\Delta T}\right] \quad (8)$$

The simplest model, a linear fit with minimal parameters (α and s) was used initially with additional parameters added to successively increase the model complexity. Significant improvement in the model results was determined using log likelihood ratio tests as per Francis (1988a). Bootstrapping was used to calculate the 95% confidence intervals for the final parameter estimates. The modeling and bootstrapping were carried out using the Solver add-in in Microsoft Excel® (C. Simpfendorfer personal communication[5]).

Longevity

Three methods were used to estimate longevity. The oldest fish aged from the vertebral method provides an initial estimate of longevity, however, this value is likely to be underestimated in a fished population due to a decrease in the largest sizes. Taylor (1958) defined the life span of a teleost species as the time required to attain 95% of L_∞. The estimated age at 95% of L_∞ (= longevity in years) was calculated by solving the VBGF and Gompertz growth functions for t and replacing $L_{(t)}$ with 0.95 L_∞. For the VBGF we obtained:

$$\text{Longevity} = (1/k)\ln\left[\frac{(L_\infty - L_0)}{L_\infty(1-x)}\right] \qquad (9)$$

and for the Gompertz growth curves we obtained:

$$\text{Longevity} = (1/k)\ln\left[\frac{\ln(L_0/L_\infty)}{\ln(x)}\right] \qquad (10)$$

with $x = L(t)/L_\infty = 0.95$.

Data from recaptured shortfin makos at liberty for the longest time period in the NMFS Cooperative Shark Tagging Program were tabulated. Ages at tagging were assigned to these fish based on size at tagging, time at liberty was then added to estimate longevity.

Results

Vertebral samples from 290 shortfin makos (64–340 cm) were processed. Additionally, vertebrae from three mid-term embryos ranging in size from 42.5 to 44.7 cm were processed. Fish with vertebrae taken from the head region, with OTB FL calculated from other measurements, or of unknown sex were not included in the FL/VR analysis reducing the sample size to 236 (108 males, 128 females).

Vertebral centrum interpretation

The FL-VR data was log-transformed and fit with a linear regression (Fig. 1). There was no significant difference between the sexes for intercept (P = 0.075) or slope (P = 0.051). Therefore, we calculated the power regression for sexes combined. The back-transformed power regression was:

$$FL = 18.53\,VR^{0.8927}\; r^2 = 0.983, n = 236.$$

Examination of the four whole columns revealed that although bands were more difficult to count on smaller vertebrae, counts between vertebrae along the column never differed by more than one. This indicated that vertebrae collected from all regions along the column could be used for counts, though not for band measurements.

Shortfin mako vertebrae did not show consistent pre-birth marks; thus, the first distinct opaque band was defined as the birth band. The

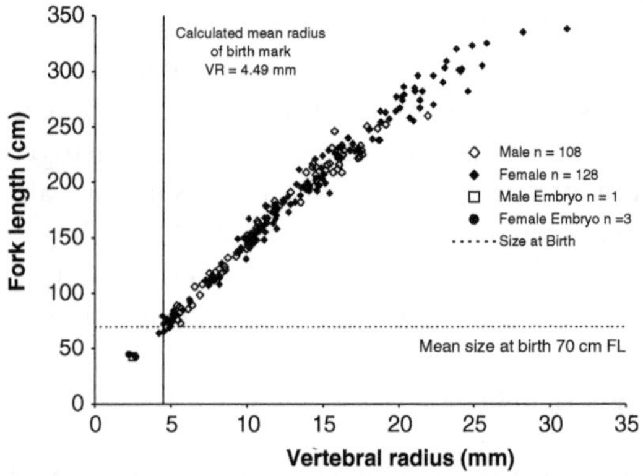

Fig. 1 Relationship between vertebral radius and fork length for male and female shortfin makos. The horizontal dotted line represents the size at birth and the vertical dotted line represents the mean radius of the birth mark

[5] Colin Simpfendorfer. Mote Marine Laboratory, 1600 City Island Park, Sarasota, FL 33577

location of the birth band (BB) coincided with a slight change in the angle of the corpus calcareum (Fig. 2). The mean BB value of the total sample (mean BB ± 95% CI = 4.49 mm ± 0.04 mm; N = 250) was lower than the mean vertebral radius (VR) of 16 YOY (64–79.5 cm; mean VR ± 95% CI = 4.89 mm ± 0.19 mm) and higher than the mean VR of three mid-term embryos (42.5–44.7 cm; mean VR ± 95% CI = 2.52 mm ± 0.32 mm) (Fig. 1). The location of the BB between the VR of the mid-term embryos and YOY indicates the birth band was identified correctly.

Vertebrae from one OTC-injected shark were returned after 1.04 years at liberty. This shark came from the east coast of South Africa; the estimated size at tagging (212 cm pre-caudal length) was the same as the measured size at recapture (240.7 cm FL OTB) thus the shark showed no growth. The vertebra from this specimen had a distinct OTC mark and one full band pair (the expected number of growth bands after 1 year at liberty) had been deposited between the time of tagging and recapture (Fig. 3). This mature male was aged at 18 years at recapture, which confirms annulus formation at this size and agrees with estimates from bomb radiocarbon dating.

Data analysis

Comparison of counts between readers at the different laboratories indicated that all readers were identifying the same bands. The coefficient of variation between NMFS and MLML (n = 53) varied about a mean of 10.8%, while those between NMFS and the University of Auckland (n = 50) varied about 9.0%. Age bias plots generated for both studies showed variation around the 1:1 plot but no systematic bias.

Comparison of the first and second counts of the first author also indicated no systematic bias (Fig. 4). The individual coefficients of variation fluctuated around the mean at 3.9% and the APE and D were 2.8%. This level of precision was much lower than for the porbeagle (15%) (Natanson et al. 2002) and was thus considered acceptable; and the second count was used for those counts that differed by one band (64%). A third count was completed on those that differed by two or more bands, if the third count did not agree with either of the first two, the vertebrae were sent to the other labs for confirmation and discarded if no consensus was reached (3%). Quality control was maintained by periodically recounting earlier samples and cross-checking the readings. Eighteen additional samples were discarded due to questionable length

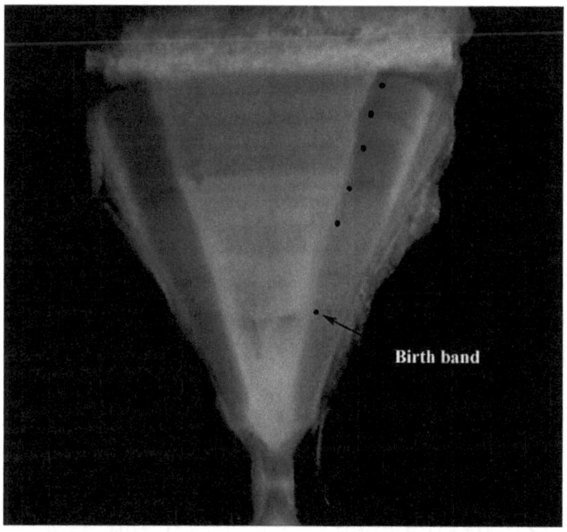

Fig. 2 Photograph of a vertebral section from a 151 cm fork length (FL) female shortfin mako estimated to be 5 years old. The birth band (BB) is indicated and band pairs are marked with a dark circle. Vertebral radius = 10.2 mm

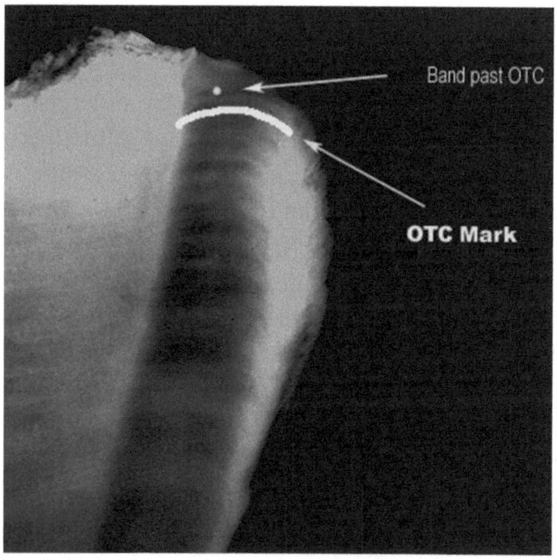

Fig. 3 Vertebra from an oxytetracycline (OTC) injected 240.7 cm 18-year-old male shortfin mako showing location of the OTC mark. Magnification = 8×

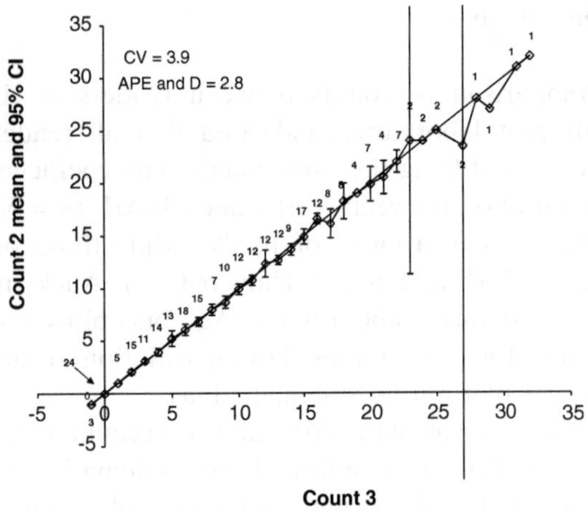

Fig. 4 Age bias graph for pair-wise comparison of 290 shortfin mako vertebral counts from two independent age readings by the first author. Each error bar represents the 95% confidence interval for the mean age assigned in reading 2 to all fish assigned a given age in reading 1. The one to one equivalence line is also presented

data or poor sectioning, the remaining 264 (118 male, 140 female and 6 of unknown sex) were used for the age analyses.

All the growth curves fit the data over the observed size ranges (Fig. 5a, b). However, comparison of the estimates from the VBGF and GGF parameters showed that the VBGF 3-parameter function provided the most reasonable estimates of maximum size and size at birth for males, while the GGF 3-parameter provided the most reasonable estimates of the parameters for females based on observed values of size at birth and maximum size (Table 1, Fig. 6). For females, estimates of maximum size and size at birth from the VBGF were substantially higher than observed values. Additionally, the low K estimates resulted in high longevity values, while the GGF predicted lower estimates of maximum size and longevity, closer to observed values. Size at birth and maximum size estimated by the 3-parameter GGF, while slightly high, were more reasonable than those estimated using the other methods. Additionally, longevity estimated using the 3-parameter GGF was consistent with the actual aged values. The largest female in our sample (366.2 cm FL) was estimated to be 20 years, which is clearly an underestimation. In addition, the shark was not measured by the authors and the length could not be confirmed, thus this datapoint was not included in the calculations (Fig. 6).

In contrast, the growth for males was better represented by the VBGF, as the GGF predicted lower values for maximum size and longevity and a higher size at birth (Table 1, Fig. 6). The L_∞ estimated for the males by all of the methods was lower than the largest known male, thus underestimating longevity. The 3-parameter VBGF fit the male data well with the exception of the largest male shark (289 cm) collected from the Gulf of Mexico (Fig. 6). The age estimate for this shark seems to fit better within data from the females (Fig. 6). This fish was not measured by us and the length is unconfirmed, thus it was not included in calculations. The data from the OTC-injected male shark fit well on the growth curve.

Length-at-age data indicate that males and females grow at similar rates until approximately 11 years (207–212 cm). After this point, which is close to the length-at-maturity of males, there is a dramatic difference between the sexes, as indicated by the lack of overlap in observed size at age data (Fig. 6). Males growth starts to level out after the size at 50% maturity (185 cm FL), whereas growth of females continues to rise. Subsequent comparisons are for the sexes separate.

Variable band-pair deposition

Growth functions for the variable band deposition scenario produced lower values of L_∞ and higher values of K than those for annual band deposition alone, showing the faster initial growth as would be expected assuming two bands per year during a portion of the lifetime (Table 1). Corresponding estimates of longevity, and size at maturity, for both sexes, were lower than those derived from annual band deposition, yet higher than the Pratt and Casey (1983) values.

Length–frequency

Data from 3374 shortfin makos (1575 female, 1799 male) obtained between 1961 and 2004 were examined for length–frequency modes (Fig. 7). No visual differences were observed in the modes between the two 20-year time periods thus all the data were combined. Though the first few modes

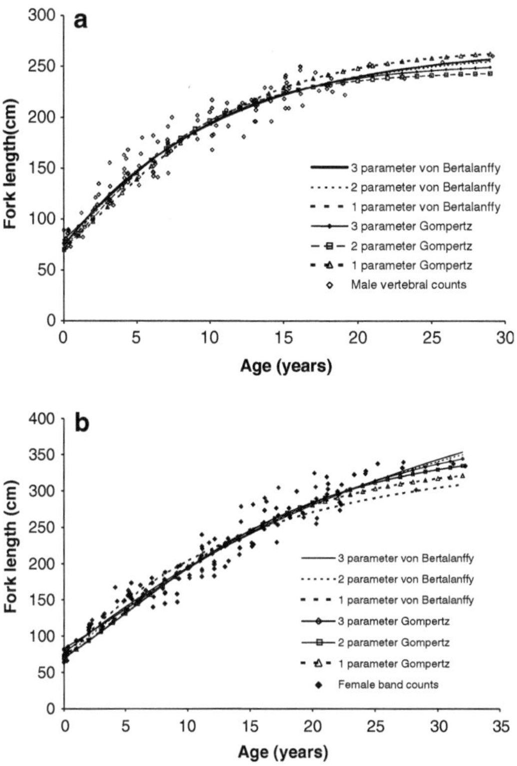

Fig. 5 Comparison of von Bertalanffy and Gompertz growth curves generated using 3, 2, and 1 parameter estimation for each model for the shortfin mako. (**a**) male, (**b**) female

in the summer months were quite clear, the later modes became obscured and it was decided to use only juveniles (up to 215 cm, higher than male maturity of 185 cm) so that the entire last mode of immature males was included in the final plots. Since juvenile growth in this species is similar, the sexes were combined. The majority of the samples were collected in June (n = 1822), and four distinct modes are visible (Fig. 7). Additionally a small mode is visible at 100 cm, but it blends into the next mode by August and becomes indistinct; thus it was not considered to be a true mode. Based on the modal analysis sharks grow 40 cm in the first year, 35 cm in the second and 30 cm in the third year.

Tagging data

A total of 22 shortfin mako was recaptured with sufficient information for tag/recapture analysis. Time at liberty ranged from 0.08 to 2.56 years and size at tagging ranged from 74 to 137 cm. Data from 14 sharks at liberty >0.9 years were used for Gulland and Holt's (1959) and Fabens' (1965) methods whereas all individuals were used for GROTAG (Francis 1988a).

None of the tag/recapture methods produced biologically reasonable values for maximum size and longevity (Table 1). The different models of the likelihood ratio tests using GROTAG (Francis 1988a) all produced the same results (Table 2). The mean annual growth rates are g_{85} = 47.5 cm year^{-1}. and g_{130} = 29.5 cm year^{-1}, corresponding to growth rates at FL = 85 cm and 130 cm, respectively. The Fabens (1965) and GROTAG results were quite similar to each other but had a lower L_∞ and a higher K than Gulland and Holt (1959) (Table 1). All of the plotted curves looked similar but GROTAG better fit the data and was considered a more reliable curve due to its use of all available data. However, as previously stated due to limited number of samples, size range and times at liberty, the growth predicted from the tag/recapture data could be questioned, particularly in light of the high confidence intervals around the parameter estimates and lack of a strong relationship between the Gulland and Holt (1959) parameters (Fig. 8).

While direct comparison of the growth curves using the vertebral, length–frequency and tag/recapture generated growth curves is misleading, it can be useful to compare the growth rates at set sizes (Francis 1988b). The tag/recapture curves and the length–frequency modes indicate a much faster growth for the young mako sharks than the vertebral growth indicates (Table 2).

Longevity

The maximum age based on vertebral band-pair counts was 29 and 32 years, for males and females, respectively. The calculated longevity estimate of 21 years for males is an underestimate and was lower than that obtained directly; while for females, the calculated longevity (38 years) was older than that obtained using vertebral counts.

The maximum ages of 21 for males and 32 for females calculated using the tag/recapture analysis are close to the ages obtained by vertebral counts. None of the long-term recaptures were

Table 1 von Bertalanffy growth function parameters and 95% confidence intervals calculated by using vertebral and tag/recapture methods

Method		L_∞	K	L_0	n	Longevity
von Bertalanffy—3 parameter	Male	253.3	0.125	71.6	118	21.3
	CI±	8.3	0.016	5.9		
	Female	432.2	0.043	81.2	140	65.8
	CI±	54.8	0.011	7.4		
von Bertalanffy—2 parameter	Male	252.1	0.128		118	20.8
	CI±	7.1	0.011			
	Female	393.1	0.054		140	51.6
	CI±	31.5	0.009			
von Bertalanffy—1 parameter	Male		0.109		118	24.8
	CI±		0.004			
	Female		0.074		140	37.3
	CI±		0.003			
Gompertz—3 parameter	Male	241.9	0.191	76.7	118	16.3
	CI±		0.020	5.3		
	Female	365.6	0.087	88.4	140	38.1
	CI±		0.013	6.6		
Gompertz—2 parameter	Male	238.3	0.212		118	15.0
	CI±		0.013			
	Female	331.3	0.118		140	28.9
	CI±		0.010			
Gompertz—1 parameter	Male		0.159		118	20.5
	CI±		0.007			
	Female		0.113		140	30.2
	CI±		0.004			
Tag/recapture						
Gulland and Holt (1959)	Combined	281.7	0.24		14	11.4
	CI±	95.7	0.05			
Fabens (1965)	Combined	199.41	0.58		14	4.5
	CI±	46.65	0.51			
GROTAG	Combined	203.9	0.51		22	5.1

measured, however, the estimated lengths combined with the times at liberty gives information on longevity. Three recaptured sharks were at liberty longer than 10 years. The male with the longest time at liberty (12.7 year) was estimated at 183 cm (corresponding to 7.9 years) at tagging and 21 years at recapture. The longest time at liberty for a female was 12.4 years. The shark was estimated at 100 cm (3 year) at tagging and 15.4 years at recapture. Another female, at liberty 10.5 years was estimated to be 274 cm at tagging corresponding to 21.3 years and 31.8 years at recapture.

Discussion

In 1983, Pratt and Casey, in aging the shortfin mako in the North Atlantic Ocean, put forth the hypothesis that this species deposits vertebral bands twice per year. Although the two-band-per-year hypothesis was brought into doubt several times since its inception (Cailliet and Bedford 1983; Natanson 2001; Natanson et al. 2002), there has been no evidence that this hypothesis is erroneous until recently. Recent advances in the validation of the periodicity of bands using bomb carbon techniques have been used to confirm that bands in this species are deposited annually (Campana et al. 2002; Ardizzone et al. this volume; Cailliet and Goldman 2004). Additionally, the results from one recaptured male after OTC-injection supports annual band periodicity at 18 years. Thus, the previous ages for the shortfin mako using the two-band-per year hypothesis (Pratt and Casey 1983) were underestimated. Biannual band-pair deposition has also been suggested in other shark species, including the sand

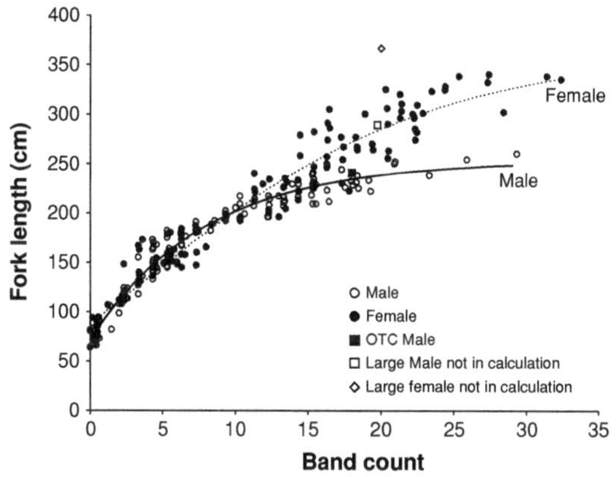

Fig. 6 Shortfin mako growth data based on vertebral band counts. Growth curves have been fitted to the data by sex the 3-parameter Von Bertalanffy growth functions (VBGF) for males and the 3-parameter Gompertz for females. Open circles are males, closed circles are females. Closed square is oxytetracycline (OTC) injected male, open square is largest male [289 cm fork length (FL)] not included in the growth curve calculation, open triangle is largest female (366 cm FL) not included in the growth curve calculation. Modes from length–frequency analysis are also presented for comparison

tiger (Branstetter and Musick 1994), basking shark (Parker and Stott 1965), and scalloped hammerhead (Chen et al. 1990; Anislado-Tolentino and Robinson-Mendoza 2001). Goldman et al. (this volume) validated annual periodicity in the sand tiger, using OTC. The data for the assumption of biannual deposition for the other two species were inconclusive and the studies lacked the appropriate supporting evidence. There are therefore no validated studies proving biannual band-pair deposition for any species.

In the present study, there was no difference in band-pair counts in vertebrae along the vertebral column. Bishop (2004) also found no difference using vertebral columns of three shortfin makos of varying size. This factor is important in the present study as many samples were taken from the head region from fish that were processed for sale. However, Hsu (2003) found a distinct difference in the vertebral counts along the column. Band counts are expected to vary between techniques, oceans and readers. The authors in the present study intercalibrated with Bishop (2004) and coordinated efforts to use the same methodology for processing, measuring and counting, thus it is not surprising that the results of these studies agreed. The discrepancy with the Hsu (2003) study could be attributed to the use of whole vertebrae versus sections and thus technique. Counts on whole vertebrae often underestimate those in sections in because of compression of the last bands. In a smaller vertebra, such as those at the extreme head or tail, it would be difficult to distinguish between bands in a whole vertebra which might be clear in a section. Although it has yet to be investigated, it is also possible that the differences between the Hsu (2003) and other studies represent a real difference in the populations of the species.

Fig. 7 Length–frequency modes for juvenile shortfin mako caught in June

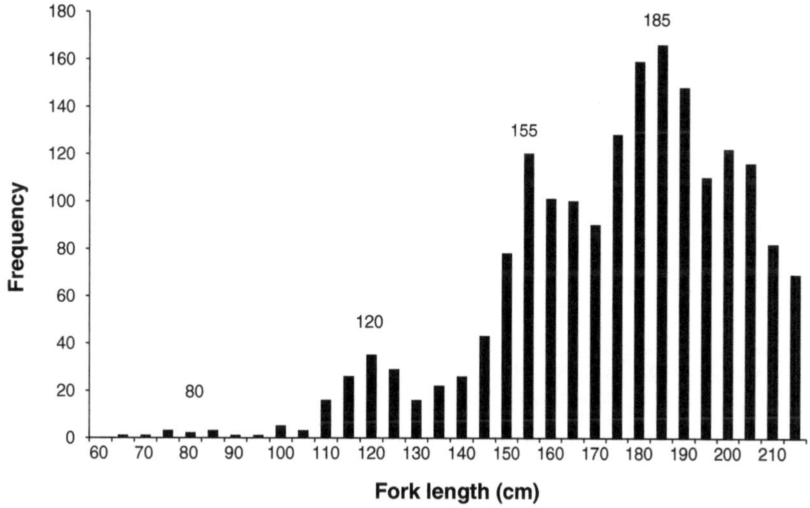

Table 2 Log-likelihood function value and parameter estimates for Growth rates from tag–recapture analysis (GROTAG) (Francis 1988a) growth model fitted to shortfin mako tagging data

Parameter	Symbol (unit)	Values	Growth vertebral	Growth length–frequency
Log likelihood	λ	–86.89		
Mean growth rates	g_{85}(cm/year)	47.5	17.4	42.5
	g_{130} (cm/year)	29.5	13.1	36.1
Measurement error	s (cm)	11.01		

Vertebral growth based on male growth curve

The band counts from Natanson (2001) support Pratt and Casey's (1983) original band counts on sharks greater than 150–160 cm FL but disagree with their assigned age estimates. Conversely, band counts from Natanson (2001) on sharks less than 150 cm FL disagree with the bands counts from the 1983 study but agree with the assigned age estimate and also correspond to the growth estimates based on Pratt and Casey's (1983) other three methods. Direct comparison of age-length and length-increment-derived growth curves has been shown to be inappropriate since the publication of the Pratt and Casey (1983) study (Francis 1988b). In the present study, we also used juvenile length–frequency and tag/recapture data to independently estimate growth. The current results based on increased sample sizes and improved data analysis techniques are very similar to the Pratt and Casey (1983) results in that the growth rates predicted by the juvenile length–frequency modes and the tag/recapture data are higher than that of the vertebral analysis. In the present study, however, the vertebral bands pairs are validated from age two to 31 (Ardizonne et al. this volume) and the vertebral growth curve cannot be manipulated to agree with the other methods; rather those methods need to be examined.

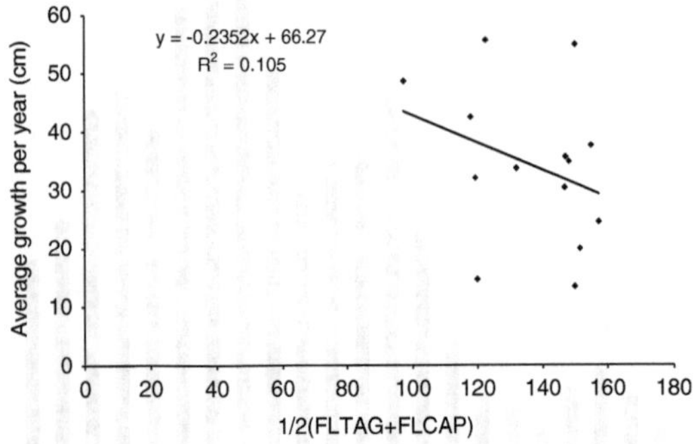

Fig. 8 Gulland and Holt (1959) tag–recapture graph for 14 recaptured shortfin makos. The negative slope equals estimated K and the x-intercept equals L_∞

The high estimates for L_∞ and K from GROTAG can be attributed to the different derivation of the VBGF parameters, small sample size and the absence of older recaptured sharks in the sample. Additionally, the tag/recapture growth data examined graphically using Gulland and Holt (1959) shows no relationship and a large variation in growth rate of these individuals (Fig. 8). The length–frequency modes are primarily applicable to the young size classes and may be a better indicator of growth for those sizes. Both tag–recapture and length–frequency show similar fast growth in the younger fish as compared to the vertebral method. However, the data from these methods is only slightly above the length at age data from the vertebrae (Fig. 6). The fast initial growth represented by the data is not reflected in the growth curves as they are unable to handle rapid initial growth. The tag–recapture growth curves approach L_∞ faster than those derived from vertebral analyses and asymptote at a lower size than biologically reasonable due to the dependence on the younger size classes. The difference between these methods is not entirely unexpected and was also seen in the shortfin mako in Bishop (2004). Size-mode analysis in the porbeagle was also quite distinct in the first year but also showed a much faster growth rate than the vertebral growth curve (Natanson et al. 2002). Since the assigned ages in this study were based on validated counts, the more rapid growth, as seen in modal analysis, is difficult to interpret. Perhaps the modes coalesce at younger ages than was previously presumed due to the large variation of size at age combined with the rapidity of the growth. This effect was seen in one mode, which appeared in July at 100 cm and blended into the next by August.

The vertebral results provide an estimate of growth for the entire size range of the species and thus are more likely to reflect species growth than methods based on juveniles alone. Male

Table 3 Growth function parameters and 95% confidence intervals calculated by using vertebral ages adjusted for deposition rate of two band pairs per year until 160 cm FL followed by one band-pair per-year deposition

Method		L_{max}	K	L_0	n	Longevity
von Bertalanffy—3 parameter	Male	240.2	0.18	84.8	118	13.9
	CI±	8.11	0.03	6.64		
	Female	388.9	0.06	90.7	140	43.7
	CI±	37.07	0.01	7.00		
von Bertalanffy—2 parameter	Male	232.5	0.24		118	11.2
	CI±	5.99	0.02			
	Female	347.0	0.90		140	30.9
	CI±	20.48	0.01			
Gompertz—3 parameter	Male	234.4	0.25	89.8	118	11.7
	CI±		0.04	6.33		
	Female	388.9	0.11	97.3	140	29.5
	CI±		0.02	6.47		
Gompertz—2 parameter	Male	225.0	0.37		118	8.5
	CI±		0.03			
	Female	315.3	0.16		140	20.4
	CI±		0.02			

and female growth past age 11 was markedly different and the sexes required different growth functions to best describe them. The female data do not appear to reach an asymptote while the male curve did. For the female, the VBGF produced an unrealistically high estimate of size

Table 4 Growth function parameters for shortfin makos from seven studies

Study		Size range cm FL	L_∞ cm FL	K	t_0	n	Location	Age at maturity	Oldest aged	Bands/ year	Longevity years
Pratt and Casey (1983)**	Male	69–328*	302	0.266	–1	49	Western North Atlantic	3	4.5	2	10
	Female		345	0.203	–1	54	Western North Atlantic	7	11.5	2	14
Cailliet and Bedford (1983)		80.6–293	292.8	0.072	–3.75	44	Pacific, California	7–8	17	1	38
Ribot-Carballal et al. (2005)		68.6–264	375.4	0.05	–4.7	109	Pacific, Baja, CA	7 m; 15 f	18	1	55
Bishop et al. (2006)	Male	100–347*	302.2	0.052	–9.04	145	Pacific, New Zealand	7–9	29	1	48
	Female		820.1	0.013	–11.3	111	Pacific, New Zealand	19–21	28	1	219
Hsu (2003)	Male	72.6–250.9	321.8	0.049	–6.07	133	China	13–14	23.6	1	
	Female	72.6–314.9	403.62	0.040	–5.27	174	China	18–19	30.6	1	
Chan (2001)**	Male	66–274	267.0	0.312	–0.95	24	Pacific, Australia	Not reported	7	2	9
	Female	74–314	349.0	0.155	–1.97	52	Pacific, Australia	Not reported	10	2	17
Natanson et al. (this volume)***	Male	72–260	253.3	0.125	71.6	118	Western North Atlantic	8	29	1	21
Gompertz	Female	64–340	365.6	0.087	88.4	140	Western North Atlantic	18	32	1	38

All lengths are fork length, those originally presented in total length were converted using equations published in the study or conversions from Bishop (2004) for the Pacific and Kohler et al. (1996) for the Atlantic. 95% Longevity is calculated

* Only total size range listed

** 2 Band/year assumed

*** Validated

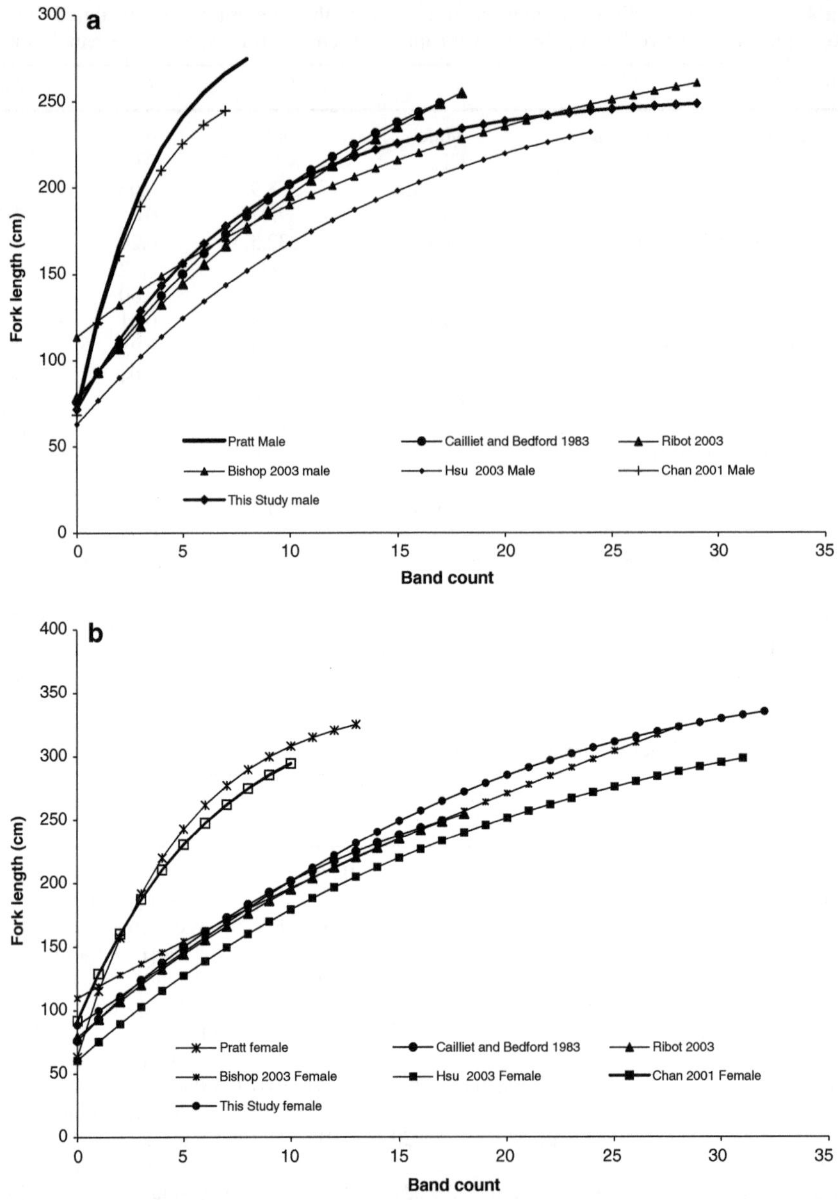

Fig. 9 Growth curves generated from vertebral data for male (**a**) and female (**b**) shortfin makos, included for comparison are the von Bertalanffy growth curves of other studies

at birth, maximum size and longevity, while the GGF provided reasonable estimates for these parameters. Bishop (2004) and Bishop et al. (2006) had similar findings with females such as no asymptote and high estimates of size at birth from their von Bertalanffy (1938) equation and felt that the Schnute (1981) four parameter function better described growth. The best fit for the female, in this study, was the 3-parameter GGF. The VBGF 3-parameter function was the best fit for males providing reasonable estimates for all parameters.

One other possibility we explored was one in which the band deposition rate changes with size and thus with age. If we accepted that Pratt and Casey (1983) were correct that biannual band-pair deposition occurred in the size range they used (less than or equal to 155–160 cm FL), and that the current validation of the annual periodicity in the older sharks is also correct, one can construct a growth curve that reflects this change. As expected, the values for growth rate, age at maturity and longevity fall between the current values and Pratt and Casey's (1983) values

(Table 3). Without direct OTC-validation of the youngest size classes, we could not be certain that the band-pair deposition rates do not change. We feel, however, that this scenario is unlikely and that band-pair deposition is annual throughout the life of the shortfin mako.

The change from the two-band-per-year to the one-band-per-year interpretation results in a difference in growth rate, age at maturity and longevity, which has far-reaching implications for this species. The only previous study of this species in the western North Atlantic Ocean resulted in age at maturity of approximately three and seven for males and females, and a longevity estimate for females of 11–12 years (Pratt and Casey 1983). Data from the current study show an age at 50% maturity of 8 and 18 years (185 and 275 cm) for males and females respectively, and longevity estimates of 29 and 36.

The longevity estimates derived from the growth curves for males is an underestimation, while the estimate of female longevity was reasonable when compared with that obtained from vertebral analysis. The estimates provided fit the maximum size of makos in the WNA for the past 40 years (N. Kohler unpublished data) and assume that the largest-aged female is in the range of the maximum size in the current population. In an unfished population, this maximum age and corresponding size was perhaps higher. Longevity estimates calculated from other studies of mako sharks are higher in those studies using calculations based on VBGF parameters (Cailliet and Bedford 1983; Hsu 2003; Ribot-Carballal et al. 2005; Bishop et al. 2006) and lower for those based on two band pairs per year (Pratt and Casey 1983; Chan 2001) (Table 4).

Other studies have also shown the difference in growth between male and female shortfin makos. Bishop et al. (2006) had a high L_∞ (2.1 times higher than her largest measured sample [347 cm FL]) and low K for the females using a 3-parameter VBGF. Although their curve fit the data in the size range of their samples they chose to use the Schnute (1981) curve instead of both the Gompertz and VBGF. Neither Cailliet et al. (1983b) or Ribot-Carballal et al. (2005) had sufficient samples to analyze their data by sex and also lacked large females. Chan (2001) had only nine females over approximately 250 cm and only two over 300 cm, thus the larger sizes were underrepresented; this may account for the lower L_∞ and higher K in his study (Fig. 9).

The occurrence of sexual differences in growth is well documented in elasmobranchs, with females usually growing larger than males. In the shortfin mako there is ample evidence that the female attains a larger size than the male. Maximum size male and female specimens in this study, 260 and 340 cm, respectively, represent the largest reliably measured shortfin makos from the North Atlantic Ocean.

Using validated vertebral band counts it has been shown that the shortfin mako grows slower, matures later and lives a longer life than previously thought in the North Atlantic Ocean. We believe that the validated vertebral interpretations generated in this study provide robust estimates of age and growth for the shortfin mako.

Acknowledgements We would like to thank the fishermen who allowed us to sample their catches and all the tournament officials who gave us the opportunity to sample at their events. We would also like to thank the fishermen who voluntarily tag and return tags to us to make much of our work possible. Brian Gervelis helped in locating and sectioning, Silver Bishop and Malcolm Francis were vital in the intercalibration portion of the study. We would also like to acknowledge the support of the Apex Predators Investigation staff and particularly Wes Pratt and Jack Casey for laying the groundwork for this study. This manuscript was greatly improved with the input of reviews by Sarah Gaichas and two anonymous reviewers. We would also like to thank Ken Goldman and John Carlson for all their overwhelming effort in organizing the age and growth symposium and editing the proceedings. This study was supported in part by funds from NOAA/NMFS to the Pacific Shark Research Center at Moss Landing Marine Laboratories, as part of the National Shark Research Consortium. This publication was supported in part by the National Sea Grant College Program of the U.S. Department of Commerce's National Oceanic and Atmospheric Administration under NOAA Grant # NA06RG0142, project # R/F-190, through the California Sea Grant College Program; and in part by the California State Resources Agency. The views expressed herein do not necessarily reflect the views of any of those organizations.

References

Anislado-Tolentina V, Robinson-Mendoza C (2001) Age and growth of the scalloped hammerhead shark, *Sphyrna lewini* (Griffith and Smith, 1834), along the central Pacific coast of Mexico. Ciencias Marinas 27:501–520

Ardizzone D, Cailliet GM, Natanson LJ, Andrews AH, Kerr LA, Brown TA (this volume) Application of bomb radiocarbon chronologies to shortfin mako (*Isurus oxyrinchus*) age validation. In: Carlson JK, Goldman KJ (eds) Age and growth of chondrichthyan fishes: new methods, techniques, and analyses. Environmental Biology of Fishes

Bishop SDH (2004) Age determination and life history characteristics of the shortfin mako (*Isurus oxyrinchus*) in New Zealand waters. Master's Thesis. University of Auckland, Auckland, NZ, p 96

Bishop SDH, Francis MP, Duffy C, Montgomery JC (2006) Age, growth, maturity, longevity and natural mortality of the shortfin mako (*Isurus oxyrinchus*) in New Zealand waters. Mar Freshwater Res 57:143–154

Branstetter S (1987) Age, growth and reproductive biology of the silky shark, *Carcharhinus falciformis*, and the scalloped hammerhead, *Sphyrna lewini*, from the northwestern Gulf of Mexico. Environ Biol Fish 19(3):161–173

Branstetter S, Musick JA (1994) Age and growth estimates for the sand tiger in the northwestern Atlantic Ocean. Trans Am Fisher Soc 123:242–254

Cailliet GM, Bedford DW (1983) The biology of three pelagic sharks from California waters and their emerging fisheries: a review. CalCOFI Rep. vol. XXIV

Cailliet GM, Goldman KJ (2004) Age determination and validation in Chondrichthyan fishes, Chapter 14. In: Carrier J, Musick JA, Heithaus MR (eds) Biology of sharks and their relatives. CRC Press, LLC, Boca Raton, Florida, pp 399–447

Cailliet GM, Martin LK, Kusher D, Wolf P, Welden BA (1983a) Techniques for enhancing vertebral bands in age estimation of California elasmobranchs. In: Prince ED, Pulos LM (eds) Proceedings international workshop on age determination of Oceanic Pelagic fishes: tunas, billfishes, sharks. NOAA Tech. Rep. NMFS 8:157–165

Cailliet GM, Martin LK, Harvey JT, Kusher D, Welden BA (1983b) Preliminary studies on the age and growth of blue, *Prionace glauca*, common thresher, *Alopias vulpinus*, and shortfin mako, *Isurus oxyrinchus*, sharks from California waters. In: Prince ED, Pulos LM (eds) Proceedings of the international workshop on age determination of Oceanic Pelagic fishes: tunas, billfishes, and sharks. USDOC Tech. Rep. NMFS 8:179–188

Cailliet GM, Natanson LJ, Welden BA, Ebert DA (1985) Preliminary studies on the age and growth of the white shark, *Carcharodon carcharias*, using vertebral bands. Memoirs South Calif Acad Sci 9:49–60

Campana SE, Annand MC, McMillan JI (1995) Graphical and statistical methods for determining the consistency of age determinations. Trans Am Fisher Soc 124:131–138

Campana SE, Natanson LJ, Myklevoll S (2002) Bomb dating and age determination of a large pelagic shark. Can J Fisher Aquat Sci 59:450–455

Chan RWK (2001) Biological studies on sharks caught off the coast of New South Wales. PhD Thesis, University of New South Wales, Sydney, Australia, p 323

Chen CT, Leu TC, Joung SJ, Lo NCH (1990) Age and growth of the scalloped hammerhead, *Sphyrna lewini*, in Northeastern Taiwan waters. Pacific Sci 44(2):156–170

Fabens AJ (1965) Properties and fitting of the von Bertalanffy growth curve. Growth 29:265–289

Francis RICC (1988a) Maximum likelihood estimation of growth and growth variability from tagging data. NZ J Mar Freshwater Res 22:43–51

Francis RICC (1988b) Are growth parameters estimated from tagging and age-length data comparable? Can J Fisher Aquat Sci 45:936–942

Goldman KJ, Branstetter S, Musick JA (this volume) A re-examination of the age and growth of sand tiger sharks, *Carcharias taurus*, in the western North Atlantic Ocean using improved ageing and back-calculation techniques. In: Carlson JK, Goldman KJ (eds) Age and growth of chondrichthyan fishes: new methods, techniques, and analyses. Environmental Biology of Fishes

Gulland JA, Holt SJ (1959) Estimation of growth parameters for data at unequal time intervals. J Cons Int Explor Mer 25:47–49

Hsu HH (2003) Age, growth, and reproduction of shortfin mako, *Isurus oxyrinchus* in the northwestern Pacific. MS thesis, National Taiwan Ocean Univ., Keelung, Taiwan, pp 107 [In Chinese]

Kohler NE, Casey JG, Turner PA (1996) Length–length and length–weight relationships for 13 species of sharks from the western North Atlantic. NOAA Tech. Memo. NMFS-NE-110, p 22

Martin LK, Cailliet GM (1988) Age and growth of the bat ray, *Myliobatis californica*, off central California. Copeia 1988(3):762–773

Mollet HF, Cliff G, Pratt HL Jr, Stevens JD (2000) Reproductive biology of the female shortfin mako, *Isurus oxyrinchus* Rafinesque, 1810, with comments on the embryonic development of lamnoids. Fish Bull 98:299–318

Natanson LJ (2001) Preliminary investigations into the age and growth of the shortfin mako, *Isurus oxyrinchus*, white shark, *Carcharodon carcharias*, and thresher shark, *Alopias vulpinus*, in the Western North Atlantic Ocean. ICCAT Working Document SCRS/01/66

Natanson LJ, Mello JJ, Campana SE (2002) Validated age and growth of the porbeagle shark, *Lamna nasus*, in the western North Atlantic. Fish Bull 100:266–278

Neter J, Wasserman W (1974) Applied linear statistical models. Richard D. Irwin Inc. Homewood, Illinois

Parker HW, Stott FC (1965) Age, size and vertebral calcification in the basking shark, *Cetorhinus maximus* (Gunnerus). Zoologische Mededelingen (LEIDEN) 40(34):305–319

Pratt HL Jr, Casey JG (1983) Age and growth of the shortfin mako, *Isurus oxyrinchus*, using four methods. Can J Fisher Aquat Sci 40(11):1944–1957

Ribot-Carballal MC, Galván-Magan~a F, Quin~ónez-Velázqeuz C (2005) Age and growth of the mako shark, *Isurus oxyrinchus*, from the western coast of Baja California Sur, Mexico. FishRes 76:14–21
Ricker WE (1979) Growth rates and models. In: Hoar WS, Randall DJ, Brett JR (eds) Fish physiology, vol. VIII, bioenergetics and growth. Academic Press, pp 677–743
Schnute J (1981) A versatile growth model with statistically stable parameters. Can J Fisher Aquat Sci 38:1128–1140
Schwartz FJ (1983) Shark ageing methods and age estimation of scalloped hammerhead, *Sphyrna lewini*, and dusky, *Carcharhinus obscurus*, sharks based on vertebral ring counts. NOAA Tech Rep NMFS 8:167–174
Taylor CC (1958) Cod growth and temperature. J Conseil Int Explor de la Mer 23:366–370
von Bertalanffy L (1938) A quantitative theory of organic growth (inquiries on growth laws II). Hum Biol 10:181–213
Wintner SP, Cliff G (1999) Age and growth determination of the white shark, *Carcharodon carcharias*, from the east coast of South Africa. Fish Bull 97(1):153–169

Environ Biol Fish (2006) 77:385–400
DOI 10.1007/s10641-006-9126-0

SI-ELASMOBRANCHS

Validated age and growth of the sandbar shark, *Carcharhinus plumbeus* (Nardo 1827) in the waters off Western Australia

Rory B. McAuley · Colin A. Simpfendorfer · Glenn A. Hyndes · Rick R. Allison · Justin A. Chidlow · Stephen J. Newman · Rod C. J. Lenanton

Received: 12 June 2006 / Accepted: 21 July 2006 / Published online: 16 August 2006
© Springer Science+Business Media B.V. 2006

Abstract Sandbar sharks, *Carcharhinus plumbeus*, collected from commercial shark fisheries in Western Australia were aged by examination of sectioned vertebrae and analysis of tag-recapture data. Growth curves were derived from consensus counts of growth bands from the vertebrae of 238 individuals ranging in size between 47 and 154 cm fork length (FL). The annual periodicity of growth band formation was validated using vertebrae from tagged sharks, which were injected with oxytetracycline (n = 9) and calcein (n = 23) and were at liberty for up to 8.1 years. The oldest female was estimated to be 25 years of age and the oldest male was 19 years. The ages at which 50% of female and male sharks were mature were estimated to be 16.2 and 13.8 years, respectively. Growth increment data from 104 tagged *C. plumbeus*, which were at liberty for up to 7.4 years, were used to construct growth curves for comparison with those derived from vertebral analysis. The two methods yielded noticeably different results. Based on a known size at birth of 42.5 cm FL, von Bertalanffy parameters estimated using length at age data from vertebral analysis were: $K = 0.039$ year^{-1} and $L_\infty = 245.8$ cm; $K = 0.044$ year^{-1} and $L_\infty = 226.3$ cm; and $K = 0.040$ year^{-1} and $L_\infty = 239.6$ cm for females, males and both sexes combined. The von Bertalanffy parameters derived from tag-recapture data were: $K = 0.153$ year^{-1} and $L_\infty = 142.0$ cm for combined sexes. However, as sharks longer than 142.0 cm were commonly encountered during sampling, these estimates appear to be biologically unrealistic. Also, given the high variability in growth rates of tagged sharks, compared to those derived from the larger vertebral analysis dataset, vertebral ageing was concluded to provide a better description of age and growth in this study. These results confirm that *C. plumbeus* is a slow-growing and late maturing species and thus recovery times from periods of overexploitation would be considerable.

Keywords Age · Growth · *Carcharhinus plumbeus* · Validation · Calcein

R. B. McAuley (✉) · R. R. Allison ·
J. A. Chidlow · S. J. Newman · R. C. J. Lenanton
Western Australian Fisheries and Marine Research Laboratories, Department of Fisheries, Government of Western Australia, PO Box 20, North Beach, WA 6920, Australia
e-mail: rmcauley@fish.wa.gov.au

C. A. Simpfendorfer
Mote Marine Laboratory, Center for Shark Research, Ken Thompson Parkway, 1600, Sarasota, FL 34236, USA

G. A. Hyndes · R. B. McAuley
Centre for Ecosystem Management, School of Natural Sciences, Edith Cowan University, 100 Joondalup Drive, Joondalup, WA 6027, Australia

Introduction

The sandbar shark, *Carcharhinus plumbeus* (Nardo 1827), is a medium-sized carcharhinid shark, which occurs globally in tropical and temperate coastal waters (Compagno 1984). Due to their cosmopolitan distribution, abundance in coastal waters, vulnerability to multiple gear-types, high quality flesh and large fins, *C. plumbeus* stocks support significant commercial target fisheries throughout the species' range, particularly on the east coast of the United States[1] (Sminkey and Musick 1996), east coast of Mexico (Bonfil 1994), East China Sea (Taniuchi 1971; Joung and Chen 1995) and on the west coast of Australia[2,3,4] However, due to their slow growth, late onset of maturity, small litter sizes and long reproductive cycle, sandbar sharks are highly susceptible to overfishing and the effects of commercial exploitation need to be well understood and carefully managed to avoid overfishing[1] (Sminkey and Musick 1996; Smith et al. 1998; Castro et al. 1999; Brewster-Geisz and Miller 2000). Furthermore, where overfishing of *C. plumbeus* stocks has occurred, e.g. in the north-west Atlantic Ocean (Sminkey and Musick 1995, 1996; Castro et al. 1999), stock restoration is likely to take many years (Bonfil 1994; Musick et al. 2000). Reliable age and growth estimates can provide important information for assessing the status of exploited stocks and for predicting their capacity for recovering from periods of overfishing.

In Australia, *C. plumbeus* occurs off both the east and west coasts and to a lesser extent off the north and south coasts (Last and Stevens 1994). In Western Australia, sandbar sharks are primarily distributed between Cape Leveque (16°30′S, 123°E) on the north coast and Albany (118°E) on the south coast. Throughout this range, they are caught by a number of fisheries that take sharks as both targeted catch and by-catch. Between 1995 and 2004, reported catches of *C. plumbeus* in Western Australia increased nearly fivefold, from 83 to 402 tonnes live weight[3,4] (R.B. McAuley, unpublished data). Sandbar sharks are now the main component of the catch in the west coast region of the temperate shark fishery (between latitudes 26°30′S and 33°S), where they are primarily caught by demersal gillnets. The developing northern shark fishery (which operates north and east of 22°78′S, 114°06′E) also targets *C. plumbeus* using demersal longlines. Both the temperate and northern shark fisheries are limited entry, with effort in the temperate fishery controlled by restrictions on the amount of gear that can be used[3,4] (Simpfendorfer and Donohue 1998).

There are currently no published age and growth data for *C. plumbeus* from Australian waters and age and growth estimates for other populations vary considerably between studies (Springer 1960; Wass 1973; Casey et al. 1985; Casey and Natanson 1992; Sminkey and Musick 1995; Joung et al. 2004). Although separate *C. plumbeus* populations do exhibit markedly different biological characteristics (Springer 1960; Bass et al. 1973; Wass 1973; Cliff et al. 1988; Joung and Chen 1995; Sminkey and Musick 1995; Joung et al. 2004), the magnitude of the differences in previously reported growth curves are too large to confidently be attributed to either population density or environmental effects. An alternative explanation for some of these differences is that tagging data, which have previously been used to estimate growth rates and verify the annual formation of vertebral growth bands (Casey et al. 1985; Casey and Natanson 1992), can provide an unreliable basis for growth rate analyses. Because the periodicity of vertebral growth band formation has never been comprehensively validated for this species, it is difficult to resolve whether the

[1] Hoff TB (1990) Conservation and management of the western North Atlantic shark resource based on the life history strategy limitation of sandbar harks. PhD dissertation, University of Delaware, Newark, 149 pp.

[2] McAuley R, Simpfendorfer C (2003) Catch composition of the Western Australian temperate demersal gillnet and demersal longline fisheries, 1994–1999. Fisheries Research Report No. 146. Department of Fisheries, Western Australia, 78 pp.

[3] Gaughan D, Chidlow J (2005) Demersal gillnet and longline fisheries status report. In: Penn JW, Fletcher WJ, Head F (eds) State of the Fisheries Report 2003/04. Department of Fisheries, Western Australia, pp.186–191.

[4] Gaughan D, Chidlow J (2005b) Northern shark fisheries status report. In: Penn JW, Fletcher WJ, Head F (eds) State of the Fisheries Report 2003/04. Department of Fisheries, Western Australia, pp.146–150.

reported differences in growth indicate real variability within and between *C. plumbeus* populations or whether they were caused by different ageing methods or inconsistent interpretation of growth bands between studies.

Due to its demonstrated vulnerability to overfishing and the rapid increases in commercial landings in Western Australia, a clear understanding of the biology and status of the *C. plumbeus* stock is essential to ensure sustainable levels of exploitation. In order to develop an age-structured population assessment of this stock, accurate and locally derived estimates of size-at-age and growth rates are necessary. The primary aim of the present study was, therefore, to determine age and growth parameters for *C. plumbeus* in Western Australian waters. Secondly, as the periodicity of growth band formation in *C. plumbeus* vertebrae has not previously been validated in wild sharks, this study also aimed to determine whether the growth bands visible in sectioned vertebrae form annually. The final aim was to examine whether growth rates derived from tagging data vary from those derived from vertebral analysis and whether the former approach could have contributed to the variation in previously reported growth rates of this species.

Materials and methods

Vertebral sample collection

Vertebral samples from 680 *C. plumbeus*, which ranged in size from 47 to 166 cm fork length (FL), were collected from both commercial and fishery-independent sources between 6 April 1999 and 28 May 2002. Sharks obtained from commercial catches were caught by demersal gillnets (n = 379) and longlines (n = 263). A smaller number of sharks were caught during fishery-independent sampling using demersal longlines (n = 20) and drumlines (n = 22). Gillnets were constructed of 0.9–1.0 mm diameter monofilament meshes of between 165 mm (6.5″) and 178 mm (7″) stretched mesh sizes with either a 15 or 20 mesh drop. Both commercial and fishery-independent longlines comprised size 12/0 J-shaped hooks, attached to the main line via approximately 2 m metal snoods. Longline hooks were baited with mullet, *Mugil cephalus*, or mackerel (family Scombridae). Drumlines consisted of size 14/0 Mustad shark hooks, baited with Australian salmon, *Arripis truttaceus*, and attached to anchored 20 l plastic drums via lengths of chain, so that hooks floated approximately 1 m below the surface. Gillnets and longlines were set demersally in depth ranges of 9–121 and 14–157 m, respectively, and drumlines were fished in depths of between 54 and 100 m. Vertebral sampling was conducted between Eighty Mile Beach (20°S, 120°E) on the north coast and east of Cape Leeuwin (35°S, 115°E) in the south west of the state (Fig. 1).

Each shark was sexed and FL measured to the nearest centimetre, as a straight line from the tip of the snout to the fork of the caudal fin. For comparison with results of other studies, the relationships between FL and total length (TL) in Western Australian *C. plumbeus* are described by the equations:

$$\mathrm{TL} = 1.118(\mathrm{FL}){+}6.302 \quad (\mathrm{n} = 878, r^2 = 0.984) \text{ for males,}$$

$$\mathrm{TL} = 1.113(\mathrm{FL}){+}5.819 \quad (\mathrm{n} = 895, r^2 = 0.986) \text{ for females and}$$

$$\mathrm{TL} = 1.122(\mathrm{FL}){+}6.004 \quad (\mathrm{n} = 1{,}773, r^2 = 0.985) \text{ for both sexes combined.}$$

The relationships for males and females were significantly different (ANCOVA, F = 5.53; df = 2, 1,769; P = 0.041).

As vertebral samples were largely obtained from commercial catches, centra were removed from the vertebral column anterior to the origin of the first dorsal fin and stored frozen until being processed. Reproductive condition was also recorded for subsamples of male (n = 65) and female (n = 48) sharks, according to the following definitions. Males were defined relative to the degree of clasper calcification: immature (uncalcified, where claspers were small and could be easily bent

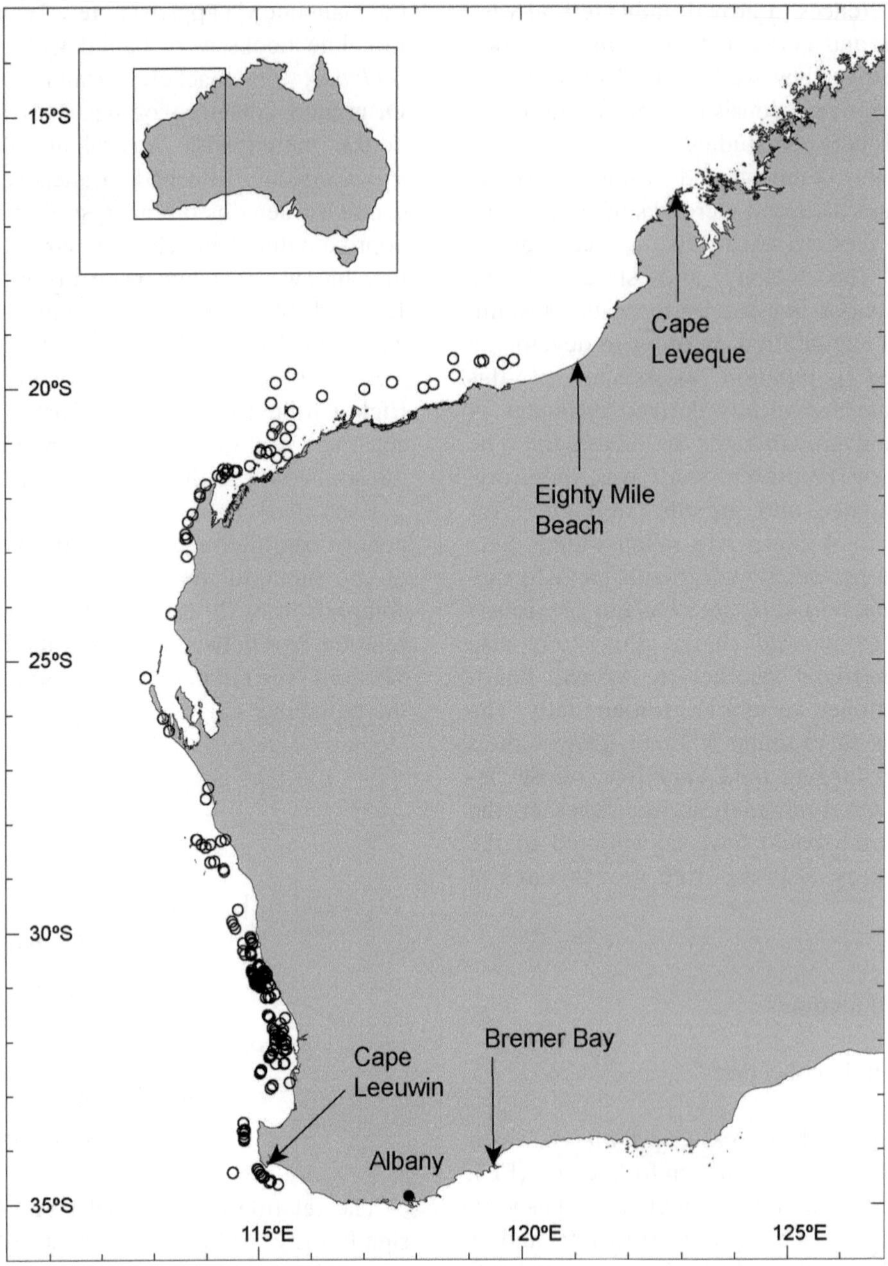

Fig. 1 Sampling locations of 680 *C. plumbeus* from waters off the west coast of Australia, which were used for vertebral analysis of age and growth

along their entire length), maturing (partially calcified, where claspers had begun to elongate and calcify but could still be bent along most or all of their length) and mature (calcified, where claspers were elongate and could not be bent at all). Female maturity was defined by a combination of uterine and ovarian development: stage 1 (immature), uterus very thin along its entire length and empty and ovaries indistinguishable from epigonal organ; stage 2 (maturing), uterus very thin along most of its length but enlarged posteriorly and empty, ovaries difficult to distinguish from epigonal organ; stage 3 (mature, not pregnant), uterus enlarged along its entire length but empty and functional ovary clearly distinguishable from epigonal organ, with differentiated ovarian follicles or developing yolky ova; stage 4 (mature, ovulatory and post-ovulatory), uterus containing yolky eggs but no visible embryos on eggs; stage 5 (mature, pregnant), uterus containing visible embryos; stage 6 (mature, post-partum), uterus enlarged and flaccid, appearing to have just given birth, possibly with visible placental scars.

Vertebral processing and analysis

After defrosting, the neural arch, transverse processes and excess tissue were excised from each vertebral sample and individual centra were separated. Centra were soaked in a 5–10% sodium hypochlorite solution for up to 60 min, depending on their size and quantity and the age of the solution, until all remaining tissue had been removed. Cleaned centra were thoroughly rinsed in fresh water and dried in an oven at 50°C for periods of up to 24 h. Three centra from each shark were embedded in polyester casting resin and longitudinal cross-sections of 170 μm thickness were taken from as close to the focus of each centrum as possible, using a variable speed linear precision saw. Sections were mounted on microscope slides with casting resin and digitally photographed through a dissecting microscope under reflected light.

Images were viewed and minor adjustments to their brightness and contrast were made using Microsoft Photo Editor. Growth bands (defined as a narrow translucent band and adjacent wide opaque band pair) were independently counted by three readers, without knowledge of the size, sex or previous results for any shark. Two readers had experience in ageing sharks, while the third had no experience in ageing sharks but was experienced in ageing teleosts. Counts commenced after the birth-mark, which was identified by a change of angle on the outer edge of the corpus calcareum and an associated translucent band. The readability of each section was scored according to the following definitions: 0, unreadable; 1, bands visible but difficult to interpret; 2, bands visible but the majority difficult to interpret accurately; 3, bands visible but a minority difficult to interpret accurately; and 4, all bands unambiguous. Sections with a readability score of 0 were excluded from further analysis.

A consensus was determined for each reader's counts of the three centra from each shark using the following criteria: (i) where at least two counts matched, the matching count was adopted; (ii) where no counts matched but two counts varied by 1, the count with the higher readability was adopted; (iii) where no counts matched but two counts varied by 1 and readability was equal, the final reading was adopted, since this was made with greater experience in the interpretation of band formation. Where a consensus could not be reached, that specimen was excluded from further analysis of that reader's results. Final consensus of the number of growth bands for each specimen was determined among readers by taking the count that matched in at least two of the consensus counts from each reader.

The index of average percentage error (IAPE) was calculated for each reader's counts and for the consensus counts according to the method described by Beamish and Fournier (1981):

$$\mathrm{IAPE} = \frac{1}{N}\sum_{j=1}^{N}\left(\frac{1}{R}\sum_{i=1}^{R}\frac{|X_{ij} - X_j|}{X_j}\right) \times 100,$$

where N is the number of animals aged, R is the number of readings, X_{ij} is the count from the jth animal at the ith reading and X_j is the mean age of the jth animal from i readings.

A form of the von Bertalanffy growth equation that fits the curve to a known size at birth (Simpfendorfer et al. 2000) was fitted to the resulting length at age data:

$$L_T = L_0 + (L_\infty - L_0)(1 - \mathrm{e}^{-KT}),$$

where L_0 is the mean size at birth (42.5 cm FL for both sexes, R.B. McAuley, unpublished data), L_T is the mean length at time T, L_∞ is the mean asymptotic length and K is the growth coefficient. The equation was fitted using the non-linear regression function of Sigmaplot 9.01 (Systat 2004). Ninety-five percent confidence intervals were estimated for von Bertalanffy parameters by re-sampling the length-at-age data from vertebral analysis, to create 1,000 new datasets (of the same size as the original data) and re-fitting the growth curve to each through the known size at birth of 42.5 cm FL. Values of t_0 were derived from each of the resulting estimates using the standard definition of the von Bertalanffy curve:

$$L_t = L_\infty\left[1 - e^{-K(t-t_0)}\right].$$

Validation of growth band periodicity

A total of 2,107 sharks, caught by a variety of fishery-dependent and -independent methods, were tagged between 22 March 1994 and 15 June 2004 in waters between Cape Leveque (16°30′S, 123°E) on the north coast and Bremer Bay (34°S, 120°E) on the south coast. Prior to release, sharks were sexed, measured (FL) and tagged with Jumbo Rototags (Dalton I.D. Systems, Dalton Group Ltd, Dalton, UK) in the first dorsal fin. Subsamples of tagged sharks were injected with either oxytetracycline (OTC, prior to December 2000, n = 151) or calcein (post-December 2000, n = 725) to mark their vertebral centra for age validation. Both OTC and calcein were injected as an aqueous solution (25 mg ml^{-1}) into the dorsal musculature anterior to the first dorsal fin and adjacent to the vertebral column at doses of 25 and 3–5 mg kg^{-1} body weight, respectively. Due to concerns about the potential toxicity of low doses of calcein in sharks (Gelsleichter et al. 1997), three calcein-injected sharks were held in captivity for observation over 12 months. The two females (69 and 72 cm FL) and one male (72 cm FL) were caught by demersal longlines during fishery-independent surveys in August 2002 and were tagged and injected following capture according to the methods described above. Captive sharks were held in a shaded circular outdoor tank of 5 m diameter and 1.5 m depth with open seawater circulation. The tank was subject to natural photoperiod and temperature regimes and sharks were fed to satiation 2–3 times weekly with pilchards *Sardinops neopilchardus*.

Marked vertebrae were prepared according to the previously described methods and digitally photographed via a dissecting microscope under both normal reflected light and then under fluorescent light with an ultraviolet filter. Fluorescent images were superimposed on their non-fluorescent counterparts using Adobe Illustrator 10.0 software. The transparency of the fluorescent layer of the composite image was then adjusted so that the fluorescing mark was visible while banding patterns from the non-fluorescent layer could be easily distinguished. The number of complete growth bands after the mark were then counted and plotted against time at liberty. The slope of the linear regression between post-injection band counts and time at liberty equates to the number of bands formed per year.

Growth rate estimation using tagging data

Recapture data were collected for 104 tagged *C. plumbeus*, which were at liberty for between 1 and 2,723 days (up to 7.4 years). Capture information, including date, location and FL were collected by commercial fishers and scientific observers during fishery-dependent and -independent surveys. To assist in the collection of comprehensive and accurate tag capture data, commercial fishers and observers were provided with measuring tapes, standardised tag-recapture reporting forms and training in measuring sharks and in the collection of vertebral samples.

Growth rates were calculated from tag length-increment data using the Francis (1988a) maximum likelihood method. This method estimates growth (ΔL_i) of tagged fish based on growth rates, g_α and g_β, at two arbitrary lengths of α (70 cm FL) and β (110 cm FL), which were chosen as they represented the range of most of the empirical data, so that:

$$\Delta L_i = \left[\frac{\beta g_\alpha - \alpha g_\beta}{g_\alpha - g_\beta} - L_i\right]\left[1 - \left(1 + \frac{g_\alpha - g_\beta}{\alpha - \beta}\right)^{\Delta T_i}\right],$$

where L_i is the length at release and ΔT_i is the period at liberty.

Variability in growth rates (v), measurement error (m, s) and the probability of incorrectly recorded length data, referred to as the contamination probability (p), were estimated by maximising the likelihood function:

$$\lambda = \sum_i \log[(1-p)\lambda_i + p/R],$$

where

$$\lambda_i = \frac{\exp(-0.5(\Delta L_i - \mu_i - m)^2/(\sigma_i^2 + s^2)}{\left[2\pi(\sigma_i^2 + s^2)\right]^{0.5}},$$

R is the range of observed growth increments and subscript i refers to the ith fish. This method assumes that v is normally distributed with a mean of μ and a standard deviation σ and that σ is proportional to μ, such that $\sigma = v\mu$. Net measurement error at release and recapture is also assumed to be normally distributed with a mean of m and a standard deviation s.

Confidence intervals of parameter estimates were calculated by refitting the model to 500 'bootstrapped' length-increments. Bootstrapped data were generated by randomly selecting (with replacement) from a normal distribution with a mean equal to the predicted growth increment and a standard deviation of $v\mu$. Bootstrapped measurement error data were generated by randomly selecting from a normal distribution with a mean equal to m and a standard deviation of s.

Resulting values of g_α and g_β were used to estimate the von Bertalanffy growth curve parameters by:

$$K = -\ln(1 + ((g\alpha - g\beta)/(\alpha - \beta))), \text{ and}$$

$$L_\infty = (\beta g\alpha - \alpha g\beta)/(g\alpha - g\beta).$$

Results

Validation of growth band periodicity

Apart from an orange discolouration of their teeth for a period of approximately 3 weeks, captive sharks showed no immediate or longer-term signs of being adversely affected by calcein injection or tagging. One shark (72 cm FL female) was released after 12 months due to worsening abrasions on its snout and body caused by impacts with the tank wall or circulation pipes. The remaining sharks were released 2 weeks later, suffering no apparent ill effects from their captivity.

Vertebral samples from nine OTC-injected and 26 calcein-injected sharks were either collected by researchers or returned after recapture by commercial fishers. However, one OTC-marked sample and three calcein-marked samples were returned with insufficient capture data (capture date or length) and were omitted from further analysis. Fluorescing marks were visible in all calcein-marked sections and in all but two sections from OTC-injected sharks. Times at liberty for the sharks from which the 29 remaining samples were collected varied between 16 and 2,963 days (up to 8.1 years). There was a significant linear relationship between the number of complete growth bands counted after the injection marks and the periods for which the corresponding sharks were at liberty (ANOVA, $r^2 = 0.972$; $F = 945$; $P < 0.0001$; Fig. 2). The slope of this regression was 1.02 (SE = 0.033), demonstrating that growth bands were formed annually.

Vertebral analysis

Vertebrae from 680 *C. plumbeus*, ranging in size between 47 and 166 cm FL (Fig. 3), were sectioned and read. Mean band counts, mean readability, IAPE and the percentage of readings by each reader that were in agreement with final consensus counts were similar for all three readers (Table 1), although readers A and B obtained slightly more individual consensus counts than reader C. Reader C provided both the highest individual growth band count (27 years) and the highest individual consensus age (26 years). Final consensus counts between readers were obtained for 238 specimens, ranging in size from 47 to 154 cm FL.

The oldest female and male sharks for which consensus ages were agreed were 25 years (152 cm FL) and 19 years (142 cm FL), respectively. The oldest immature female and male sharks were both 12 years. Maturing females were aged at between 12 and 16 years and a single maturing male was aged at 15 years. The youngest mature female and male sharks were aged at 14 and 13 years, respectively. On the basis of these data, age at maturity is estimated to be 12–16 years for females and 12–15 years for males. Based on the known sizes at which 50% of Western Australian *C. plumbeus* are mature ($L_{0.5} = 135.9$ cm FL for females and 126.9 cm FL for males, R.B. McAuley, unpublished data), ages at maturity were calculated from these curves as 16.2 and 13.8 years for females and males,

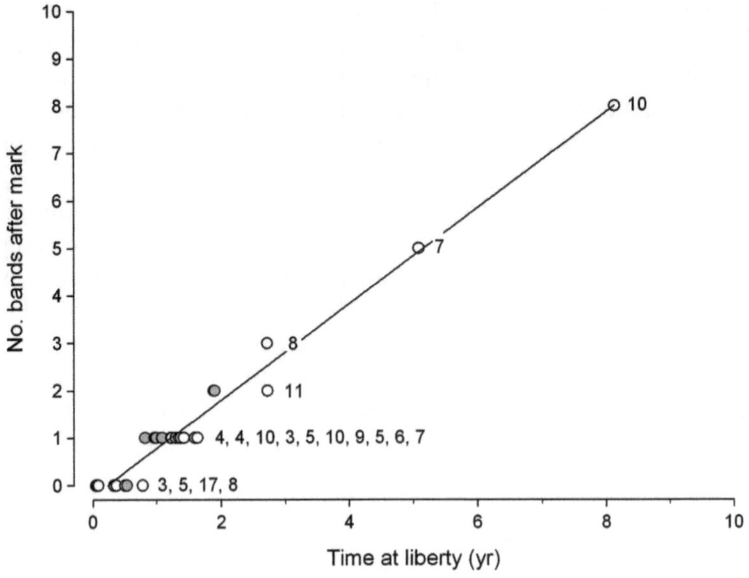

Fig. 2 Number of complete growth bands counted after OTC or calcein mark in 29 sectioned *C. plumbeus* centra. *Numbers beside datapoints* are the estimated ages of sharks at capture, shown by *white circles*. Centra for which absolute age estimates could not be determined are shown as *grey circles*

respectively. These estimates concur with the ranges determined from aged sharks with known reproductive condition.

The modified form of the von Bertalanffy equation provided adequate descriptions of the male, female and combined length-at-age data derived from vertebral analysis ($r^2 = 0.92$, 0.94 and 0.93, respectively). Although the fitted growth curves suggested that males grew slightly faster and reached a smaller asymptotic length than females (Fig. 4; Table 2), these differences were not significant (Kimura 1980; Likelihood Ratio Test, $\chi^2 = 0.350$, $df = 2$, $P = 0.839$). The growth rate for both sexes combined was predicted to slow gradually from 7.8 cm year^{-1} in the first year, to 2.4 cm year^{-1} at the projected age of 30.

Based on the maximum observed sizes of *C. plumbeus* in Western Australia (165 cm for males and 166 cm FL for females, R.B. McAuley, unpublished data), maximum ages were estimated from these parameters to be 23.9 and 25.3 years for females and males, respectively. Using the maximum reported size of *C. plumbeus* in Australian waters [240 cm stretched TL (Last and Stevens 1994), which equals 196 cm FL, according to the relationship given in Stevens and McLoughlin (1991)], these parameters gave maximum age estimates of 36.4 and 40.9 years for females and males, respectively.

Tagging data

The von Bertalanffy parameter estimates derived from analysis of tagging data by the Francis (1988a) model were noticeably different to and less precise than those calculated from vertebral analysis (Fig. 4d; Table 3). With the observed data, the model predicted that L_∞ was considerably lower (142.0 cm) and K was higher (0.153 year^{-1}) than their values estimated by vertebral analysis. This combination of parameters indicated that initial growth was more rapid but that *C. plumbeus* reached a far lower asymptotic length than estimated by vertebral analysis. However, sharks longer than the theoretical maximum size (L_∞) estimated from tagging data were commonly encountered during sampling (Fig. 3) and this estimate was considerably less than the maximum observed size in this population (166 cm FL, R.B. McAuley, unpublished data).

The observed growth rates of tagged sharks were highly variable, especially for those sharks captured after less than 3 months at liberty (Fig. 5). The model estimated that growth rate variability (v) amounted to 26.4% of the annual growth predicted by the model (with 95% confidence intervals of 1.6–38.4%). Although the mean net measurement error estimate ($m = -0.322$ cm) was reasonably low, the standard

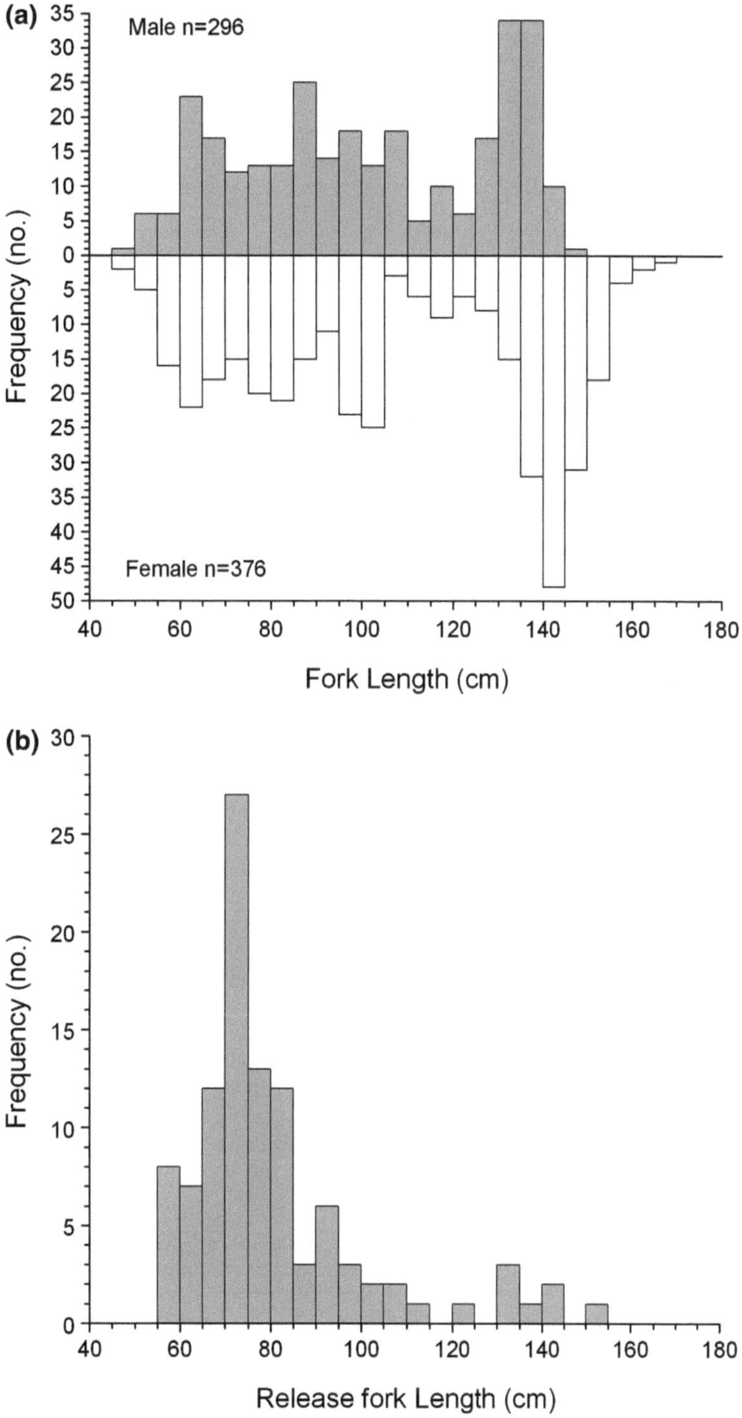

Fig. 3 Size frequency distributions of **a** 672 sexed *C. plumbeus* specimens used for vertebral analysis (NB. vertebrae from a further eight unsexed specimens were also aged) and **b** release lengths of 104 tagged *C. plumbeus* used for analysis of growth increment data

deviation of measurement error (s = 2.510 cm) was high, accounting for ± 24.6 and ± 55.8% of the predicted annual growth at lengths α (70 cm) and β (110 cm), respectively. The bootstrapped data suggested that the range of measurement error was even higher with a median estimate of s = 3.643 cm and 95% confidence intervals of 2.383–5.162 cm (i.e. 23.4–50.6% of predicted

Table 1 Growth band count statistics from the vertebrae of 680 *C. plumbeus* from waters off the west coast of Australia, for three readers and for final consensus count

Reader	Mean readability	Mean band count	IAPE	No individual consensus counts	Agreement with final band count (%)
A	1.8	10.9	10.0	528	49.8
B	2.1	10.2	14.0	522	44.8
C	1.9	10.6	12.6	482	47.3
Final	–	8.7	11.7	238	–

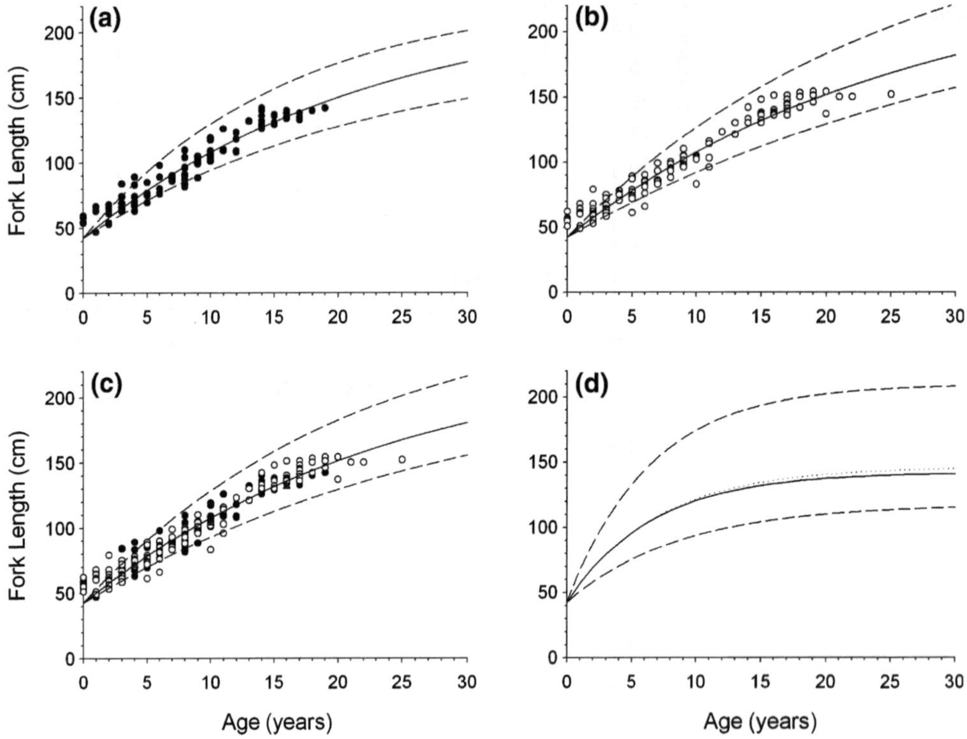

Fig. 4 Length at age of *C. plumbeus*, determined by: vertebral analysis of **a** males (n = 105, *filled circles*), **b** females (n = 130, *open circles*), **c** sexes combined (n = 238, unknown sexes shown as *grey triangles*) and **d** growth increments of tagged sharks (n = 104). *Solid lines* are the fitted von Bertalanffy growth curves for each dataset, *dashed lines* indicate 95% confidence intervals and in (**d**) *dotted line* is the growth curve derived from median estimates of von Bertalanffy growth parameters

Table 2 Summary of von Bertalanffy parameters, estimated by fitting the growth curve to length-at-age data derived from vertebral analysis of 238 *C. plumbeus* from waters off the west coast of Australia

Sex	N	K (year^{-1})	L_∞ (cm FL)	t_0	r^2
Male	105	0.044 (0.044, 0.063)	226.3 (188.6, 229.2)	–4.7 (–4.0, –4.7)	0.925
Female	130	0.039 (0.029, 0.040)	245.8 (241.3, 297.9)	–4.9 (–4.8, –5.4)	0.940
Combined	238	0.040 (0.033, 0.047)	239.6 (222.2, 273.0)	–4.9 (–4.5, –5.1)	0.934

Values in parentheses are 95% confidence intervals of bootstrapped estimates

annual growth at length α and 53.0–114.7% of growth at length β). The model also estimated a contamination probability of 6.4% in the observed data, which increased to 8.9% in the bootstrapped data (with 95% confidence limits of 1.9–22.7%), indicating that data from between

Table 3 Growth and growth variability parameter estimates for *C. plumbeus* from waters off the west coast of Australia, with 95% confidence intervals, calculated from tagging data by the Francis (1988a) model

	g_α (cm year^{-1})	g_β (cm year^{-1})	v	m (cm)	s (cm)	p	L_∞ (cm FL)	K (year^{-1})	t_0 (years)
Fitted	10.2	4.5	0.264	–0.322	2.510	0.064	142.0	0.153	–2.3
Median bootstrap	10.2	4.85	0.095	–0.428	3.643	0.089	146.2	0.144	2.4
95% confidence intervals	6.8, 15.2	1.1, 10.8	0.016, 0.384	–4.581, 3.017	2.383, 5.162	0.019, 0.227	117.5, 209.8	0.115, 0.155	–1.5, –3.9

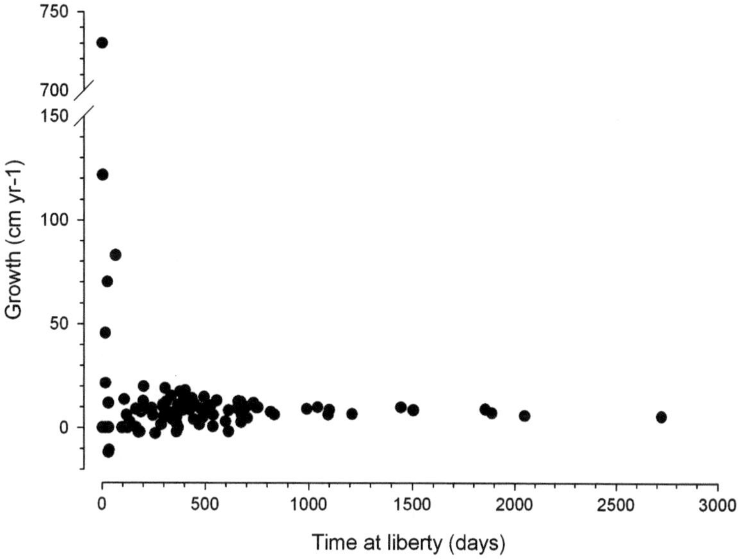

Fig. 5 Growth rate versus time at liberty of 104 recaptured *C. plumbeus*, which were tagged off the west coast of Australia between March 1994 and October 2003

seven and nine sharks were likely to exert undue influence in the model fitting procedures.

Discussion

This study validated the annual nature of vertebral growth band formation in sandbar sharks up to 17 years of age, from waters off the west coast of Australia. As such, it is the first to validate the annual periodicity of growth band formation for this commercially valuable species. Although Branstetter (1987) did previously provide validation based on two captive juveniles, other age and growth studies have had to rely on the indirect methods of marginal increment analysis (Casey et al. 1985; Sminkey and Musick 1995; Joung et al. 2004) or comparison of vertebral results with growth rates of tagged sharks (Casey et al. 1985; Casey and Natanson 1992). This study is also one of only a few that have demonstrated annual band formation in the genus *Carcharhinus* in the wild (Branstetter 1987; Brown and Gruber 1988; Simpfendorfer et al. 2002). However, unlike these previous studies, annual growth bands were validated for a relatively wide range of age classes (3–17 years), which were at liberty for periods of up to 8.1 years.

This is also believed to be the first study of elasmobranch age and growth that has used calcein to mark the centra of wild sharks. These results confirm the conclusions of Walker et al.[5] and Gelsleichter et al. (1997), that calcein produces distinct fluorescent marks in vertebrae from

[5] Walker TI, Officer RA, Clement JG, Brown LP (1995) Southern shark age validation: Part 1—project overview, vertebral structure and formation of growth-increment bands used for age determination. Final report to Fisheries Research and Development Corporation (FRDC Project 91/037). Department of Conservation and Natural Resources, Queenscliff, Vic., Australia.

captive sharks. Although the former study reported no deleterious effects on the health of school sharks, *Galeorhinus galeus*, or gummy sharks, *Mustelus antarcticus*, injected with calcein, the latter study found that doses of 25 mg kg^{-1}, which are typically used in validating the age and growth of teleosts (e.g. Monaghan 1993), had a toxic effect on nurse sharks, *Ginglymostoma cirratum*. Gelsleichter et al. (1997) therefore suggested administering calcein at a dose of 5 mg kg^{-1}, as this produced suitable marks for age validation without inducing mortality. The results of the current study indicate that even lower doses of calcein might be adequate for marking centra, as the 3–5 mg kg^{-1} doses used in this study produced highly visible fluorescing marks in all of the 26 calcein-marked vertebrae. This was in contrast to the centra obtained from the nine OTC-injected sharks, in which no marks could be detected in centra from two sharks and could only be faintly seen in centra from one. Although previous studies have generally not reported any failure to detect OTC marks in the vertebrae of injected sharks, marks could not be detected in nearly a third of the samples in the present study. However, failure to mineralise OTC was also reported by Kusher et al. (1992), who found 'sufficient OTC uptake' in only 35% of injected leopard sharks, *Triakis semifasciata*, while Smith et al. (2002) reported a rapid decay in OTC marks in vertebrae of leopard sharks that had been at liberty for ca. 20 years. Results from the present study therefore suggest that even at low doses, calcein may be a more reliable chemical marker than OTC for age validation studies in which vertebrae from wild sharks are used.

The von Bertalanffy growth curves estimated by vertebral analysis for *C. plumbeus* in Western Australian waters were dissimilar to those previously reported from the Pacific (Wass 1973), western North Atlantic (Casey et al. 1985; Casey and Natanson 1992; Sminkey and Musick 1995) and more recently from northeastern Taiwanese waters (Joung et al. 2004; Table 4). Allowing for the fact that *C. plumbeus* in the western North Atlantic population are born at a larger size and attain a greater maximum length than those from Australia's west coast[2] (Springer 1960), the von Bertalanffy growth curves estimated in this study most closely resembled those previously estimated by Sminkey and Musick (1995). Age at maturity in Western Australian sandbar sharks was similar to previous estimates of 15 for males and 13 for females by Casey et al. (1985) and 15–16 for males and females by Sminkey and Musick (1995) but was much younger than the late 20 s suggested by Casey and Natanson (1995) and much older than the 5 years derived from the results of Wass (1973) and 7.5–8.2 years reported by Joung et al. (2004). Even though Sminkey and Musick (1995) demonstrated a significant difference between growth rates of juveniles in the western North Atlantic before and after a period of stock depletion, it seems highly unlikely that even disparate populations of *C. plumbeus* that have experienced differing levels of exploitation, could realistically exhibit such dramatic differences in ages at maturity. Nor can density-dependent effects adequately account for the magnitude of the discrepancies between growth rates determined by vertebral analysis (Casey et al. 1985; Sminkey and Musick 1995) and those derived from tagging data (Casey and Natanson 1992) in the western North Atlantic stock.

Whilst it was possible to estimate age at maturity in Western Australian *C. plumbeus* precisely, maximum age could not be conclusively determined due to the scarcity of samples and lack of consensus readings from the largest size classes. Although the largest aged female (25 years) equalled the maximum observed size in Western Australia, only three vertebral samples were obtained from sharks larger than 160 cm FL and a consensus age could only be agreed for one of these. The largest male shark for which a consensus age was agreed (19 years) was 23 cm smaller than the maximum observed size in this population. Although vertebrae from seven larger male sharks (143–147 cm FL) were examined, consensus ages could not be agreed for any of these. Despite extensive sampling throughout the range of *C. plumbeus* in Western Australia, it was impossible to obtain large sample sizes from the very oldest sharks due to their inherent scarcity. Furthermore, the compression of growth bands towards the margin of the corpus calcareum, which has been noted in several species (e.g. Casey et al. 1985; Simpfendorfer 1993;

Table 4 Comparison of published and derived age and growth parameters for *C. plumbeus*

Region	Female					Male					Method(s)	Reference
	L_∞ (cm)	K (year^{-1})	t_0 (years)	t_{mat} (years)	t_{max} (years)	L_∞ (cm)	K (year^{-1})	t_0 (years)	t_{max} (years)	t_{max} (years)		
Western North Atlantic	299.0 FL	0.04	–4.9	13	21	257.0 FL	0.05	–4.5	12	15	Vertebral analysis and tagging	Casey et al. (1985)
Western North Atlantic[a]	186.0 FL	0.046	–6.5	Late 20 s	>50						Tagging	Casey and Natanson (1992)
Western North Atlantic	263.3 TL	0.059	–4.8	15–16		245.9 TL	0.059	–5.4	15–16		Vertebral analysis	Sminkey and Musick (1995); 1980–1981 sample
Western North Atlantic	220.5 TL	0.086	–3.9	15–16		221.8 TL	0.087	–3.8	15–16		Vertebral analysis	Sminkey and Musick (1995); 1991–1992 sample
Hawaii	193.9 TL	0.001	–0.4	6.5		181.7 TL	0.001	–0.4	5.38		Captive growth rates	Wass (1973)
East China Sea	223.0 TL	0.1	–4.5	8.2	19.8	200.0 TL	0.14	–4	7.5–8.2	20.8	Vertebral analysis	Joung et al. (2004)
Western Australia	245.8 FL (281.0 TL)	0.039	–4.9	16.2	25–36	226.3 FL (259.2 TL)	0.044	–4.7	13.8	24–41	Vertebral analysis	Present study

L_∞, K, and t_0 are parameters of the von Bertalanffy growth curve. Natural TLs given in parentheses were derived from length relationships presented in the current study

t_{mat} indicates age at maturity; t_{max} is the maximum age

[a] Denotes estimates based on growth curve for combined sexes

Natanson and Kohler 1996; Smith et al. 2002), could have caused less consistent readings of the vertebrae from older sharks. The relatively small number of length-at-age estimates for older sharks is likely to have biased the fitted von Bertalanffy growth curves and caused estimates of L_∞ to exceed the maximum observed lengths in this population. It is therefore difficult to estimate maximum size and hence maximum age with any certainty. However, when extrapolated to the maximum reported size of *C. plumbeus* in Australia (Last and Stevens 1994), these data suggest that maximum age could be as high as 41 years for males and 36 years for females.

Francis (1988a, b) demonstrated that, because age-based growth data determined from vertebral analysis are fundamentally different to length-based growth data estimated from tagging data, growth increments of tagged fish are not comparable to and therefore should not be used to verify, length-at-age data. Nonetheless, this approach has been used for several shark species (Natanson et al. 1999; Smith et al. 2002; Wintner et al. 2002; Skomal and Natanson 2003). The von Bertalanffy growth curves derived from the tag-return data in the present study bore little resemblance to those estimated from vertebral analysis, thereby confirming some of the problems associated with such comparisons. However, the analysis of tag-recapture data using the Francis (1988a) model did provide some useful insights into and quantification of the limitations of using tagging data to describe growth rates of slow-growing species such as *C. plumbeus*. For example, although the estimated growth variability ($v = 0.26$) was lower than has been determined for other chondrichthyan species (Francis and Francis 1992; Francis 1997), including the related and co-occurring dusky shark, *Carcharhinus obscurus* ($v = 0.34$, Simpfendorfer 2000), this level of natural variability in the growth of individual sharks suggests that in some situations, tagging data provide an unreliable basis for the verification of growth rates of the population. These situations include when most of the data come from short-term recaptures (when growth-increments are most variable; Fig. 5) and when sample sizes are small relative to the amount of variation.

The growth curve derived from the tagging data was clearly unrepresentative of the larger length-at-age dataset (Fig. 4). The estimated levels of measurement error in the tagging data appear to have been sufficiently large to obscure the growth rates predicted by the model, particularly in larger sharks. This would result in unreliable estimates of the von Bertalanffy parameters. Also, despite every reasonable attempt to ensure the provision of accurate tag release and recapture information, the model predicted a 6.4% probability of data contamination. In other words, six tag returns were either misrecorded or represented observations of atypical individuals. As growth rates of tagged sharks have typically been derived from relatively small sample sizes, even at this level of contamination there is considerable potential for outliers in the data to seriously bias results. It is therefore suggested that the use of growth rates determined from the length-increments of tagged sharks should not be regarded as a substitute for the direct validation of the frequency of growth band formation in vertebrae.

The results of the present study confirm that *C. plumbeus* from waters off the west coast of Australia are slow-growing and take many years to reach maturity. In conjunction with their low fecundity (R.B. McAuley, unpublished data), this stock is likely to be particularly sensitive to exploitation pressure and consequently fishery managers need to be responsive to its intrinsic vulnerability to overharvesting. Based on these results and the known size composition of catches in the temperate Western Australian demersal gillnet fishery[2], *C. plumbeus* catches in the southern half of this stock's range primarily comprise all but the first few juvenile age classes and adults. In conjunction with recent increases in catches of adult-sized sharks in the Western Australian north coast shark fishery[3,4] (R.B. McAuley, unpublished data), sandbar sharks appear to be subject to fishing at all but the very youngest and oldest ages. The highly mobile, increasingly efficient and wide ranging capacity of fishing vessels within the shark fleets of Western Australia indicate that complex management arrangements are required to maintain sustainable levels of breeding stock. It has previously been argued that limiting fishing mortality to only a few age classes is a desirable approach to sustainable management of long-lived shark species such as this (Stevens et al. 1997; Walker 1998; Simpfendorfer and Donohue 1998; Prince 2005). Fishery managers should therefore consider low frequency or low intensity harvest strategies in conjunction with targeted spatial or temporal closures to restrict fishing mortality to a minimum of juvenile age-classes, in order to protect the breeding stock biomass of these sharks. As depletion of the breeding biomass of this stock will result in long population recovery times, the economic loss associated with recovering and rebuilding this stock may persist longer than short-term economic losses from adoption of such management strategies. To avoid such depletion, more detailed assessment of the effects of current age-specific harvest levels of the Western Australian *C. plumbeus* stock is urgently required, for which these results will be essential.

Acknowledgements This research was funded by a grant from the Fisheries Research and Development Corporation. Our study would not have been possible without the considerable assistance of the skippers and crews of the Western Australian temperate demersal gillnet and longline and northern shark fisheries, on whose vessels most of these data were collected and their help is gratefully acknowledged. We also wish to thank all the staff of the WA Department of Fisheries who contributed to the collection of data, especially Ryan Ashworth, Dennyse Newbound, Ben Sale and the skipper and crew of the Research Vessels Flinders and Naturaliste. Also, many thanks are due to the staff at the Western Australian Fisheries and Marine Research Laboratory who provided valuable advice on this manuscript.

References

Bass AJ, D'Aubrey JD, Kistnasamy N (1973) Sharks of the east coast of southern Africa. I. The genus Carcharhinus (Carcharhinidae). Oceanogr Res Inst (Durban) Investig Rep 33:168

Beamish RJ, Fournier DA (1981) A method for comparing the precision of a set of age determinations. Can J Fish Aquat Sci 38:982–983

Bonfil R (1994) Overview of world elasmobranch fisheries. FAO Fisheries Technical Paper No. 341. FAO, Rome, pp. 119

Branstetter S (1987) Age and growth validation of newborn sharks held in laboratory aquaria, with comments on the life history of the Atlantic Sharpnose shark, *Rhizoprionodon terraenovae*. Copeia 1987:291–300

Brewster-Geisz KK, Miller TJ (2000) Management of the sandbar shark, *Carcharhinus plumbeus*: implications of a stage-based model. Fish Bull 98:236–249

Brown CA, Gruber SH (1988) Age assessment of the lemon shark, *Negaprion brevirostris*, using tetracycline validated vertebral centra. Copeia 1988:747–753

Casey JG, Natanson LJ (1992) Revised estimates of age and growth of the sandbar shark (*Carcharhinus plumbeus*) from the western North Atlantic. Can J Fish Aquat Sci 49:1474–1477

Casey JG, Pratt HL Jr, Stillwell CE (1985) Age and growth of the sandbar shark (*Carcharhinus plumbeus*) from the western North Atlantic. Can J Fish Aquat Sci 42:963–975

Castro JI, Woodley CM, Brudek RL (1999) A preliminary evaluation of the status of shark species. FAO Fisheries Technical Paper No. 380. FAO, Rome, pp. 72

Cliff G, Dudley SFJ, Davis B (1988) Sharks caught in the protective gill nets off Natal, South Africa. 1. The sandbar shark, Carcharhinus plumbeus (Nardo). S Afr J Mar Sci 7:255–265

Compagno LJV (1984) FAO species catalogue, vol. 4. Sharks of the World. An annotated and illustrated catalogue of sharks known to date. FAO Fisheries Synopsis No. 125

Francis MP (1997) Spatial and temporal variation in the growth rate of elephantfish (*Callorhincus milii*). NZ J Mar Freshw Res 31:9–24

Francis RICC (1988a) Maximum likelihood estimation of growth and growth variability from tagging data. NZ J Mar Freshw Res 22:42–51

Francis RICC (1988b) Are growth parameters estimated from tagging and age-length data comparable? Can J Fish Aquat Sci 45:936–942

Francis MP, Francis RICC (1992) Growth rate estimates of New Zealand rig (*Mustelus lenticulatus*). Aust J Mar Freshw Res 43:1157–1176

Gelsleichter J, Cortés E, Manire CA, Hueter RE (1997) Use of calcein as a fluorescent marker for elasmobranch vertebral cartilage. Trans Am Fish Soc 126:862–865

Joung SJ, Chen CT (1995) Reproduction in the Sandbar Shark, *Carcharhinus plumbeus*, in the Waters off northeastern Taiwan. Copeia 3:659–665

Joung SJ, Liao YY, Chen CT (2004) Age and growth of sandbar shark, *Carcharhinus plumbeus*, in northeastern Taiwan waters. Fish Res 70:83–96

Kimura DK (1980) Likelihood methods for the von Bertalanffy growth curve. Fish Bull 77:765–775

Kusher DI, Smith SE, Cailliet GM (1992) Validated age and growth of the leopard shark, *Triakis semifasciata*, with comments on reproduction. Environ Biol Fish 35:187–203

Last PR, Stevens JD (1994) Sharks and rays of Australia. CSIRO, Melbourne, Australia, pp. 513

Monaghan JP Jr (1993) Comparison of calcein and tetracycline as chemical markers in summer flounder. Trans Am Fish Soc 122:298–301

Musick JA, Burgess G, Cailliet G, Camhi M, Fordham S (2000) Management of sharks and their relatives (Elasmobranchii). Fisheries 25(3):9–13

Natanson LJ, Kohler NE (1996) A preliminary estimate of the age and growth of the dusky shark *Carcharhinus obscurus* from the south-west Indian Ocean, with comparisons to the western North Atlantic population. S Afr J Mar Sci 17:217–224

Natanson LJ, Gasey JG, Kohler NE, Colket T IV (1999) Growth of the tiger shark, *Galeocerdo cuvier*, in the western North Atlantic based on tag returns and length frequencies; and a note on the effects of tagging. Fish Bull 97:944–953

Prince JD (2005) Gauntlet fisheries for elasmobranchs—the secret of sustainable shark fisheries. J Northwest Atl Fish Sci 35:article 29

SigmaPlot for Windows version 9.01 (2004) Systat Software Inc

Simpfendorfer CA (1993) Age and growth of the Australian sharpnose shark, *Rhizoprionodon taylori*, from north Queensland, Australia. Environ Biol Fish 36:233–241

Simpfendorfer CA (2000) Growth rates of juvenile dusky sharks, *Carcharhinus obscurus* (Lesueur, 1818), from southwestern Australia estimated from tag-recapture data. Fish Bull 98:811–822

Simpfendorfer CA, Donohue K (1998) Keeping the fish in 'fish and chips': research and management of the Western Australian shark fishery. Mar Freshw Res 49:593–600

Simpfendorfer CA, Childlow J, McAuley R, Unsworth P (2000) Age and growth of the whisikery shark, *Furgaleus Mackiki*, from southwestern Australia. Environ Biol Fishes 58:335–343

Simpfendorfer CA, McAuley RB, Chidlow J, Unsworth P (2002) Validated age and growth of the dusky shark, *Carcharhinus obscurus*, from Western Australian waters. Mar Freshw Res 53:567–573

Skomal GB, Natanson LJ (2003) Age and growth of the blue shark (*Prionace glauca*) in the North Atlantic Ocean. Fish Bull 101:627–639

Sminkey T, Musick J (1995) Age and Growth of the Sandbar Shark, *Carcharhinus plumbeus*, before and after population depletion. Copeia 4:871–883

Sminkey T, Musick J (1996) Demographic analysis of the sandbar shark, *Carcharhinus plumbeus*, in the Western North Atlantic. Fish Bull 94:341–347

Smith SE, Au DW, Show C (1998) Intrinsic rebound potentials of 26 species of Pacific sharks. Mar Freshw Res 49:663–678

Smith SE, Mitchell RA, Fuller D (2002) Age-validation of a leopard shark (*Triakis semifasciata*) recaptured after 20 years. Fish Bull 101:194–198

Springer S (1960) Natural history of the sandbar shark, *Eulamia milberti*. US Fish Wildl Serv Fish Bull 61:1–38

Stevens JD, McLoughlin KJ (1991) Distribution, size and sex composition, reproductive biology and diet of sharks from northern Australia. Aust J Mar Freshw Res 42:151–199

Stevens JD, Walker TI, Simpfendorfer CA (1997) Are southern Australian shark fisheries sustainable? In: Hancock DA, Smith DC, Grant A, Beumer JP (eds) Developing and sustaining world fisheries resources. The state of science and management. Proceedings of the second World fisheries congress. CSIRO, Melbourne, pp. 62–66

Taniuchi T (1971) Reproduction of the sandbar shark, *Carcharhinus milberti*, in the East China Sea. Jpn J Ichthyol 18:94–98

Walker TI (1998) Can shark resources be harvested sustainably? A question revisited with a review of shark fisheries. Mar Freshw Res 49:553–572

Wass RC (1973) Size, growth and reproduction of the sandbar shark, *Carcharhinus milberti*, in Hawaii. Pac Sci 27:305–318

Wintner SP, Dudley SFJ, Kistnasamy N, Everett B (2002) Age and growth estimates for the Zambesi shark, *Carcharhinus leucas*, from the east coast of South Africa. Mar Freshw Res 53:557–566

Environ Biol Fish (2006) 77:401–406
DOI 10.1007/s10641-006-9108-2

ORIGINAL PAPER

Analysis of variability in vertebral morphology and growth ring counts in two Carcharhinid sharks

**Andrew N. Piercy · Travis S. Ford ·
Laura M. Levy · Franklin F. Snelson Jr**

Received: 2 June 2006 / Accepted: 23 June 2006 / Published online: 15 August 2006
© Springer Science+Business Media B.V. 2006

Abstract Inter- and intra-regional variations in vertebrae morphology and growth increment counts (band counts) were analyzed for two carcharhinid shark species, *Carcharhinus plumbeus* (n = 10) and *C. limbatus* (n = 11). Five sequential vertebrae were removed from the cervical region, above the branchial chamber and posterior to the chondrocrainium, and thoracic region, below the first dorsal fin. Dorsal–ventral height, medial–lateral breadth, and caudal–cranial length were measured for each sampled vertebra. Results indicate no significant difference in vertebral morphology within a sampled region of the vertebral column. However, a significant difference in vertebral morphology was noted between regions for both shark species, with thoracic vertebrae consistently larger than cervical vertebrae. A sub-set of three vertebrae was taken from each sampled region of each shark for sectioning and counting of growth increments. Analyses of growth increment counts by two readers indicated no significant difference in band counts within and between sampled regions.

A. N. Piercy (✉) · T. S. Ford · F. F. Snelson Jr
Florida Program for Shark Research, Florida Museum of Natural History, University of Florida, Gainesville, FL 32611, USA
e-mail: apiercy@flmnh.ufl.edu

L. M. Levy
Department of Zoology, University of Florida, Gainesville, FL 32611, USA

Keywords Shark · Growth · Vertebrae · Age · *Carcharhinus plumbeus* · *Carcharhinus limbatus*

Introduction

Accurate knowledge of the age and growth of fishes is paramount for successful management of fish populations (Ricker 1975; Cailliet et al. 1986). Shark age and growth studies have traditionally relied on the quantification of growth of a shark hard structure, such as dorsal fin spines or vertebrae, as a proxy for somatic growth (Cailliet and Goldman 2004). The enumeration of growth marks or increments present on the shark hard structure is used to determine the age of a shark. A defined relationship, usually linear, between hard structure growth and somatic growth is needed to justify the use of a proxy.

The majority of shark age and growth studies have utilized vertebrae taken below the first dorsal fin of the shark, herein termed thoracic region (e.g. Casey et al. 1985; Sminkey and Musick 1995). Vertebrae from the thoracic region are relatively homogeneous and possess a large radius and clearer growth bands (Cailliet et al. 1983; Joung et al. 2004). However, fishery dependent sampling often precludes removal of thoracic vertebrae, as it will damage the harvested meat. Alternatively, sampling of the vertebrae just posterior to the chondrocranium above the branchial region, herein termed cervical region, can be done without damaging harvestable meat and has been used in several shark age and growth studies (e.g. Branstetter 1987; Natanson et al. 2002). However, before vertebrae other than thoracic are considered for shark age and growth measurements, differences in the growth of these structures must be assessed. Specifically, are there region specific differences in vertebral growth? Will cervical vertebral growth reflect the same growth pattern as thoracic vertebrae? To this end, the goal of this study was to assess inter- and intra-regional variation in shark vertebrae morphology and growth increment counts (band counts).

Materials and methods

Sample collection

Specimens for this study were collected during fishery independent sampling of sharks conducted by the Florida Program for Shark Research, Florida Museum of Natural History, University of Florida. Long-line fishing gear was used to catch large coastal sharks off the East and West coasts of Florida. Upon capture, sharks were sexed, measured for fork length, and species identification was determined. Samples of the vertebral column containing five or more sequential centra were removed from two regions of the body; the cervical region, dorsal to the branchial chamber, and the thoracic region, ventral to the first dorsal fin. Upon removal, the anterior end of the vertebral columns was marked with a rubber band to facilitate correct labeling of vertebrae sequence. Vertebrae samples were then placed on ice and transported to the lab. Shark vertebrae samples were manually and chemically cleaned of tissue using scalpels and 6% sodium hypochlorite solution. Once cleaned, vertebrae were washed under a running tap water rinse to remove residual sodium hypochlorite. Cleaned and dried vertebrae were given a number relative to their sequence along the extracted portion of the vertebral column (e.g. the first whole anterior cervical vertebrae was numbered C-1).

Vertebrae measuring and sectioning

The morphology of five whole cervical and thoracic vertebrae was measured from each specimen. Measurements of caudal–cranial length (CCL), medial–lateral breadth (MLB), and dorsal–ventral height (DVH) were taken for each vertebra using Ultra-Cal IV digital calipers (Fred V Fowler Co. Inc, Newton, MA) (Fig. 1).

Three centra from each vertebral region (C1, C3, C5, and T1, T3, T5) were used for growth band analysis. Sagittal sections, 0.6 mm in thickness, were cut from selected vertebrae using a Buehler 82 Isomet low-speed saw. Each section was mounted on a glass microscope slide with clear resin (Cytoseal 60, Fisher Scientific, Pittsburgh,

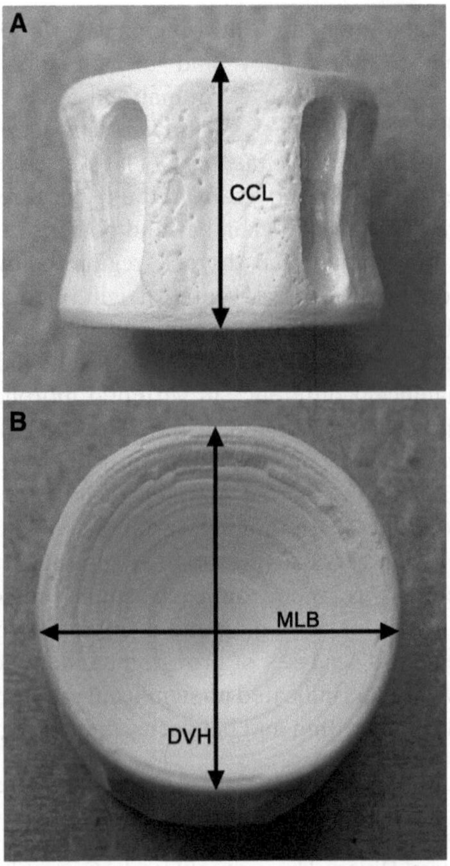

Fig. 1 (**A**) Ventral view of shark vertebrae with haemal arches removed. Line represents caudal–cranial length (CCL) measurement. (**B**) Saggital view of shark vertebrae with neural and haemal arches removed. Lines represent medial–lateral breadth (MLB) and dorsal–ventral height (DVH)

Pennsylvania, USA), randomized, and examined using a dissecting microscope under transmitted light. Opaque bands representing summer growth and translucent bands representing winter growth were identified following the description and terminology in Cailliet and Goldman (2004). A band was counted if it was present across the corpus calcarum and intermedialia (Fig. 2). Winter growth bands were enumerated by two readers for each vertebral section.

Data analysis

Variation in vertebral morphology was examined using an analysis of covariance test (ANCOVA) (SAS Institute Inc., Cary, NC, USA). Coefficients of variation were calculated to assess inter- and intra-regional variation in growth band counts. Additionally, differences in median cervical and thoracic growth band counts were assessed through a correlation test and ANCOVA.

Results

Vertebrae morphology

The morphology of cervical and thoracic vertebrae was examined for eleven sandbar sharks, *Carcharhinus plumbeus*, and ten blacktip sharks, *C. limbatus*, ranging in sizes from 127 to 163 cm fork length and 95 to 156 cm fork length,

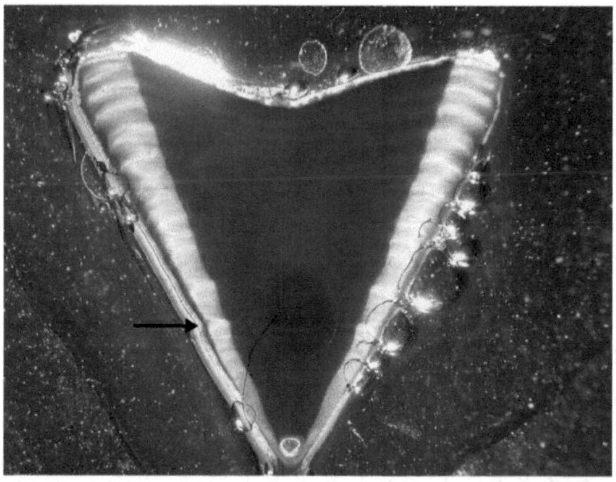

Fig. 2 Sagittal cross section of a sandbar shark vertebra prepared for age estimation. Arrow indicates a counted ring

respectively. Measurements of CCL, MLB, and DVH were recorded to a hundredth of a millimeter. Analysis of covariance for *C. plumbeus* and *C. limbatus* specimens showed that mean CCL, MLB, and DVH measurements were not significantly different within a vertebral region (cervical or thoracic) (Table 1). However, there were significant differences in morphology between vertebral regions for both shark species (Table 1). Cervical vertebrae of both sharks were significantly smaller than thoracic vertebrae for all three measurements (Table 2).

Vertebral growth increments

Growth rings from the vertebrae of nine *C. plumbeus* and ten *C. limbatus* were analyzed. Average coefficients of variation within and between regions were relatively homogenous and low for both shark species (Table 3). Median growth band counts of *C. plumbeus* cervical vertebrae were significantly correlated with thoracic vertebrae counts for both reader 1 ($P < 0.001$) and reader 2 ($P = 0.002$) counts. Further, the relationship between cervical and thoracic growth band counts was shown to be linear with a slope not significantly different than one (ANCOVA Reader 1: t-value = 0.15, $P = 0.885$; Reader 2: t-value = –0.70, $P = 0.495$) indicating that cervical counts are not significantly different than thoracic counts for both readers. Median growth band counts of *C. limbatus* cervical vertebrae were also significantly correlated with thoracic vertebrae counts for both reader 1 ($P < 0.0001$) and reader 2 ($P < 0.0001$). As seen in *C. plumbeus*, the relationship between cervical and thoracic growth band counts for *C. limbatus* vertebrae was linear with a slope not significantly different than one (ANCOVA Reader 1: $t = -0.39$, $P = 0.703$; Reader 2: $t = -1.21$, $P = 0.245$).

Discussion

Cervical and thoracic vertebrae are typically used for shark age and growth studies as they are generally considered to be homogeneous (Joung et al. 2004) and are favored over caudal vertebrae as the smaller caudal vertebrae may

Table 1 Analysis of covariance results for vertebrae morphological measurements of two Carcharhinid sharks

Carcharhinus plumbeus				*Carcharhinus limbatus*			
Caudal Cranial Length				Caudal Cranial Length			
ANCOVA model: $R^2 = 0.82$	F value = 46.49	$p < 0.0001$		ANCOVA model: $R^2 = 0.89$	F value = 72.27	$p < 0.0001$	
Parameters	Degrees of Freedom	F value	P value	Parameters	Degrees of Freedom	F value	P value
Fork Length	1	215.43	< 0.0001	Fork Length	1	517.23	< 0.0001
Vertebrae Region	1	243.73	< 0.0001	Vertebrae Region	1	200.47	< 0.0001
Vertebrae Sequence (Region)	8	0.71	0.6785	Vertebrae Sequence (Region)	8	0.62	0.7602
Medical Lateral Breadth				Medical Lateral Breadth			
ANCOVA model: $R^2 = 0.77$	F value = 33.80	$p < 0.0001$		ANCOVA model: $R^2 = 0.94$	F value = 139.45	$p < 0.0001$	
Parameters	Degrees of Freedom	F value	P value	Parameters	Degrees of Freedom	F value	P value
Fork Length	1	269.98	< 0.0001	Fork Length	1	1374.80	< 0.0001
Vertebrae Region	1	67.52	< 0.0001	Vertebrae Region	1	15.95	< 0.0001
Vertebrae Sequence (Region)	8	0.06	0.9999	Vertebrae Sequence (Region)	8	0.47	0.8734
Dorsal Ventral Height				Dorsal Ventral Height			
ANCOVA model: $R^2 = 0.78$	F value = 35.25	$p < 0.0001$		ANCOVA model: $R^2 = 0.91$	F value = 91.91	$p < 0.0001$	
Parameters	Degrees of Freedom	F value	P value	Parameters	Degrees of Freedom	F value	P value
Fork Length	1	257.93	< 0.0001	Fork Length	1	781.08	< 0.0001
Vertebrae Region	1	89.63	< 0.0001	Vertebrae Region	1	136.94	< 0.0001
Vertebrae Sequence (Region)	8	0.61	0.7653	Vertebrae Sequence (Region)	8	0.14	0.9974

Table 2 Least square means values for morphological measurements of two Carcharhinid sharks

Carcharhinus plumbeus			*Carcharhinus limbatus*		
Vertebral region	Least Square Means	*P*-value	Vertebral region	Least Square Means	*P*-value
Caudal-cranial length					
Cervical	8.94	< 0.0001	Cervical	7.90	< 0.0001
Thoracic	10.45		Thoracic	9.21	
Medial-lateral breadth					
Cervical	17.33	< 0.0001	Cervical	18.27	< 0.0001
Thoracic	18.64		Thoracic	18.83	
Dorsal-ventral height					
Cervical	16.79	< 0.0001	Cervical	16.32	< 0.0001
Thoracic	18.53		Thoracic	18.17	

lack some growth bands (Cailliet and Goldman 2004). Our results indicate that vertebral morphology is homogeneous within a region but heterogeneous between regions with thoracic vertebrae being significantly larger than cervical vertebrae in *C. plumbeus* and *C. limbatus* sharks. This regional difference in vertebrae morphology precludes the combined use of cervical and thoracic vertebrae in back-calculation methods often used in age and growth studies, as the vertebral radius and shark length relationship may differ by vertebral region. If back calculations are required, then separate vertebral radius and shark length regressions should be used for each sampled region.

The results of this study show there to be no inter- or intra-regional differences in the number of growth increments present in the vertebrae of *C. plumbeus* and *C. limbatus* sharks. These findings are reinforced by Joung et al. (2004), though not presenting any data, reporting that *C. plumbeus* sharks in Taiwan waters also exhibit no regional differences in vertebral growth increment numbers between caudal and cervical vertebrae. Officer et al. (1996) reported no intra-regional differences in growth increments in the vertebrae of *Mustelus antarticus* and *Galeorhinus galeus*. These multi-study results indicate that homogeneity in vertebral growth increments within a region is likely common in sharks. However, Officer et al. (1996) did report significant inter-regional differences in the number of growth increments present in the vertebrae of *M. antarticus* and *G. galeus*. Thoracic vertebrae were shown to exhibit greater numbers of growth increments than caudal and cervical vertebrae of larger specimens of *M. antarticus* and all sizes of *G. galeus*. Variation in ageing techniques may influence the readability of the vertebrae as seen in the Officer et al. (1996) study. With this in mind we utilized the most commonly used ageing technique of analyzing sagittal vertebral sections.

Table 3 Average coefficient of variation for growth band counts. Values for cervical versus thoracic column are based on median counts

	Cervical	Thoracic	Cervical vs. Thoracic
Carcharhinus plumbeus			
Reader 1	4.87	3.65	4.16
Reader 2	6.05	8.80	5.26
Carcharhinus limbatus			
Reader 1	4.67	4.62	2.89
Reader 2	7.82	6.98	7.03

The absence of regional differences in vertebral growth increments in our study, contrasted with the presence of such differences in the sharks of Officer et al.'s 1996 study, promotes the need for species-specific examination of possible regional differences in vertebral growth. The possibility of inter-regional differences in vertebral growth increments reinforces the need for validation of age estimates.

Acknowledgements The authors thank Alexia Morgan, Taylor Chapple, and Peter Cooper for assistance in collecting shark specimens for this study. We also thank Mike Allen for statistical guidance. Funding for this project came from the National Marine Fisheries Service's Highly Migratory Species Division through the Florida Program for Shark research involvement in the National Shark Research Consortium.

References

Branstetter S (1987) Age and growth estimates for blacktip, *Carcharhinus limbatus*, and spinner, *C. brevipinna*, sharks from the Northwestern Gulf of Mexico. Copeia 1987:964–974

Cailliet GM & KJ Goldman (2004) Age determination and validation in chondrichthyan fishes. In: Carrier JC, Musick JA, Heithaus MR (eds) Biology of sharks and their relatives. CRC Press, pp 399–447

Cailliet GM, Martin LK, Kusher D, Wolf P, Welden BA (1983) Techniques for enhancing vertebral bands in age estimation of California elasmobranchs. In: Prince ED, Pulos LM (eds) Proceedings of the international workshop on age determination of ocean pelagic fishes: tunas, billfishes, and sharks. NOAA Tech Rep NMFS 8, pp 157–165

Cailliet GM, Radtke RL, Welden BA (1986) Elasmobranch age determination and verification: a review. In: Uyeno T, Arai R, Taniuchi T, Matsuura K (eds) Indo-Pacific Fish Biology: Proceedings of the second international conference on Indo-Pacific Fishes. Ichthyological Society of Japan, Tokyo, pp 345–359

Casey JG, Pratt HL Jr, Stillwell CE (1985) Age and growth of the sandbar shark (*Carcharhinus plumbeus*) from the Western North Atlantic. Can J Fish Aquat Sci 42:963–974

Joung SJ, Liao YY, Chen CT (2004) Age and growth of sandbar shark, *Carcharhinus plumbeus*, in northeastern Tawian waters. Fisheries Research 70:83–96

Natanson LJ, Mello JJ, Campana SE (2002) Validated age and growth of the porbeagle shark (*Lamna nasus*) in the western North Atlantic Ocean. Fish Bull 100:266–278

Officer RA, Gason AS, Walker TI, Clement JG (1996) Sources of variation in counts of growth increments in vertebrae from gummy shark, *Mustelus antarcticus*, and school shark, *Galeorhinus galeus*: implications for age determination. Can J Fish Aquat Sci 53:1765–1777

Ricker WE (1975) Computational and interpretation of biological statistics of fish populations. Bull Fish Res Board Can 191:1–382

Sminkey TR, Musick JA (1995) Age and growth of the sandbar shark, *Carcharhinus plumbeus*, before and after population depletion. Copeia 1995:871–885

Environ Biol Fish (2006) 77:407–421
DOI 10.1007/s10641-006-9109-1

ORIGINAL PAPER

Morphometric minefields—towards a measurement standard for chondrichthyan fishes

Malcolm P. Francis

Received: 5 June 2006 / Accepted: 20 June 2006 / Published online: 22 August 2006
© Springer Science+Business Media B.V. 2006

Abstract Size measurements are crucial for studies on the growth, maturation, maximum size, and population structure of cartilaginous fishes. However, researchers use a variety of measurement techniques even when working on the same species. Accurate comparison of results among studies is only possible if the measurement technique used is adequately defined and, if different techniques are used, a conversion equation can be derived. These conditions have not always been met, leading to invalid comparisons and incorrect conclusions. This paper reviews methods used for measuring chondrichthyans, and summarises the variety of constraints that influence the choice of a measurement technique. Estimates of the variability present in some measurement techniques are derived for shortfin mako shark, *Isurus oxyrinchus*, porbeagle shark, *Lamna nasus*, blue shark, *Prionace glauca*, Antarctic thorny skate, *Amblyraja georgiana*, and Pacific electric ray, *Torpedo californica*. Total length measured with the tail in the natural position (sharks) and disc widths (batoids) have higher variability than other methods, and are not recommended. Instead, the longest longitudinal axis should be measured where possible and practical; i.e., flexed total length for sharks, total length for batoids (excluding suborder Myliobatoidei), pelvic length for batoids of the suborder Myliobatoidei, and chimaera length (snout to posterior end of supracaudal fin) for chimaeroids (except for *Callorhinchus*, for which fork length should be measured from the anterior edge of the snout protuberance). Straight-line measurements are preferred to measurements over the curve of the body. Importantly, measurement methods must be clearly defined, giving information on the anterior reference point, the posterior reference point, and how the measurement was made between these two. Measurements using at least two different methods are recommended on at least a subsample of the fish in order to develop conversion regression relationships.

Keywords Measurement · Chondrichthyans · Sharks · Rays · Skates · Chimaeras

M. P. Francis (✉)
Inshore and Pelagic Fishes, National Institute of Water and Atmospheric Research Ltd, 301 Evans Bay Parade, Greta Point, Private Bag 14901, Wellington 6000, New Zealand
e-mail: m.francis@niwa.co.nz

Introduction

Fish biologists and ecologists usually measure the animals that they sample during field or aquarium studies. Typically, a fish's length, width or weight is used as a measure of its 'size', which is a crucial variable in studies of growth, sexual maturation, maximum size, and population structure. Errors in fish size measurements could bias estimates of growth rate and growth curve parameters, so accurate and precise measurements are a prerequisite for age and growth studies.

Measuring fish is now such a common and fundamental practice that we often adopt a measurement technique without clearly defining it, and without considering whether it is appropriate, repeatable, or comparable. In this paper, I explore these issues for chondrichthyan (cartilaginous) fishes, a group for which a large array of measurement methods have been applied, and for which measurement problems seem to be particularly common. However many of the conclusions and recommendations apply equally to teleosts.

In recent decades, the volume of research published on chondrichthyans has increased dramatically, and the geographical scope or impact of many studies is worldwide. It is now possible (and often essential) to compare our results with those from studies made previously on the same population (temporal comparisons) or on the same species elsewhere (spatial comparisons). But accurate comparisons among studies are only possible if (a) the measurement techniques used in the comparison have been adequately defined, and (b) the same measurement technique has been used in both studies, or a regression equation exists for converting between two (or more) different techniques. These conditions have not always been met, leading to invalid comparisons and incorrect conclusions.

It is clear that we cannot standardise measurement techniques across all chondrichthyans, and for reasons discussed later it may not be possible to standardise measurement techniques within a species. But it is equally clear that there is great scope for clarifying and simplifying current practice. The plethora of measurement methods in use worldwide, and the poor definition of methods by many authors, are obstacles to good science, and possibly to good fisheries management. The aim of this study is to review the measurement techniques used most frequently for chondrichthyans, consider their merits and drawbacks, and make recommendations for improvements to current practice. I do not address the use of measurements by taxonomists, who require a larger suite of precisely defined and accurately measured morphometrics than is generally required by fish biologists and ecologists.

Measurement techniques

A glance through any well-illustrated guide to chondrichthyans (e.g. Compagno 1984; Last and Stevens 1994; Compagno et al. 2005) reveals a huge diversity of body morphology. Diversity in the snout, tail and wing tips has resulted in a large number of different measurement techniques, but most techniques vary in the way the tail is treated. The snout and wing tip reference points are usually unambiguous, though there are some notable exceptions, such as mobulid rays, *Manta birostris* and *Mobula* spp., and cownose rays, *Rhinoptera* spp., which all have bilobed, anterio-lateral extensions on the head, and *Callorhinchus* spp., which have a fleshy protuberance on the snout. Further variation is introduced by differences in the way the measurement is taken—whether in a straight line or over the curve of the body, and whether along the ventral, dorsal or lateral surface.

The main measurement methods currently in use, and some less common methods, are summarised below. However the list is not comprehensive. Acronyms used throughout this paper are given in parentheses and italics.

Sharks

Flexed total length (TL_{flex})—tip of the snout to the posterior tip of the tail, with the tail flexed down so that the upper lobe lies along the body midline. Also commonly called stretched total length.

Natural total length (TL_{nat})—tip of the snout to the posterior tip of the tail, with the tail in the natural position. This measurement is taken along the body midline to a point intersected by a perpendicular dropped from the posterior tip of the upper lobe of the tail.

Calculated total length (TL_{calc})—sum of the precaudal length (see below) and tail length. The latter is estimated as a fixed fraction of the length of the upper lobe of the tail; the magnitude of the fraction depends on the tail angle. This method was developed by Bass (1973) and Bass et al. (1975) to overcome the subjectivity associated with determining the natural tail position, and the difficulty experienced in dropping a perpendicular from the tail tip to the midline. It has been used rarely, and then mainly in South Africa for comparative purposes (e.g., Cliff et al. 1990; Cliff and Dudley 1992). However Gilmore et al. (1983) used it in their study of *Odontaspis taurus*.

Fork length (*FL*)—tip of the snout to the fork in the tail.

Precaudal length (*PCL*)—tip of the snout to the origin of the upper lobe of the caudal fin. In species with an upper precaudal pit, the measurement is usually taken to the anterior edge of the pit; in other species it is taken to the point where the upper caudal lobe becomes clearly distinct from the caudal peduncle.

TL_{flex}, TL_{nat}, *FL* and *PCL* are all commonly used in shark studies. *PCL* is most frequently used in species having a precaudal pit, and *FL* is restricted to species having a distinct tail fork.

All of the above methods can be made in a straight line (point to point) (*SL*) or over the curve of the body (*OTB*). *SL* measurements may be made with a measuring board, callipers, or a tape measure laid along the ground, whereas *OTB* measurements are made with a flexible measuring tape draped over the body. *SL* measurements are commonly used worldwide, whereas *OTB* measurements are uncommon except in the north-west Atlantic Ocean. The Northeast Fisheries Science Center (NEFSC) of the U.S. National Marine Fisheries Service (NMFS) has been using *OTB* measurements since 1965 (H.L. Pratt, Mote Marine Laboratory, Key West, Florida, USA, personal communication; L.J. Natanson, NMFS, Narragansett, Rhode Island, USA, personal communication). Before 1965, *SL* measurements were taken with callipers (J.G. Casey, personal communication to L.J. Natanson). Most of the shark data in the large database held by NEFSC were collected since 1965. Some important and often-cited publications from NEFSC incorrectly stated that their measurements were *SL* (e.g., Pratt 1979; Kohler et al. 1995); other publications (e.g., Casey et al. 1985; Natanson et al. 1999) imply the same by citing Bigelow and Schroeder (1948), who clearly stated (p. 61) that they used *SL* measurements. Scientists and observers in other parts of north-eastern USA and eastern Canada also use *OTB* measurements (MacNeil and Campana 2002; Skomal and Natanson 2003; Campana et al. 2005).

All of the above methods may be taken with the shark lying on its side or belly, except for TL_{nat} which can only be measured with the shark lying on its side. When a shark is lying on its belly, the natural position of the tail is difficult to simulate and measure. For species in which the upper lobe of the caudal fin deviates only slightly from the body midline (e.g., scyliorhinid catsharks), TL_{nat} and TL_{flex} differ little.

Batoids

Total length (*TL*)—tip of the snout to the tip of the tail, with the tail straightened and aligned with the anterio-posterior axis of the body.

Disc length (*DL*)—tip of the snout to the posterior edge of the disc. Hubbs and Ishiyama (1968) defined the posterior reference point as the posterior margin of the pectoral fins, but few authors since then have defined it. Because the posterior reference point is ambiguous (it could be either the posterior edge of the pectoral fins or the posterior edge of the pelvic fins), *DL* should not be used without qualification. I propose the following definitions and acronyms that explicitly specify the posterior reference point:

Disc length pectoral (DL_{pec})—tip of the snout to the posterior edge of the pectoral fins;

Disc length pelvic (DL_{pel})—tip of the snout to the posterior edge of the pelvic fins, excluding the anterior pelvic lobes or 'legs' (which sometimes project beyond the posterior pelvic lobes) and claspers.

Disc width (*DW*)—distance between the wing tips.

Pre-dorsal length (*PD1* or *PD2*)—tip of the snout to the origin of the first or second dorsal fin.

TL is the commonest measurement method used for skates (Suborder Rajoidei), but DL_{pec}, DL_{pel} and *DW* have been used occasionally (Ishiyama 1951; Natanson 1993; Francis et al. 2001). *DW* is the commonest method used for rays (Suborder Myliobatoidei). *PD1* and *PD2* have been used in batoids with dorsal fins, though not as the main measurement method (J. Neer, NMFS, Panama City, Florida, USA personal communication). All batoid measurements may be made *SL* or *OTB*, and are mainly if not always made with the ventral surface downwards.

Chimaeroids

Callorhinchidae

FL or *PCL*—as defined above for sharks. In *Callorhinchus*, the only genus in this family, the tail is shark-like, so in theory any of the methods used for sharks are possible. However, the tip of the tail is often missing, making TL_{nat} and TL_{flex} impractical. *Callorhinchus* has a fleshy protuberance on the snout, and some researchers have measured from the anterior tip of the protuberance (Sullivan 1977; Francis 1997), whereas others have squashed the snout flat on a measuring board, and used the forehead as the anterior reference point (Coakley 1973; Freer and Griffiths 1993).

Chimaeridae and Rhinochimaeridae

Chimaera length (*CL*)—tip of the snout to the posterior edge of the supracaudal fin, excluding the caudal filament.

Precaudal length (*PCL*)—tip of the snout to the anterior edge of the supracaudal fin (see Didier 2002). This measurement is often called *pre-supracaudal fin length* but as it is analogous to *PCL* in sharks, I prefer the latter term for consistency and simplicity.

Snout to vent length (*SVL*)—tip of the snout to the anterior edge of the vent.

For *Callorhinchus*, *FL* and *PCL* have been used recently, though with two different anterior reference points (Freer and Griffiths 1993; Di Giácomo and Perier 1994; Francis 1997). For chimaerids and rhinochimaerids, *CL* is the standard measurement technique in New Zealand fisheries research, whereas *PCL* and *SVL* have been used elsewhere (Johnson and Horton 1972; Luchetti 2004; Moura et al. 2004; Calis et al. 2005). All chimaeroid measurements may be made *SL* or *OTB*, and may be taken with the fish lying on its side or belly.

Constraints on measurement techniques

Unfortunately scientists do not have complete freedom to choose a measurement method. Many factors constrain how this seemingly simple task is carried out:

1. A method may not be possible on a particular species (e.g. *FL* on a species without a forked tail).
2. A method may not be possible on all individuals of a species; e.g. any form of total length in species with a tail that is readily damaged (e.g. chimaerids and dasyatid rays), or that dries out in the sun after capture (e.g. sharks landed at fishing tournaments). A special case is presented by embryos: their morphology may prevent the use of a method that is suitable for post-natal individuals of the same species. For example, the caudal fin lobes of embryonic lamnids are curved inwards towards each other (Francis and Stevens 2000), making measurement of TL_{nat} and TL_{flex} problematic; the former probably shrinks after birth as the tail lobes splay apart, but the latter may increase as the upper caudal lobe straightens. In species that hatch from deposited egg cases, the embryonic tail is often modified to pump water through the egg case; for example *Callorhinchus* spp. have a very elongated upper caudal lobe and no lower caudal lobe or tail fork until near hatching (Didier 1995), making *FL* impossible to measure.
3. A method may not be logistically feasible for routine use in the field (e.g. *TL* in very large skates or TL_{nat} and TL_{flex} in large sharks, especially *Alopias* spp.).
4. Some methods are more time-consuming to use than others, which can be an issue for large samples.
5. A method may be subjective and suffer from large measurement error (e.g. *PCL* in species without a precaudal pit).
6. A method may be specified in legislation (especially minimum length regulations), and scientific sampling needs to be consistent.

7. A method may be chosen to enable direct comparison with a previous study on the same or similar species.

Therefore a measurement method is usually chosen after considering a variety of factors. Some constraints may not be consistently present in space or time: legislative requirements may be specific to nations or states; constraints on measuring TL_{nat} and TL_{flex} in large sharks may apply aboard small vessels but not aboard large vessels or ashore; partial processing of the catch at sea may prevent some measurements from being taken when ashore; a study that does not involve embryo measurements will not be constrained by embryo morphology. The end result is that scientists have adopted an extraordinary array of measurement techniques for chondrichthyan fishes.

Measurement variability

Notwithstanding the constraints listed in the previous section, scientists often have a choice of two or more methods for measuring a particular chondrichthyan. In those situations, an objective tool for assessing the relative merits of each method is desirable. One useful criterion is the amount of measurement error in each method: all else being equal, a method with low measurement error is better than one with high measurement error.

Methods

I explored the magnitude of measurement error for some commonly-used techniques. For this purpose I used pairwise comparisons of multiple measurement techniques applied to the same animals. For each data set, I fitted ordinary least squares (OLS) Y-on-X and X-on-Y linear regressions to each pair of measurements using the SAS statistical package (SAS Institute, Cary, North Carolina, USA):

$$Y = a + bX$$

where a is the intercept and b is the slope. In theory, geometric mean regression (GM; also called functional regression) is preferable, because both X and Y variables have significant and similar measurement error (Ricker 1973; Mollet and Cailliet 1996). However, when the correlation between variables is high enough, the GM and OLS regressions are very similar and OLS can be used safely (McArdle 1988 suggested $r > 0.9$, i.e., $R^2 > 0.81$).

Model fit was assessed by inspecting the regression fit and the raw residuals. If there was any non-linearity in the data, as indicated by the pattern of the residuals, log–log regressions were used instead of linear regressions:

$$\mathrm{Log}_{10}Y = \acute{a} + b\mathrm{Log}_{10}X$$

where $\acute{a} = \mathrm{Log}_{10}a$.

The Root Mean Square Error (RMSE) is a measure of variability derived from the squared residuals:

$$\mathrm{RMSE} = \sqrt{\frac{\sum_i (y_i - \hat{y}_i)^2}{n}}$$

where y_i represents the ith observation of the dependent variable, $\hat{y}_i$ represents the ith predicted value, $y_i - \hat{y}_i$ is the residual, and n is the sample size. However, for measurement data the residuals (and therefore RMSE) are expected to increase on average with body size. To enable direct comparison among populations and species having different body sizes, measurement variability was expressed as a *Relative* Root Mean Square Error (RRMSE), in which the residuals are scaled by the value of the dependent variable:

$$\mathrm{RRMSE} = \sqrt{\frac{\sum_i \left(\frac{y_i - \hat{y}_i}{\hat{y}_i}\right)^2}{n}} \qquad (1)$$

Using this formula, an RRMSE of, for example, 0.02 in a regression of fork length on total length means that, on average, the error in the predicted fork length is expected to be 2%.

Equation (1) was used to calculate RRMSE for linear regressions. For log–log regressions, the RMSE becomes a relative error when converted back to natural space (C. Francis, NIWA, Wellington, New Zealand, personal communication). Therefore RRMSE was calculated as the antilog of RMSE. For example, a RMSE of 0.0166 becomes a relative error of 1.039, or 3.9%, when back-transformed.

Regression lines having Y-axis intercepts significantly different from zero indicate allometry (Gould 1966; Mollet and Cailliet 1996). However, for all of the datasets considered, there were few small fish, so estimates of the intercepts were considerable extrapolations beyond the data, and significant results are probably often spurious. Allometry occurring outside the range of the data is not an issue for this study, and is not considered further in this paper. Both SL and OTB measurements are included in the following analyses. Subscripts are used to explicitly identify these measurements, e.g. FL_{SL} and FL_{OTB}.

Six datasets were analysed. Details of the measurements taken on each fish are given in the Results.

1. Shortfin mako sharks, *Isurus oxyrinchus*, caught in New Zealand recreational fishing competitions were measured on land, lying on their sides, by scientists using metal measuring tapes (M.P. Francis, C. Duffy and S. Bishop, unpublished data).
2. Shortfin mako sharks caught by commercial tuna longline vessels in New Zealand waters were measured at sea, lying on their sides, by Ministry of Fisheries observers using callipers (M.P. Francis, unpublished data).
3. Porbeagle sharks, *Lamna nasus*, caught by longline in the north-west Atlantic Ocean off eastern Canada were measured at sea or on land, lying on their sides, by scientists using metal measuring tapes (S.E. Campana, Bedford Institute of Oceanography, Dartmouth, Canada, unpublished data).
4. Blue sharks, *Prionace glauca*, caught by longline in the north-west Atlantic Ocean off eastern Canada were measured at sea or on land, lying on their sides, by scientists using metal measuring tapes (S.E. Campana, unpublished data).
5. Antarctic thorny skates, *Amblyraja georgiana*, caught by commercial toothfish longliners in the Ross Sea, Antarctica were measured at sea, lying on their bellies, by observers using measuring boards (M.P. Francis, unpublished data).
6. Pacific electric rays, *Torpedo californica*, collected off California were measured on land, lying on their bellies, by scientists using a measuring board (J. Neer, unpublished data). Most of the adult specimens were frozen and thawed before measuring.

Results

Shortfin mako sharks, dataset 1

This dataset of shortfin makos was small, especially for FL_{OTB} regressions (Table 1). A subset of the pairwise regressions with FL_{SL} as the independent variable is shown in Fig. 1. Data from eight near-term embryos (Duffy and Francis 2001) fell on or near the regression lines (Fig. 1), but we have no data from earlier embryos that might demonstrate an effect of the inward curving of the caudal fin lobes.

Comparison of regression slopes showed that FL_{OTB} was about 3% greater than FL_{SL} (Table 1), although the samples size for this comparison was only 30 and the result needs confirmation. TL_{flexSL} was almost 6% greater than TL_{natSL} (Table 1).

Residuals from the linear regressions of PCL_{SL}, FL_{OTB}, and TL_{flexSL} against FL_{SL} were uniformly distributed and small (mostly less than 5 cm) (Fig. 2). By contrast, residuals from TL_{natSL} versus FL_{SL} were larger (many values over 5 cm and some over 10 cm), and the variability increased with increasing FL_{SL}. TL_{natSL} showed the highest RRMSEs (1.3–1.8%) of all pairwise regressions (Table 1). The other measurement methods (TL_{flexSL}, FL_{OTB}, FL_{SL} and PCL_{SL}) produced low RRMSEs (0.9–1.2%) except when regressed against TL_{natSL}. These results suggest that measurement error is highest (though still relatively small at less than 2%) for TL_{nat}, and low for the other four methods.

Shortfin mako sharks, dataset 2

Dataset 2 for shortfin makos was much larger than dataset 1, but it contained only three

Table 1 Pairwise linear regression results among five measurement methods for shortfin mako shark (datasets 1 and 2)

Dep. var.	Indep. var.	a	SE_a	b	SE_b	N	Adj R^2	Range dep. var.	Range indep. var.	RRMSE (%)
Scientists (dataset 1)										
TL_{flexSL}	TL_{natSL}	–4.27	1.61	1.056	0.007	62	0.9975	97–394	95–379	**1.3**
	FL_{OTB}	1.77	1.91	1.097	0.010	30	0.9978	127–324	117–294	1.2
	FL_{SL}	–0.15	1.05	1.137	0.005	61	0.9989	97–394	85–346	1.0
	PCL_{SL}	2.96	1.24	1.244	0.006	61	0.9984	97–394	76–314	1.2
TL_{natSL}	TL_{flexSL}	4.59	1.50	0.944	0.006	62	0.9975	95–379	97–394	**1.3**
	FL_{OTB}	7.10	2.73	1.032	0.014	30	0.9949	125–309	117–294	**1.7**
	FL_{SL}	3.17	1.38	1.085	0.007	130	0.9953	95–379	85–346	**1.7**
	PCL_{SL}	4.43	1.38	1.196	0.007	136	0.9949	95–379	76–314	**1.8**
FL_{OTB}	TL_{flexSL}	–1.20	1.75	0.910	0.008	30	0.9978	117–294	127–324	1.2
	TL_{natSL}	–5.90	2.72	0.964	0.013	30	0.9949	117–294	125–309	**1.7**
	FL_{SL}	–0.79	1.27	1.032	0.007	30	0.9988	117–294	113–287	0.9
	PCL_{SL}	–0.23	1.62	1.143	0.009	30	0.9981	117–294	102–256	1.0
FL_{SL}	TL_{flexSL}	0.36	0.92	0.878	0.004	61	0.9989	85–346	97–394	1.0
	TL_{natSL}	–1.96	1.28	0.918	0.006	130	0.9953	85–346	95–379	**1.7**
	FL_{OTB}	0.97	1.23	0.968	0.006	30	0.9988	113–287	117–294	0.9
	PCL_{SL}	2.09	0.63	1.097	0.003	129	0.9988	85–346	76–314	0.9
PCL_{SL}	TL_{flexSL}	–2.09	1.00	0.803	0.004	61	0.9984	76–314	97–394	1.2
	TL_{natSL}	–2.77	1.17	0.832	0.005	136	0.9949	76–314	95–379	**1.8**
	FL_{OTB}	0.51	1.41	0.874	0.007	30	0.9981	102–256	117–294	1.0
	FL_{SL}	–1.69	0.58	0.911	0.003	129	0.9988	76–314	85–346	0.9
Observers (dataset 2)										
TL_{natSL}	FL_{SL}	1.97	1.15	1.077	0.007	253	0.9901	80–330	70–297	**2.7**
FL_{SL}	TL_{natSL}	–0.19	1.07	0.919	0.006	253	0.9901	70–297	80–330	**2.7**
	PCL_{SL}	1.06	0.36	1.097	0.002	854	0.9966	61–333	55–304	**1.6**
PCL_{SL}	FL_{SL}	–0.43	0.33	0.908	0.002	854	0.9966	55–304	61–333	**1.6**

Bold values indicate relative root mean square errors (RRMSE) ≥ 1.3%. Measurements were straight line (*SL*) or over the body (*OTB*). Dep. var., dependent variable; Indep. var., independent variable; a, Y-intercept; b, slope; SE_a and SE_b, standard errors for a and b; N, sample size; Adj R^2, coefficient of determination adjusted for degrees of freedom. Ranges are in centimeters

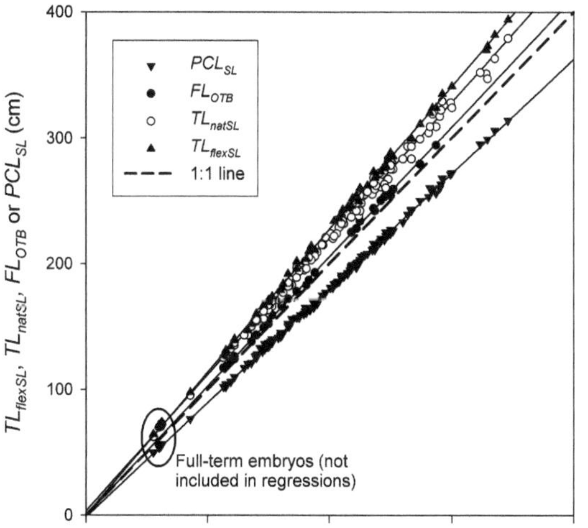

Fig. 1 Pairwise linear regressions among five measurement methods for shortfin mako shark (dataset 1). Embryo data were not included in the fitted regressions

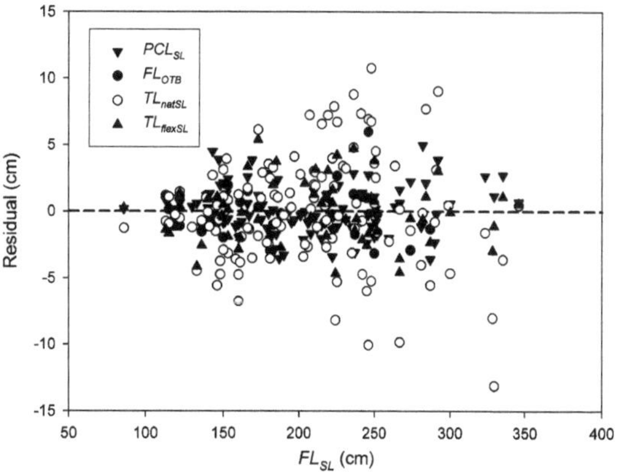

Fig. 2 Residuals from pairwise linear regressions (see Fig. 1) among five measurement methods for shortfin mako shark (dataset 1)

measurement methods: TL_{natSL}, FL_{SL} and PCL_{SL}, and only two pairwise comparisons were possible (PCL_{SL} and TL_{natSL} were never measured on the same sharks) (Table 1). RRMSEs were substantially larger than for their counterparts in dataset 1: 2.7% for TL_{natSL} versus FL_{SL} (cf. 1.7%) and 1.6% for PCL_{SL} versus FL_{SL} (cf. 0.9%), suggesting that multiple observers working at sea are not able to measure sharks as precisely as a few scientists working ashore. Nevertheless the ranking of the pairwise comparisons was the same for both datasets, with comparisons involving TL_{natSL} being more variable.

Porbeagle sharks, dataset 3

Sample sizes were moderate to large (Table 2). The highest RRMSEs (≥2.0%) were obtained for regressions involving TL_{natOTB}, and the lowest RRMSEs (0.8%) were obtained for the pairwise regression between FL_{OTB} and PCL_{OTB}. Other comparisons had intermediate values. Thus FL_{OTB} and PCL_{OTB} had the lowest co-variation, possibly reflecting the fact that the two measurements can easily be made sequentially on a shark with minimal adjustment of the measuring tape.

Blue sharks, dataset 4

Sample sizes varied from small to large (Table 3). The highest RRMSEs (1.8–2.0%) all involved TL_{natSL} or TL_{natOTB}, but interestingly the regressions between these two methods had low RRMSEs (1.3%). This might result from the two TL_{nat} measurements being taken without altering the position of the tail, thus ensuring good correlation between them. FL_{OTB} and FL_{SL} regressions produced similar, intermediate RRMSEs in regressions with other methods.

Antarctic thorny skates, dataset 5

This dataset was moderately large (Table 4). Preliminary linear regressions showed that residuals were not uniformly distributed, indicating non-linear behaviour. Data were therefore $\log_{10}$ transformed before re-fitting the linear regressions. A homogeneity of slopes test showed that males and females were significantly different: for a given TL_{SL}, males had a longer, narrower disc than females (DL_{pelSL} versus TL_{SL}, $P < 0.0001$; DW_{SL} versus TL_{SL}, $P = 0.0009$). Consequently, the two sexes were separated before final analysis.

Log–log regressions for Antarctic thorny skates are shown in Table 4. A subset of the female regressions with TL_{SL} as the independent variable is shown in Fig. 3 (males were omitted for clarity). The highest RRMSEs all involved DW_{SL} (2.7–3.3%). Regressions between TL_{SL} and DL_{pelSL} were less variable (RRMSEs 2.3–2.5%). This suggests that measurement error is highest for DW_{SL}.

Table 2 Pairwise linear regression results among four measurement methods for porbeagle shark (dataset 3)

Dep. var.	Indep. var.	a	SE_a	b	SE_b	N	Adj R^2	Range dep. var.	Range indep. var.	RRMSE (%)
TL_{natOTB}	FL_{OTB}	−0.51	0.44	1.121	0.003	940	0.9951	95–293	85–264	1.7
	FL_{SL}	−1.49	1.41	1.169	0.009	142	0.9916	105–285	91–253	**2.3**
	PCL_{OTB}	1.67	0.52	1.255	0.003	918	0.9932	95–293	77–234	**2.0**
FL_{OTB}	TL_{natOTB}	1.27	0.39	0.888	0.002	940	0.9951	85–264	95–293	1.7
	FL_{SL}	−0.32	0.78	1.045	0.005	173	0.9959	86–264	83–253	1.8
	PCL_{OTB}	1.84	0.18	1.121	0.001	924	0.9989	85–264	77–234	0.8
FL_{SL}	TL_{natOTB}	2.52	1.19	0.848	0.007	142	0.9916	91–253	105–285	**2.3**
	FL_{OTB}	0.90	0.74	0.953	0.005	173	0.9959	83–253	86–264	1.8
	PCL_{OTB}	2.56	0.93	1.065	0.006	141	0.9948	91–253	83–234	1.8
PCL_{OTB}	TL_{natOTB}	−0.33	0.41	0.791	0.002	918	0.9932	77–234	95–293	**2.0**
	FL_{OTB}	−1.48	0.16	0.892	0.001	924	0.9989	77–234	85–264	0.8
	FL_{SL}	−1.68	0.88	0.934	0.006	141	0.9948	83–234	91–253	1.9

Bold values indicate relative root mean square errors (RRMSE) ≥ 2.0%. Measurements were straight line (*SL*) or over the body (*OTB*). Dep. var., dependent variable; Indep. var., independent variable; *a*, Y-intercept; *b*, slope; SE_a and SE_b, standard errors for *a* and *b*; *N*, sample size; Adj R^2, coefficient of determination adjusted for degrees of freedom. Ranges are in centimetres

Table 3 Pairwise linear regression results among four measurement methods for blue shark (dataset 4)

Dep. var.	Indep. var.	a	SE_a	b	SE_b	N	Adj R^2	Range dep. var.	Range indep. var.	RRMSE (%)
TL_{natOTB}	TL_{natSL}	5.33	2.20	0.986	0.010	86	0.9909	154–294	153–293	1.3
	FL_{OTB}	6.04	2.71	1.181	0.015	86	0.9860	154–294	129–238	1.6
	FL_{SL}	3.52	3.46	1.219	0.020	86	0.9780	154–294	123–233	**1.9**
TL_{natSL}	TL_{natOTB}	–3.44	2.26	1.005	0.010	86	0.9909	153–293	154–294	1.3
	FL_{OTB}	3.53	0.85	1.175	0.005	777	0.9885	153–351	129–292	**1.9**
	FL_{SL}	5.53	0.86	1.194	0.005	777	0.9881	153–351	123–286	**1.8**
FL_{OTB}	TL_{natOTB}	–2.61	2.33	0.835	0.011	86	0.9860	129–238	154–294	1.7
	TL_{natSL}	–0.85	0.73	0.841	0.003	777	0.9885	129–292	153–351	**1.9**
	FL_{SL}	2.33	0.58	1.013	0.003	783	0.9924	129–292	123–286	1.5
FL_{SL}	TL_{natOTB}	0.94	2.82	0.803	0.013	86	0.9780	123–233	154–294	**2.0**
	TL_{natSL}	–2.44	0.73	0.827	0.003	777	0.9881	123–286	153–351	**1.9**
	FL_{OTB}	–0.92	0.57	0.980	0.003	783	0.9924	123–286	129–292	1.6

Bold values indicate relative root mean square errors (RRMSE) ≥ 1.8%. Measurements were straight line (*SL*) or over the body (*OTB*). Dep. var., dependent variable; Indep. var., independent variable; *a*, Y-intercept; *b*, slope; SE_a and SE_b, standard errors for *a* and *b*; *N*, sample size; Adj R^2, coefficient of determination adjusted for degrees of freedom. Ranges are in centimetres

Table 4 Pairwise log–log regression results among three measurement methods for Antarctic thorny skates (dataset 5)

Sex	Dep. var.	Indep. var.	á	$SE_á$	b	SE_b	N	Adj R^2	Range dep. var.	Range indep. var.	RRMSE (%)
Female	TL_{SL}	DL_{pelSL}	0.27	0.01	0.940	0.004	679	0.9866	42–115	29–79	2.3
		DW_{SL}	0.09	0.02	1.001	0.010	255	0.9764	42–111	35–88	**3.1**
	DL_{pelSL}	TL_{SL}	–0.26	0.01	1.050	0.005	679	0.9866	29–79	42–115	2.4
		DW_{SL}	–0.18	0.02	1.065	0.009	256	0.9822	29–77	35–88	**2.9**
	DW_{SL}	TL_{SL}	–0.05	0.02	0.975	0.010	255	0.9764	35–88	42–111	**3.1**
		DL_{pelSL}	0.20	0.01	0.923	0.008	256	0.9822	35–88	29–77	**2.7**
Male	TL_{SL}	DL_{pelSL}	0.31	0.01	0.915	0.004	501	0.9890	39–121	26–83	2.3
		DW_{SL}	0.01	0.02	1.051	0.009	262	0.9803	39–117	30–90	**3.3**
	DL_{pelSL}	TL_{SL}	–0.31	0.01	1.080	0.005	501	0.9890	26–83	39–121	2.5
		DW_{SL}	–0.29	0.02	1.133	0.009	266	0.9850	26–83	30–90	**3.1**
	DW_{SL}	TL_{SL}	0.02	0.02	0.933	0.008	262	0.9803	30–90	39–117	**3.1**
		DL_{pelSL}	0.28	0.01	0.869	0.007	266	0.9850	30–90	26–83	**2.7**

Bold values indicate relative root mean square errors (RRMSE) > 2.5%. All measurements were straight line (*SL*). Dep. var., dependent variable; Indep. var., independent variable; *á*, Y-intercept; *b*, slope; $SE_á$ and SE_b, standard errors for *á* and *b*; *N*, sample size; Adj R^2, coefficient of determination adjusted for degrees of freedom. Ranges are in centimetres

Pacific electric rays, dataset 6

Sample sizes were small. Homogeneity of slopes tests showed that males and females were significantly different for the regressions $PD1_{SL}$ versus TL_{SL} ($P = 0.0014$) and DW_{SL} versus TL_{SL} ($P = 0.0033$). The two sexes were separated before final analysis for all pairwise comparisons (Table 5). A subset of the regressions with TL_{SL} as the independent variable is shown in Fig. 4. Five notable outliers appear to represent recording errors rather than measurement variability, so those fish were omitted from all regressions to avoid inflation of the RRMSEs.

Regressions involving DW_{SL} had very high RRMSEs over 8% (Table 5). The lowest RRMSEs came from regressions involving $PD1_{SL}$ and $PD2_{SL}$, and intermediate RRMSEs occurred for regressions involving TL_{SL} (all values ≤ 3.1%). Measurement error is clearly highest for DW_{SL}, probably inflated by combining fresh and frozen specimens.

Discussion

The index of regression variability used in the present study, RRMSE, incorporates two sources

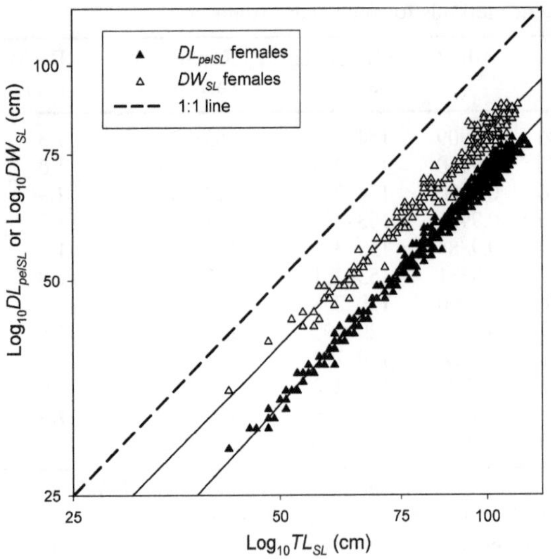

Fig. 3 Pairwise log–log regressions among three measurement methods for female Antarctic thorny skate (dataset 5)

of variation—natural variability and measurement variability. The latter source is the main focus of this study, but it could not be estimated separately (such estimation would require repeated length measurements on the same individual). Nevertheless, there is some evidence to suggest that measurement variability is a major contributor to RRMSE: for shortfin mako sharks, dataset 1, variability in comparisons involving TL_{flexSL} was substantially lower than in those involving TL_{natSL} (Table 1), even though both measurements used the same anterior and posterior reference points. If the relationship between total length and other partial lengths was inherently variable, then both TL_{flexSL} and TL_{natSL} would be expected to have similar RRMSEs. When developing regression equations to inter-convert different measurement

Table 5 Pairwise linear regression results among four measurement methods for Pacific electric rays (dataset 6)

Sex	Dep. var.	Indep. var.	a	SE_a	b	SE_b	N	Adj R^2	Range dep. var.	Range indep. var.	RRMSE (%)
Female	TL_{SL}	$PD2_{SL}$	0.47	0.30	1.332	0.008	76	0.9975	19–102	14–75	2.2
		$PD1_{SL}$	0.09	0.35	1.551	0.011	76	0.9965	19–102	12–64	3.0
		DW_{SL}	2.86	0.86	1.365	0.023	79	0.9783	19–102	11–72	**9.4**
	$PD2_{SL}$	TL_{SL}	–0.27	0.23	0.749	0.004	76	0.9975	14–75	19–102	2.2
		$PD1_{SL}$	–0.28	0.14	1.164	0.004	77	0.9990	14–75	12–64	2.0
		DW_{SL}	2.07	0.53	1.007	0.015	77	0.9843	14–75	11–72	**9.0**
	$PD1_{SL}$	TL_{SL}	0.04	0.23	0.643	0.004	76	0.9965	12–64	19–102	3.1
		$PD2_{SL}$	0.27	0.12	0.858	0.003	77	0.9990	12–64	14–75	2.0
		DW_{SL}	2.00	0.42	0.866	0.012	77	0.9865	12–64	11–72	**8.3**
	DW_{SL}	TL_{SL}	–1.37	0.65	0.717	0.012	79	0.9783	11–72	19–102	**10.3**
		$PD2_{SL}$	–1.54	0.54	0.977	0.014	77	0.9843	11–72	14–75	**9.9**
		$PD1_{SL}$	–1.87	0.51	1.139	0.015	77	0.9865	11–72	12–64	**9.2**
Male	TL_{SL}	$PD2_{SL}$	0.23	0.37	1.347	0.009	102	0.9955	18–84	14–63	2.5
		$PD1_{SL}$	–0.46	0.43	1.600	0.012	102	0.9939	18–84	12–53	3.0
		DW_{SL}	0.78	0.81	1.455	0.022	112	0.9762	18–84	10–60	**9.2**
	$PD2_{SL}$	TL_{SL}	–0.04	0.27	0.739	0.005	102	0.9955	14–63	18–84	2.5
		$PD1_{SL}$	–0.50	0.18	1.188	0.005	102	0.9981	14–63	12–53	2.0
		DW_{SL}	0.43	0.57	1.075	0.015	102	0.9801	14–63	10–60	**9.1**
	$PD1_{SL}$	TL_{SL}	0.47	0.27	0.621	0.005	102	0.9939	12–53	18–84	3.0
		$PD2_{SL}$	0.48	0.15	0.840	0.004	102	0.9981	12–53	14–63	2.0
		DW_{SL}	0.79	0.46	0.905	0.012	102	0.9817	12–53	10–60	**8.2**
	DW_{SL}	TL_{SL}	0.29	0.55	0.671	0.010	112	0.9762	10–60	18–84	**8.7**
		$PD2_{SL}$	0.28	0.53	0.912	0.013	102	0.9801	10–60	14–63	**8.6**
		$PD1_{SL}$	–0.24	0.51	1.085	0.015	102	0.9817	10–60	12–53	**8.1**

Bold values indicate relative root mean square errors (RRMSE) > 8.0%. All measurements were straight line (*SL*). Dep. var., dependent variable; Indep. var., independent variable; *a*, Y-intercept; *b*, slope; SE_a and SE_b, standard errors for *a* and *b*; *N*, sample size; Adj R^2, coefficient of determination adjusted for degrees of freedom. Ranges are in centimetres

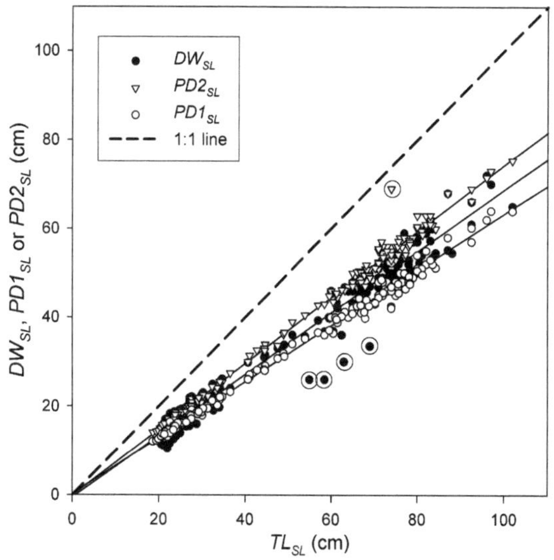

Fig. 4 Pairwise linear regressions among four measurement methods for Pacific electric ray (dataset 6). For clarity, regression lines are shown for both sexes combined, but analyses were carried out separately by sex. Circled outliers were omitted from regressions and analyses

methods, the source of the variability is immaterial—variability of both types affects the precision of length predictions made from such regressions.

For dataset 1, the RRMSEs from shortfin mako shark regressions were, at best, about 1% (Table 1). For a shark of about 300 cm FL_{SL}, this means that 95% of the animals would fall within about ±2% of 300 cm, or ±6 cm, of the regression line. Using a regression of TL_{natSL} against FL_{SL}, the RRMSE of 1.7% would generate an absolute variation of about ±10 cm. This magnitude of variation may be acceptable for some purposes but not for others. For example, the power to distinguish a difference in length at maturity between two populations depends not only on the measurement variability, but also on the difference in mean length at maturity between the two populations. If measurement variability is low, the power to detect small between-population differences is greater.

Published studies frequently do not define their measurement methods adequately, so it is easy to inadvertently apply an incorrect regression equation, or even to decide incorrectly that no conversion is necessary. This generates a bias in the converted length, which might accentuate, nullify or even reverse a true population difference. A simple example, using 11 regression equations between fork length and total length for shortfin mako sharks populations, illustrates this point (Table 6). The fork length predicted for a 300 cm total length shark varied between 264 and 283 cm, a range of 19 cm, or 7% of the lower value. Methods involving FL_{OTB} produced the highest predicted fork lengths, and TL_{flex} produced the lowest predicted fork lengths. Interestingly, TL_{flex} has rarely been used in shortfin mako studies worldwide. Perhaps this reflects the perceived rigidity of the caudal fin of shortfin makos, making TL_{flex} difficult to measure; the corollary of this is that TL_{nat} should be less variable than TL_{flex} for this species. However, the evidence from this study is that TL_{flex} is less variable than TL_{nat}, and my personal experience is that the tail is easily flexed down to the midline for the former method.

An important conclusion from this study is that two commonly used measurement methods, TL_{nat} and DW, performed worse than other methods. The reason for the poor performance of TL_{nat} is probably the difficulty in judging a "natural" tail position—this judgement is apparently not reproducible across sharks, or across measurers. The reason for the poor performance of DW is not clear, though Notarbartolo-di-sciara (1987) provided some clues in relation to measuring mobulid rays: "Disc width is the largest reliable dimension on a ray's body, and therefore it affords a good degree of accuracy and precision when fresh and well preserved specimens are measured. However, it may be a difficult measurement to obtain on curled specimens, or it may be an unreliable reference index in very old specimens in which the texture of the pectoral fin has become loose".

It remains to be seen whether the results of the present study apply to a wide range of chondrichthyans—the six datasets examined here cover only a small fraction of the species and families of chondrichthyans, and each dataset provided information on only a subset of the possible measurement techniques. Nevertheless, the following recommendations are offered in an attempt to improve the quality of length data

Table 6 Calculated fork length at a total length of 300 cm using 11 different regression equations for shortfin mako sharks

Source	Region	Measurement method			
		Fork length	Total length	Sample size	Calc. FL for TL = 300 cm
Present study	SW Pacific	FL_{OTB}	TL_{natSL}	30	283
Campana et al. (2005)	NW Atlantic	FL_{OTB}	TL_{natOTB}	13	282
Kohler et al. (1995)	NW Atlantic	FL_{OTB}	TL_{natOTB}	199	277
Present study	SW Pacific	FL_{OTB}	TL_{flexSL}	30	272
Present study	SW Pacific	FL_{SL}	TL_{natSL}	130	273
Chan (2001)	SW Pacific	FL_{SL}	TL_{natSL}	50	272
E. Acuña (unpubl. data)	SE Pacific	FL_{SL}	TL_{natSL}	421	271
V. Buencuerpo (unpubl. data)	NE Atlantic	FL_{SL}	TL_{natSL}	1,764	269
Joung and Hsu (2005)	NW Pacific	FL_{SL}	TL_{natSL}	1,236	268
R. Rasmussen (unpubl. data)	NE Pacific	FL_{SL}	TL_{natSL}	3,976	266
Present study	SW Pacific	FL_{SL}	TL_{flexSL}	61	264

Rows are arranged by fork length measurement method and descending order of calculated fork length. Measurements were straight line (SL) or over the body (OTB)

collected from chondrichthyans, and to reduce the number of measurement techniques in use:

1. The ideal chondrichthyan measuring technique for biologists and ecologists dealing with moderate to large samples of fish should be (a) a good indicator of the size of an animal; (b) simple to apply to a high proportion of individuals in a population; (c) accurate and precise; and (d) unambiguous.
2. Measure the longest longitudinal axis consistent with other constraints. Intuitively, this dimension is the best index of 'size'. I recommend the use of TL_{flex} for sharks, TL for batoids of all suborders except Myliobatoidei, DL_{pel} for batoids of the suborder Myliobatoidei,[1] and CL for chimaeroids (except for *Callorhinchus* for which FL measured from the anterior edge of the snout protuberance is most appropriate). DL_{pel} is preferred over DL_{pec} for myliobatoids because the former is longer, and because it is easier to measure when a ray is placed on a measuring board (the posterior ends of the pectoral fins are often displaced from the midline and overlie the pelvic fins). Concerns about gender differences in the size and shape of batoid pelvic fins can be addressed by recording the sex of all animals and treating the sexes separately. This is important anyway because many batoids have sexually dimorphic disc shapes (D. Ebert, Moss Landing Marine Laboratories, Moss Landing, California, USA, personal communication). If the recommended method is not possible or practical, use the next longest longitudinal body dimension (FL for sharks, DL_{pel} or DL_{pec} for batoids, and PCL for chimaeroids including *Callorhinchus*). For species that may increase in girth (sharks) or thickness (batoids) with age, girth or weight may also be informative and useful in developing growth models.
3. Avoid the use of TL_{nat} and DW because they appear to have higher measurement error than other methods.
4. Use SL methods in preference to OTB methods. Although FL_{OTB} performed as well as, or better than, FL_{SL} in shortfin mako, porbeagle and blue sharks, OTB methods are likely to be affected by pregnancy, stomach fullness, liver size and body condition.
5. Define the measurement method clearly and adequately. Each definition should include an anterior reference point, a posterior reference point, and how the distance between these two points is measured. This last item should include a description of the measuring device, the path of the measurement

[1] Although some members of the Myliobatoidei (e.g. urolophids) have a well formed caudal fin, it is often damaged (P. Kyne, University of Queensland, Australia, pers. comm.).

(whether *SL* or *OTB*), and the orientation of the animal (whether on its side, belly or other). [The orientation of the animal has not been addressed here, and its effect on measurements is unknown.] Journal reviewers and editors should insist that such information is provided in published papers and reports.

6. For greater utility, make two or more different measurement methods on each animal. If done routinely, this potentially allows direct comparison with other studies and avoidance of the need to calculate conversion regressions. It also provides a means for detecting gross recording errors, which appear as outliers on bivariate or multivariate plots. If it is not possible to make multiple measurements on every individual, this should at least be done on a moderate subset of animals to allow the development of study-specific conversion regressions; this avoids the need to use a regression equation from another study, with the associated risk that it may not be appropriate.
7. Measure a large sample size over a wide length range. Report the sample size and length range of dependent and independent variables, because these affect the applicability of the regressions. Caution should be exercised when extrapolating beyond the range of data, especially if there is evidence of allometry.
8. Check for male–female differences using a homogeneity of slopes test and analysis of covariance, and only pool the sexes if differences are non-significant. Failure to treat sexes separately when they have different underlying morphometric relationships will inflate variability and possibly introduce bias. The same applies to pooling datasets from different populations or time periods.
9. Check data for allometry (is the *Y*-intercept significantly different from zero?), non-linearity (examine residuals), and homogeneity of variances (examine residuals). Allometry is evidence of non-linearity (though the non-linear part of the relationship may fall outside the data range). Use log–log regressions where necessary to linearise data, and improve homogeneity of variances. Also, consider fitting non-linear models.
10. For length–length regressions, R^2 is usually high and it is safe to use OLS regression in place of the more correct GM regression.
11. Minimise other sources of error in measurements, or compensate for them. Chondrichthyans shrink after freezing and thawing: *Callorhinchus milii* shrinks by 2–5% depending on length (Francis 1997), and neonate *Mustelus lenticulatus* shrink by about 4% (Francis and Francis 1992), so corrections are necessary if estimates of fresh length are required. A common practice is to round lengths down to the centimetre below actual length. Any parameter estimated from such length data (e.g. median length at maturity) must be corrected for this rounding bias by adding 0.5 cm.
12. Establishment of a database in which morphometric data and conversion regressions can be archived and accessed via the worldwide web should be explored. Although such an exercise could be fraught with problems, particularly in ensuring that consistent and well-defined measurement protocols were used for deposited data, the benefits may outweigh the disadvantages, especially for species having broad geographic distributions.

Acknowledgments I thank Clinton Duffy and Silver Bishop for help in collecting the New Zealand competition shortfin mako data, and the New Zealand Ministry of Fisheries and CCAMLR observers for collecting the New Zealand tuna longline mako data and the Ross Sea Antarctic thorny skate data. Steve Campana kindly provided the porbeagle and blue shark data, and Julie Neer the Pacific electric ray data. Enzo Acuña (Universidade Catolica del Norte, Coquimbo, Chile), Valentin Buencuerpo (Universidade Complutense de Madrid, Madrid, Spain) and Rand Rasmussen (NMFS, La Jolla, California, USA) allowed me to use their unpublished shortfin mako regression equations. Chris Francis provided invaluable statistical advice. I appreciate the numerous helpful discussions with many colleagues over recent years that helped shape some of the ideas and work in this paper; they included John Carlson, Jon Dodrill, Dave Ebert, Peter Kyne, Skip McKinnell, Henry Mollet, Lisa Natanson, Julie Neer, Wes Pratt, Sue Smith, John Stevens, and William White.

References

Bass AJ (1973) Analysis and description of variation in the proportional dimensions of scyliorhinid, carcharhinid and sphyrnid sharks. Oceanographic Research Institute Investigational Report 32, 28 pp

Bass AJ, D'Aubrey JD, Kistnasamy N (1975) Sharks of the east coast of southern Africa. IV. The families Odontaspididae, Scapanorhynchidae, Isuridae, Cetorhinidae, Alopiidae, Orectolobidae and Rhiniodontidae. Oceanographic Research Institute Investigational Report 39, 102 pp

Bigelow HB, Schroeder WC (1948) Fishes of the Western North Atlantic. Part 1. Lancelets, cyclostomes, sharks. Mem Sears Found Mar Res 1:59–546

Calis E, Jackson EH, Nolan CP, Jeal F (2005) Preliminary age and growth estimates of the rabbitfish, *Chimaera monstrosa*, with implications for future resource management. J Northw Atl Fish Sci 35:15–26

Campana SE, Marks L, Joyce W (2005) The biology and fishery of shortfin mako sharks (*Isurus oxyrinchus*) in Atlantic Canadian waters. Fish Res 73:341–352

Casey JG, Pratt HL, Stillwell CE (1985) Age and growth of the sandbar shark (*Carcharhinus plumbeus*) from the western North Atlantic. Can J Fish Aquat Sci 42:963–975

Chan RWK (2001) Biological studies on sharks caught off the coast of New South Wales. PhD Thesis, University of New South Wales, Sydney, 323 pp

Cliff G, Dudley SFJ (1992) Sharks caught in the protective gill nets off Natal, South Africa. 6. The copper shark *Carcharhinus brachyurus* (Günther). S Afr J Mar Sci 12:663–674

Cliff G., Dudley SFJ, Davis B (1990) Sharks caught in the protective gill nets off Natal, South Africa. 3. The shortfin mako shark *Isurus oxyrinchus* (Rafinesque). S Afr J Mar Sci 9:115–126

Coakley A (1973) A study in the conservation of elephant fish (*Calloryncus milii*, Bory) in New Zealand. New Zealand Marine Department Fisheries Technical Report 126, 22 pp

Compagno L, Dando M, Fowler S (2005) Sharks of the world. Princeton University Press, Princeton, 368 pp

Compagno LJV (1984) Sharks of the world. An annotated and illustrated catalogue of shark species known to date. FAO Fish Synop 125(4):655

Di Giácomo EE, Perier MR (1994) Reproductive biology of the cockfish, *Callorhynchus callorhynchus* (Holocephali: Callorhynchidae), in Patagonian waters (Argentina). Fish Bull 92:531–539

Didier DA (1995) Phylogenetic systematics of extant chimaeroid fishes (Holocephali, Chimaeroidei). American Museum Novitates (3119), 86 pp

Didier DA (2002) Two new species of chimaeroid fishes from the southwestern Pacific Ocean (Holocephali, Chimaeridae). Ichthyol Res 49:299–306

Duffy C, Francis MP (2001) Evidence of summer parturition in shortfin mako (*Isurus oxyrinchus*) sharks from New Zealand waters. N Z J Mar Freshw Res 35:319–324

Francis MP (1997) Spatial and temporal variation in the growth rate of elephantfish (*Callorhinchus milii*). N Z J Mar Freshw Res 31:9–23

Francis MP, Francis RICC (1992) Growth rate estimates for New Zealand rig (*Mustelus lenticulatus*). Aust J Mar Freshw Res 43:1157–1176

Francis MP, Ó Maolagáin C, Stevens D (2001) Age, growth, maturity, and mortality of rough and smooth skates (*Dipturus nasutus* and *D. innominatus*). New Zealand Fisheries Assessment Report 2001/17, 21 pp

Francis MP, Stevens JD (2000) Reproduction, embryonic development and growth of the porbeagle shark, *Lamna nasus*, in the south-west Pacific Ocean. Fish Bull 98:41–63

Freer DWL, Griffiths CL (1993) Estimation of age and growth in the St Joseph *Callorhinchus capensis* (Dumeril). S Afr J Mar Sci 13:75–81

Gilmore RG, Dodrill JW, Linley PA (1983) Reproduction and embryonic development of the sand tiger shark, *Odontaspis taurus* (Rafinesque). Fish Bull 81:201–225

Gould SJ (1966) Allometry and size in ontogeny and phylogeny. Biol Rev Camb Philos Soc 41:587–640

Hubbs CL, Ishiyama R (1968) Methods for the taxonomic study and description of skates (Rajidae). Copeia 1968:483–491

Ishiyama R (1951) Studies on the rays and skates belonging to the family Rajidae, found in Japan and adjacent regions. 2. On the age-determination of Japanese black-skate *Raja fusca* Garman (preliminary report). Bull Jpn Soc Sci Fish 16:112–118

Johnson AG, Horton HF (1972) Length-weight relationship, food habits, parasites, and sex and age determination of the ratfish, *Hydrolagus colliei* (Lay and Bennett). Fish Bull 70:421–429

Joung S-J, Hsu H-H (2005) Reproduction and embryonic development of the shortfin mako, *Isurus oxyrinchus* Rafinesque, 1810, in the northwestern Pacific. Zool Stud 44:487–496

Kohler NE, Casey JG, Turner PA (1995) Length-weight relationships for 13 species of sharks from the western North Atlantic. Fish Bull 93:412–418

Last PR, Stevens JD (1994) Sharks and rays of Australia. CSIRO, Hobart, 513 pp

Luchetti E (2004) Aspects de la biologie de *Chimaera monstrosa* L. (Holocephali: Chimaeridae) en Atlantique Nord-Est. J Rech Oceanogr 29:40

MacNeil M.A, Campana SE (2002) Comparison of whole and sectioned vertebrae for determining the age of young blue shark (*Prionace glauca*). J Northw Atl Fish Sci 30:77–82

McArdle BH (1988) The structural relationship: regression in biology. Can J Zool 66:2329–2339

Mollet HF, Cailliet GM (1996) Using allometry to predict body mass from linear measurements of the white shark. In: Klimley AP, Ainley DG (eds) Great white sharks. The biology of *Carcharodon carcharias*. Academic Press, San Diego, pp 81–89

Moura T, Figueiredo I, Machado PB, Gordo LS (2004) Growth pattern and reproductive strategy of the holocephalan *Chimaera monstrosa* along the Portuguese continental slope. J Mar Biol Assoc UK 84:801–804

Natanson LJ, (1993) Effect of temperature on band deposition in the little skate, *Raja erinacea*. Copeia 1993:199–206

Natanson LJ, Casey JG, Kohler NE, Colket T (1999) Growth of the tiger shark, *Galeocerdo cuvier*, in the western North Atlantic based on tag returns and length frequencies; and a note on the effects of tagging. Fish Bull 97:944–953

Notarbartolo-di-Sciara G (1987) A revisionary study of the genus *Mobula* Rafinesque, 1810 (Chondrichthyes: Mobulidae) with the description of a new species. Zool J Linn Soc 91:1–91

Pratt HL (1979) Reproduction in the blue shark, *Prionace glauca*. Fish Bull 77:445–470

Ricker WE (1973) Linear regressions in fishery research. J Fish Res Board Can 30:409–434

Skomal GB, Natanson LJ (2003) Age and growth of the blue shark (*Prionace glauca*) in the North Atlantic Ocean. Fish Bull 101:627–639

Sullivan KJ (1977) Age and growth of the elephant fish *Callorhinchus milii* (Elasmobranchii: Callorhynchidae). N Z J Mar Freshw Res 11:745–753

MIX
Papier aus verantwortungsvollen Quellen
Paper from responsible sources
FSC® C105338

If you have any concerns about our products,
you can contact us on
ProductSafety@springernature.com

In case Publisher is established outside the EU,
the EU authorized representative is:
Springer Nature Customer Service Center GmbH
Europaplatz 3, 69115 Heidelberg, Germany

Printed by Libri Plureos GmbH
in Hamburg, Germany